MATHEMATICS RESEARCH DEVELOPMENTS

FRONTIERS OF COMBINATORICS AND NUMBER THEORY

VOLUME 2

MATHEMATICS RESEARCH DEVELOPMENTS

MATHEMATICS RESEARCH DEVELOPMENTS

FRONTIERS OF COMBINATORICS AND NUMBER THEORY

VOLUME 2

ZHI-WEI SUN
EDITOR

nova publishers
New York

LIBRARY OF CONGRESS CATALOGING-IN-PUBLICATION DATA

ISBN: 978-1-61942-918-5

Published by Nova Science Publishers, Inc. † New York

CONTENTS

PREFACE

This book is a collection of articles in the second volume of the Journal of Combinatorics and Number Theory. As scheduled this volume consists of three issues and it contains nineteen contributed papers from eminent experts or active young mathematicians which cover various important topics in number theory and combinatorics. (Imprint: Nova).

First the notion of r-almost s-uniform tree is introduced which includes both the case of finitely generated free groups and uniform binary trees as special cases. The goal of Chapter 1 is to study pseudorandomness of binary functions defined on r-almost s-uniform trees. The measures of pseudorandomness of binary functions are introduced; the connection between these measures is analyzed; the size of these measures for truly random binary functions is studied; binary functions with strong pseudorandom properties are constructed.

In Chapter 2 we show that there are infinitely many composite numbers N, relatively prime to 10, with the property that if any one of its digits is replaced by an arbitrary digit, then the resulting number is composite. We also show that there are infinitely many composite numbers N, relatively prime to 10, having the property that if an arbitrary digit is inserted once into the decimal representation of N, then the resulting number is composite.

In Chapter 3, let $m > 2$ be an even integer. We develop a method to construct families of prime numbers $q = mf + 1$ with f odd, and of complex nonreal cyclotomic integers $\beta_q \in \mathbb{Z}[\zeta_q]$ of degree m, with some prescribed properties, which are expressed as linear combinations, over $\mathbb{Z}$, of Gaussian periods. We apply that method in the construction of families of numbers β_q, as above, such that $\beta_q\overline{\beta_q} \in \mathbb{Z}$. Some of those families bring us families of cyclic difference sets. We also apply the method in the construction of families of cyclic polynomials of degree m whose roots are complex nonreal q-cyclotomic integers with absolute norm q. This complements results of a previous article where we show how to construct families of cyclic polynomials whose roots are real q-cyclotomic units. We give several examples for small values of m, and a MAPLE program that can be used to perform the calculations and to search for more examples.

In Chapter 4 we apply some number theoretical techniques to derive results on Boolean functions. We apply Stickelberger's theorem on 2-adic valuations of Gauss sums to the Kasami-Welch functions $\mathrm{tr}_L(x^{4^k-2^k+1})$ on $\mathbb{F}_{2^n}$, where n is odd and $(k,n) = 1$. We obtain information on the Fourier spectrum, including a characterization of the support of the

Fourier transform. One interesting feature is that the behaviour is different for different values of k. We also apply the Gross-Koblitz formula to the Gold functions $\mathrm{tr}_L(x^{2^k+1})$.

In Chapter 5 we represent the sums $\sum_{k=0}^{n-1}\binom{n}{k}^{-2}$, $\sum_{k=0}^{m}\binom{m}{k}^{-1}\binom{a}{n-k}^{-1}$, $\sum_{k=0}^{n-1}\frac{q^{-k(k-1)}}{\left[{n \atop k}\right]_q}$, and the sum of the reciprocals of the summands in Dixon's identity, each as a product of an *indefinite hypergeometric sum* times a (closed form) *hypergeometric sequence*.

In Chapter 6 we study the sumset $A + k \cdot A$ for the first non trivial case, $k = 3$, where $k \cdot A = \{k \cdot a,\ a \in A\}$. We prove that $|A + 3 \cdot A| \geq 4|A| - 4$ and that the equality holds only for $A = \{0, 1, 3\}$, $A = \{0, 1, 4\}$, $A = 3 \cdot \{0, \dots, n\} \cup (3 \cdot \{0, \dots, n\} + 1)$ and all the affine transforms of these sets.

In Chapter 7 we study the arithmetic (real) function $f = g * \mathbf{1}$, with g "essentially bounded" and supported over the integers of $[1, Q]$. In particular, we obtain non-trivial bounds, through f "correlations", for the "Selberg integral" and the "symmetry integral" of f in almost all short intervals $[x - h, x + h]$, $N \leq x \leq 2N$, beyond the "classical" level, up to level of distribution, say, $\lambda = \log Q / \log N < 2/3$ ($h \gg N^{1/3+\varepsilon}$, $\varepsilon > 0$ fixed). This time we don't apply Large Sieve inequality, as in our paper [CS]. Precisely, our method is completely elementary.

Chapter 8 describes a method to compute the Brauer-Manin obstruction for smooth cubic surfaces over $\mathbb{Q}$ such that $\mathrm{Br}(S)/\mathrm{Br}(\mathbb{Q})$ is of order two or four. This covers the vast majority of the cases when this group is non-zero. Our approach is to associate a Brauer class with every Galois invariant double-six. We show that all order two Brauer classes may be obtained in this way. We also recover Sir Peter Swinnerton-Dyer's result that $\mathrm{Br}(S)/\mathrm{Br}(\mathbb{Q})$ may take only five values.

Chapter 10 proves some identities on generating functions for values of L-functions related to the work of two papers [2, 5].

In Chapter 11 [13], P. Paillier and J. Villar make a conjecture about the malleability of an RSA modulus. In this paper we show how subtle the notion of malleability is. Specifically, we demonstrate how to factor an RSA modulus n using some information on the factorization of a concrete n' coprime to n. However, we believe the conjecture might be true, when imposing some extra conditions on the auxiliary n' allowed to be used.

In Chapter 12 we give some identities which involve the Mertens function $M(n)$. Our proofs are combinatorial with relatively prime subsets as a main tool.

In Chapter 13, we extend the combinatorial method of Sam's proof of Ehrhart's reciprocity theorem in two ways: we avoid having to triangulate the polytope in question, and the results hold for valuations invariant under translations by lattice vectors. As a consequence we obtain a simple proof of Stanley's reciprocity theorem.

Chapter 14 describes a new class of identities, which hold for certain general theta series, in two completely independent variables. It provides explicit examples of these identities involving the Dedekind eta function, Jacobi theta functions, and various theta functions of Ramanujan.

Chapter 15 shows that for any indefinite binary quadratic forms f and g the number of primitive integral lattice points in the (nonconvex and nonsmooth) plane region bounded by the four hyperbolas $|f(X,Y)| = B$ and $|g(X,Y)| = C$ is at most $32\frac{(BC)^{1/2}}{|R|^{1/4}} + 8$. Here R is the resultant of f and g. When f and g each have coprime integral coefficients, then R is also an integer and we arrive at the uniform bound $32(BC)^{1/2} + 8$, which is independent of the choice of forms f and g.

Let $\mathbb{N} = \{1, 2, 3, \cdots\}$ denote the natural numbers. Given integers $e \geq 1$ and $b \geq 2$, let $x = \sum_{i=0}^{n} a_i b^i$ with $0 \leq a_i \leq b - 1$ (thus a_i are the digits of x in base b). We define the *happy function* $S_{e,b} : \mathbb{N} \longrightarrow \mathbb{N}$ by

$$S_{e,b}(x) = \sum_{i=0}^{n} a_i^e.$$

A positive integer x is then said to be (e, b)-happy if $S_{e,b}^r(x) = 1$ for some $r \geq 0$, otherwise we say it is (e, b)-unhappy. In Chapter 16 we investigate the cycles and fixed points of the happy functions $S_{e,b}$. We give an upper bound for the size of elements belonging to the cycles of $S_{e,b}$. We also prove that the number of fixed points of $S_{2,b}$ is

$$\begin{cases} 2\tau\left(\frac{b^2+1}{2}\right) - 1 & \text{if } b \text{ is odd,} \\ \tau(b^2+1) - 1 & \text{if } b \text{ is even,} \end{cases}$$

where $\tau(m)$ denotes the number of positive divisors of m. We use our results to determine the densities of happy numbers in a small handful of cases.

Let $\overline{h}_m(n)$ be the number of non-increasing sequences of non-negative integral powers of m whose sum is n, and where the first occurrence of power of m may be overlined, such that each power of m is allowed to be used at most m times as a non-overlined part. In Chapter 17 we study this hyper m-ary overpartition function $\overline{h}_m(n)$ and obtain some congruence properties. For example, we show that: For all $m \geq 2$, a necessary and sufficient condition for $\overline{h}_m(n) \equiv 1 (\bmod 2)$ is that n can be expressed as $n = (m+1)\sum_{i=0}^{t} \varepsilon_i m^i$, where $\varepsilon_i = 0$ or 1 for each i, and $\varepsilon_t = 1$ if $t > 0$.

Chapter 18 presents a test for determining whether a real quadratic irrational has a purely periodic nearest square continued fraction expansion. This test is somewhat more explicit than the standard test and simplifies the programming of the algorithm.

In Chapter 19, let A be the generalized Cartan matrix of rank 2 Kac-Moody algebra $\mathfrak{g}$. We write $\mathfrak{g} = \mathfrak{g}(a, b)$ when A has non-diagonal entries $-a$ and $-b$. To each such A, its Weyl group and corresponding root lattice, we associate a 'Fibonacci type' integer sequence. These sequences are derived from the coordinates of the real root vectors in the root space. Each element of each sequence can be expressed as a polynomial in the non-diagonal entries of the generalized Cartan matrix, whose coefficients are shallow diagonals of Pascal's triangle. Among the Fibonacci type sequences are the bisected Lucas and Fibonacci sequences, the Kekule numbers for the benzoids, the integers whose squares are triangular numbers, Chebyshev polynomials of the second kind, as well as some other known sequences. Each sequence has an associated 'hyperbolic golden ratio' Ψ which is a unit in a real quadratic field. We show that Ψ can be obtained as the limit of ratios of areas of hyperbolic triangles spanned by a triple of adjacent real roots in the root space. For the bisected Fibonacci sequence that occurs in our setting, Ψ coincides with the classical golden ratio $\psi = 1 + \phi$ for the bisection, where $\phi = 1.61803399\ldots$. It follows that ψ is an eigenvalue for the fundamental endomorphism $w_1 w_2$ in the root system of $\mathfrak{g}(5, 1)$, and ψ^2 is an eigenvalue for $w_1 w_2$ in $\mathfrak{g}(3, 3)$. This leads to an infinite family of identities involving continued fractions. We identify the generalized Lucas and Fibonacci sequences with Chebyshev polynomials of the 2nd kind, which in turn can be expressed as products. This leads us to an arithmetic proof that Kitchloo's 'generalized binomial coefficients' occuring in the cohomology of flag varieties of rank 2 Kac-Moody groups over $\mathbb{C}$ are integers.

MA

In: Frontiers of Combinatorics and Number Theory, Volume 2 ISBN: 978-1-61942-918-5
Editor: Zhi-Wei Sun, pp. 1-25

Chapter 1

PSEUDORANDOM BINARY FUNCTIONS ON ALMOST UNIFORM TREES

Katalin Gyarmati[1,*], Pascal Hubert[2,†] and András Sárközy[1,‡]
[1]Eötvös Loránd University
Department of Algebra and Number Theory
H-1117 Budapest, Pázmány Péter sétány 1/C, Hungary
[2] Laboratorie d'Analyse, Topologie et Probabilités
Faculté des Sciences de Saint Jérôme
Avenue Escadrille Normandie-Niemen
F-13397 Marseille Cedex 20, France

Abstract

First the notion of r-almost s-uniform tree is introduced which includes both the case of finitely generated free groups and uniform binary trees as special cases. The goal of the paper is to study pseudorandomness of binary functions defined on r-almost s-uniform trees. The measures of pseudorandomness of binary functions are introduced; the connection between these measures is analyzed; the size of these measures for truly random binary functions is studied; binary functions with strong pseudorandom properties are constructed.

2000 Mathematics Subject Classification: Primary: 05C05, Secondary: 11K45.

Keywords and phrases: pseudorandom, binary function, uniform tree, correlation, Legendre symbol.

1. Introduction

Recently in a series of papers a new constructive approach has been developed to study pseudorandomness of binary sequences

$$E_N = \{e_1, \dots, e_N\} \in \{-1, 1\}^N.$$

*E-mail address: gykati@cs.elte.hu
†E-mail address: hubert@cmi.univ-mrs.fr
‡E-mail address: sarkozy@cs.elte.hu

In particular, in [25] first the following measures of pseudorandomness are introduced: the *well-distribution measure* of E_N is defined by

$$W(E_N) = \max_{a,b,t} \left| \sum_{j=0}^{t-1} e_{a+jb} \right|$$

where the maximum is taken over all $a, b, t \in \mathbb{N}$ with $1 \le a \le a + (t-1)b \le N$, the *correlation measure of order* k of E_N is defined as

$$C_k(E_N) = \max_{M,\mathbf{D}} \left| \sum_{n=1}^{M} e_{n+d_1} \dots e_{n+d_k} \right|$$

where the maximum is taken over all $\mathbf{D} = (d_1, \dots, d_k)$ and M such that $0 \le d_1 < \dots < d_k \le N - M$, and $N_k(E_N)$, the *normality measure of order* k of E_N, is defined as

$$\max_{\mathbf{X} \in \{-1,1\}^k} \max_{0<M\le N+1-k} \left| |\{n : 0 \le n < M, \ (e_{n+1}, e_{n+2}, \dots, e_{n+k}) = \mathbf{X}\}| - \frac{M}{2^k} \right|.$$

Then the sequence E_N is considered to be a "good" pseudorandom sequence if both $W(E_N)$ and $C_k(E_N)$ (at least for "small" k) are "small" in terms of N; in particular, both are $o(N)$ as $N \longrightarrow \infty$ (they showed that the normality measures can be estimated in terms of the correlation measures). Indeed, later Cassaigne, Mauduit and Sárközy [5] proved that this terminology is justified since for almost all $E_N \in \{-1,1\}^N$, both $W(E_N)$ and $C_k(E_N)$ are less than $N^{1/2}(\log N)^c$; see also [3]. It was also shown in [25] that the Legendre symbol forms a "good" pseudorandom sequence. In some other paper further sequences were tested for pseudorandomness, further constructions were given for sequences with good pseudorandom properties, and the measures of pseudorandomness were studied (in the next sections we will present some further details of these results).

Later this theory of pseudorandomness of binary sequences has been extended in various directions: pseudorandomness of binary vectors [31], binary lattices [17], [18], [19], [20], [21], [22], [24], [27], [29], [30], subsets of $\{1, 2, \dots, n\}$ [6], [7], [8], subsets of $\mathbb{Z}_n$ [9], [10], sequences of k symbols [1], [4], [13], [26], [23], etc. have been studied. In this paper our goal is to continue this work by studying pseudorandomness of *binary functions on finite almost uniform trees*. More precisely, we will study finite r-almost s-uniform trees with $r \ge 2$, $s \ge 2$:

Definition 1. If $r, s \in \mathbb{N}$ and $r \ge 2, \ s \ge 2$, then a tree is called an r-almost, s-uniform tree if the root has r children and, except for the vertices in the last row, all the other vertices have s children.

If $r = s$ then the tree is called s-uniform tree, and in the $r = s = 2$ special case the tree is called uniform binary tree.

First we show a uniform binary tree:

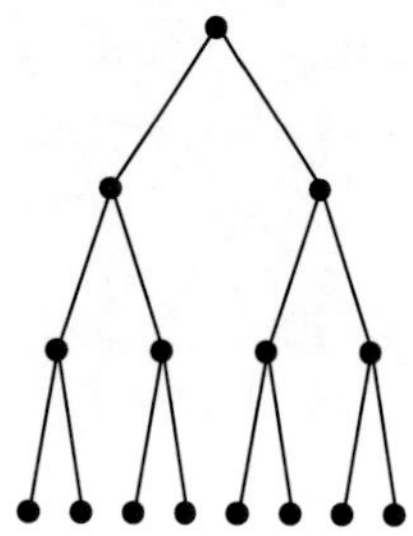

Figure 1.

Next we present a 4-almost 3-uniform tree:

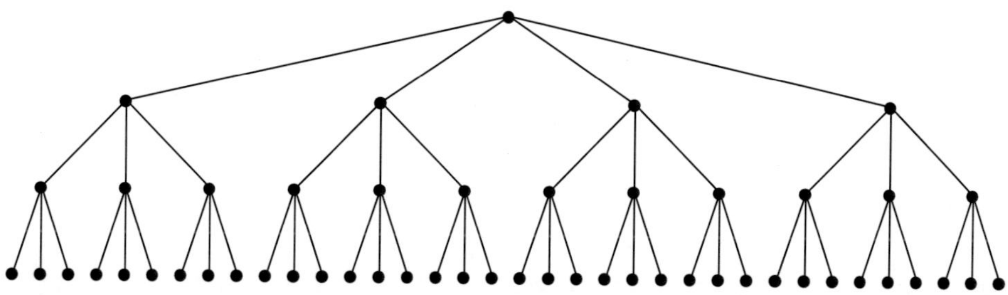

Figure 2.

These are the most important special cases: the uniform binary trees and the 4-almost 3-uniform trees. The importance of the uniform binary trees is clear, while the significance of the 4-almost 3-uniform trees is that there is a bijection between the k-letter reduced words of a free group on two generators and the paths (of length k) connecting the root of a 4-almost 3-uniform tree of "height" k (of $k+1$ rows) with one of the vertices in the last row. (See [34].) Indeed, this connection with the free groups is the reason of that we are also considering trees where the number of the children of the root is not necessarily the same as the common number of children of all other vertices.

Denote the set of the vertices of a tree T by $\mathcal{P}(T)$.

Definition 2. A *binary function on the tree* T is a function f of the type $f : \mathcal{P}(T) \to \{-1, 1\}$.

In this paper *our goal is to study pseudorandomness of binary functions defined on r-almost s-uniform trees.* We will introduce measures of pseudorandomness of binary functions of this type; we will analyze the connection between these measures; we will study the size of these measures for truly random binary functions; and we will construct binary functions with strong pseudorandom properties.

2. Notations, Terminology, Connection with Binary Sequences, the Measures of Pseudorandomness

Throughout this paper we will use the following notations:

Tree will always mean an r-almost s-uniform rooted tree for some r, s. We will use the words path, distance, height, subtree in the usual sense.

Definition 3. If T is an r-almost s-uniform tree, then a rooted subtree T' of T is called a proper subtree of T if either its root is the root of T and it is an r'-almost s-uniform tree for some $r' \leq r$, or its root is different from the root of T and it is an s-uniform tree.

We will use the word *row* in the following sense: the root forms the first row, the children of the root form the second row, and in general, the i-th row consists of the children of the vertices in the $i-1$-st row. Thus an r-almost s-uniform tree of height k has $k+1$ rows.

Consider an r-almost s-uniform tree, and draw it as in Figure 1 and Figure 2: the root is on the top, and the rows are drawn in horizontal lines. The j-th vertex from the left in the i-th row will be denoted by $P(i,j)$ (so that, starting from the left, the vertices in the i-th row are $P(i,1)$, $P(i,2)$, $P(i,3), \dots$).

From now on, we will formulate some basic facts in forms of *propositions*. It is an easy exercise to prove these facts (see [32], [35] for an approach of this type), thus we will not present the proofs.

Proposition 1. *The number of vertices in the i-th row of an r-almost s-uniform tree is 1 if* $i = 1$ *(then this single vertex is the root* $P(1,1)$*), and* rs^{i-2} *if* $i \geq 2$ *(then these vertices are denoted by* $P(i,1)$, $P(i,2), \dots, P(i, rs^{i-2})$*).*

Proposition 2. *The total number* $N = N(T)$ *of vertices of an r-almost s-uniform tree T of height* $k(\geq 1)$ *is*

$$N = N(T) = 1 + r + rs + rs^2 + \cdots + rs^{k-1} = 1 + r\frac{s^k - 1}{s-1}.$$

(The number of vertices of a tree T will be always denoted by $N = N(T)$.)

We will also use the following alternative notation for the vertices: The root is denoted by Q_1 : $Q_1 = P(1,1)$, the vertices in the second row by $Q_2, Q_3, \dots, Q_{r+1}$: $Q_2 = P(2,1)$, $Q_3 = P(2,2), \dots, Q_{r+1} = P(2,r)$; the vertices in the third row by $Q_{r+2}, Q_{r+3}, \dots, Q_{1+r(s+1)}$: $Q_{r+2} = P(3,1)$, $Q_{r+3} = P(3,2), \dots, Q_{1+r(s+1)} = P(3, rs)$, and so on; finally Q_N denotes the last vertex in the last row: $Q_N = P(k+1, rs^{k-1})$.

Proposition 3. *Define* $y(i)$, $y'(i)$ *so that* $Q_{y(i)} = P(i,1)$ *is the first vertex in the i-th row and* $Q_{y'}(i) = P(i, rs^{i-2})$ *is the last one. Then we have*

$$\begin{aligned}
y'(1) &= y(1) = 1 \\
y'(i) &= y(i) + rs^{i-2} - 1 \textit{ for } i > 1, \\
y'(i) &= y(i+1) - 1 \textit{ for } i \leq k, \\
y'(k+1) &= N, \\
y(i) &= 2 + r\frac{s^{i-2} - 1}{s-1} \textit{ for } i > 1, \\
y'(i) &= 1 + r\frac{s^{i-1} - 1}{s-1} \textit{ for } i > 1
\end{aligned}$$

and the j-th vertex in the i-th row is

$$P(i,j) = Q_{y(i)+j-1} = Q_{r\frac{s^{i-2}-1}{s-1}} \text{ for } i > 1.$$

To any binary function $f : \mathcal{P}(T) \to \{-1,1\}$ defined on an r-almost s-uniform tree T of height k one may assign the unique binary sequence

$$E_N = E_N(f,T) = (e_1, e_2, \ldots, e_N) \in \{-1,1\}^N \tag{2.1}$$

defined by

$$e_n = f(Q_n) \text{ for } n = 1, 2, \ldots, N. \tag{2.2}$$

Now we are ready to introduce the measures of pseudorandomness of binary functions defined on r-almost s-uniform trees T.

The most natural way of measuring the uniformity of the distribution of the binary function f over T relative to arithmetic progressions is to study the analogous property of $E_N(f,T)$.

Definition 4. The well-distribution measure of the binary function f over T is defined by

$$\overline{W}(f,T) = W(E_N(f,T)).$$

Note that k-ary functions on uniform trees (uniform trees coloured by k colours instead of using 2 colours as we do) and arithmetic progressions also appear in a paper of Furstenberg and Weiss [11]. However, there are two basic differences between their problem and the problem studied by us here. First, they consider arithmetic progressions along paths only, while we consider arithmetic progressions lying in (one or more) rows. Secondly, they are looking for the extreme case of monochromatic arithmetic progressions, while we are interested whether the distribution in arithmetic progression is "typical", "random type" or it is not.

The definition of correlation is not so simple. We propose the following definition:

Definition 5. For $k \geq 2$ and $\ell \geq 2$ the correlation measure $C_{k,\ell}(f,T)$ of height k and order ℓ of f over T is defined in the following way: consider ℓ different isomorphic proper subtrees $T_1, T_2, \ldots, T_\ell$ of height k of T, denote the set of their vertices by $\mathcal{P}_1, \mathcal{P}_2, \ldots, \mathcal{P}_\ell$, and for $t = 1, 2, \ldots, \ell$ let $\mathcal{P}_t = \{P_t(i,j) : i = 1, 2, \ldots, k+1,\ j = 1, 2, \ldots, q(i)\} = \{Q_{t,n} : n = 1, 2, \ldots, N(T_t)\}$ (note that both the number of vertices in the i-th row and $N(T_t)$ are independent of t by the isomorphism), and write

$$\begin{aligned} U(T_1, T_2, \ldots, T_\ell) &= \sum_{i=1}^{k+1} \sum_{j=1}^{q(i)} f(P_1(i,j)) f(P_2(i,j)) \cdots f(P_\ell(i,j)) \\ &= \sum_{n=1}^{N(T_t)} f(Q_{1,n}) f(Q_{2,n}) \ldots f(Q_{\ell,n}). \end{aligned}$$

Then

$$\overline{C}_{k,\ell}(f,T) = \max_{T_1, T_2, \ldots, T_\ell} |U(T_1, T_2, \ldots, T_\ell)|$$

where the maximum is taken over all ℓ-tuples $T_1, T_2, \ldots, T_\ell$ of proper subtrees of the type described above.

Definition 6. The universal correlation measure of order ℓ of f over T is defined by

$$\tilde{C}_\ell(f,T) = \max_k \overline{C}_{k,\ell}(f,T).$$

(The normality measure will be introduced and studied later.)

3. The Measures of f and the Associated Sequence E_N. The Measures of f in the Truly Random Case

We will prove the following theorem:

Theorem 1. *If T is an r-almost s-uniform tree of height K, f is a binary function on T, and $k,\ell \in \mathbb{N}$ $1 \le k \le K$ and $\ell \ge 2$, then*

$$\overline{C}_{k,\ell}(f,T) \le (k+1)C_\ell(E_N(f,T)). \tag{3.3}$$

Corollary 1. *For $\ell \ge 2$ we have*

$$\tilde{C}_\ell(f,T) < \left(\frac{\log N/r}{\log s} + 2\right) C_\ell(E_N(f,T)). \tag{3.4}$$

Proof of Theorem 1. Defining the isomorphic proper subtrees $T_1, T_2, \ldots, T_\ell$ as in Definition 5, we have

$$U(T_1,T_2,\ldots,T_\ell) = \sum_{i=1}^{k+1}\sum_{j=1}^{q(i)} f(P_1(i,j))f(P_2(i,j))\ldots f(P_\ell(i,j)). \tag{3.5}$$

For $1 \le t \le \ell$, let e_{m_t} denote the element of E_N assigned to $P_t(i,1)$. Then $m_1, m_2, \ldots, m_\ell$ are distinct, since $T_1, T_2, \ldots, T_\ell$ are distinct; we may assume that $m_1 < m_2 < \cdots < m_\ell$. Define d_t by $d_t = m_t - 1$ for $t = 1, 2, \ldots, \ell$. Then by Proposition 3 the absolute value of the inner sum in (3.5) is

$$\begin{aligned} \left|\sum_{j=1}^{q(i)} f(P_1(i,j))f(P_2(i,j))\ldots f(P_\ell(i,j))\right| &= \left|\sum_{j=1}^{q(i)} e_{m_1+j-1}e_{m_2+j-1}\cdots e_{m_\ell+j-1}\right| \\ &= \left|\sum_{j=1}^{q(i)} e_{j+d_1}e_{j+d_2}\cdots e_{j+d_\ell}\right| \\ &\le C_\ell(E_N(f,t)). \end{aligned} \tag{3.6}$$

It follows from (3.5) and (3.6) that

$$U(T_1,T_2,\ldots,T_\ell) \le (k+1)C_\ell(E_N(f,T))$$

which proves (3.3). □

Proof of Corollary 1. By Proposition 2 we have

$$N > rs^{K-1}$$

whence

$$K < \frac{\log n/r}{\log s} + 1$$

and using this, (3.4) follows from (3.3). □

It was shown in [5] (by using Weil's theorem [36], that for almost all $E_N \in \{-1,1\}^N$, both $W(E_N)$ and $C_\ell(E_N)$ are around $N^{1/2}$ (in a slightly sharper form); these results were sharpened in [3] where it was proved that for every $\ell \in \mathbb{N}$, $\varepsilon > 0$ there is a $\delta > 0$ such that for large N we have

$$P\left(\delta N^{1/2} < W(E_N) < \frac{1}{\delta} N^{1/2}\right) > 1 - \varepsilon \tag{3.7}$$

and

$$P\left(\delta\,(\ell N \log N)^{1/2} < C_\ell(E_N) < \frac{1}{\delta}(\ell N \log N)^{1/2}\right) > 1 - \varepsilon. \tag{3.8}$$

It follows from these results, Definition 4 and Corollary 1 that for every $\ell \in \mathbb{N}$, $\varepsilon > 0$ there is a $\delta > 0$ such for a large N and more than $(1-\varepsilon)2^N$ binary functions $f : \mathcal{P}(T) \to \{-1,1\}$ (where again $N = N(T)$) we have

$$\delta N^{1/2} < W(f,T) < \frac{1}{\delta} N^{1/2}$$

and

$$\tilde{C}_\ell(f,T) < \frac{1}{\delta}(\ell N)^{1/2}(\log N)^{3/2}. \tag{3.9}$$

On the other hand, it does not follow from the earlier results that for more than $(1-\varepsilon)2^N$ binary functions $f : \mathcal{P}(T) \to \{-1,1\}$ we have

$$\tilde{C}_\ell(f,T) > \delta N^{1/2}.$$

Thus we will prove

Theorem 2. *For every $\varepsilon > 0$, $\ell \in \mathbb{N}$, $\ell \geq 2$, $r,s \in \mathbb{N}$, $r,s \geq 2$ there are numbers $\delta = \delta(\varepsilon,\ell,r,s)$ and $N_0 = N_0(\varepsilon,\ell,r,s)$ such that if T is an r-almost s-uniform tree with $N = N(T) > N_0$, then for more than $(1-\varepsilon)2^N$ binary functions $f : \mathcal{P}(T) \to \{-1,1\}$ we have*

$$\delta N^{1/2} < \tilde{C}_\ell(f,T). \tag{3.10}$$

(Note that there is a logarithm power gap between the lower bound (3.10) and upper bound (3.9). One might like to close this gap; it is not clear which one of the two bounds is closer to the truth.)

Proof of Theorem 2. Denote the height of T by k, and let $f(Q_1) = f(P(1,1)) = e_1$, $f(Q_2) = f(P(2,1)) = e_2, \dots, f(Q_N) = f(P(k+1, rs^{k-1})) = e_N$ be independent random variables of distribution

$$P(e_i = +1) = P(e_i = -1) = \frac{1}{2}. \tag{3.11}$$

We have to show that

$$P\left(\delta N^{1/2} < \tilde{C}_\ell(f,T)\right) > 1 - \varepsilon \tag{3.12}$$

(for δ small enough and N large enough in terms of ε, ℓ, r and s).

Now define the integer i so that the i-th row is the first row of T which contains at least ℓ vertices. In other words, we have

$$i = 2, \quad \text{if } \ell \le r \tag{3.13}$$

and i is defined by

$$rs^{i-3} < \ell \le rs^{i-2} \quad \text{if } \ell > r. \tag{3.14}$$

Let $T_1, T_2, \dots, T_\ell$ denote the s-uniform subtrees of T whose roots are $P(i,1), P(i,2), \dots, P(i,\ell)$ (the first ℓ vertices of the i-th row of T) and whose height is $k - i + 1$ (so that the vertices in their last rows belong to the last row of T). Moreover, for $1 \le t \le \ell$ let T_t' denote the s-uniform subtree of T that we get from T_t by dropping its last row (so that the root of T_t' is $P(i,t)$ and its height is $k - i$). Finally, define m by

$$Q_{m+1} = P(k+1, 1)$$

(in other words, the last vertex in the k-th row of T is Q_m).

First we select the values of $e_1, e_2, \dots, e_m$ and $e_{m+s^{k-i+1}+1}, e_{m+s^{k-i+1}+2}, \dots, e_N$ (each of them are selected according to the law (3.11)). We will show that independently of the choice of these e_j's, the remaining e_j's, i.e.,

$$e_{m+1}, e_{m+2}, \dots, e_{m+s^{k-i+1}} \tag{3.15}$$

can be selected with probability $> 1 - \varepsilon$ so that the event

$$\delta N^{1/2} < \tilde{C}_\ell(f,T) \tag{3.16}$$

holds.

For $j = 1, 2, \dots, s^{k-i+1}$, write

$$g_j = \prod_{t=1}^{\ell-1} f\left(P\left(k+1, j + ts^{k-i+1}\right)\right) \left(= \prod_{t=1}^{\ell} e_{m+j+ts^{k-i+1}}\right),$$

and define h_j by

$$h_j = e_{m+j} g_j \quad \left(\text{for } j = 1, 2, \dots, s^{k-i+1}\right). \tag{3.17}$$

For every fixed choice of the e_j's not in (3.15) i.e., for any $g_1, g_2, \dots, g_{s^{k-i+1}} \in \{-1, 1\}$, the h_j's in (3.17), together with the e_{m+j}'s are distributed according to the law in (3.11) so that

$$P(h_j = -1) = P(h_j = +1) = \frac{1}{2} \quad \left(\text{for } j = 1, 2, \dots, s^{k-i+1}\right). \tag{3.18}$$

Write $S = h_1 + h_2 + \cdots + h_{s^{k-i+1}}$. By (3.18), the expectation of S is 0, its standard deviation is $\frac{1}{2}\left(s^{k-i+1}\right)^{1/2}$, thus by the central limit theorem the limit distribution of $S/\frac{1}{2}\left(s^{k-i+1}\right)^{1/2}$ for $s^{k-i+1} \to \infty$ (which follows from $N \to \infty$) is Gaussian distribution. It follows that if N is large enough and $\delta(>0)$ is small enough in terms of ε, ℓ, r and s, then the probability of the event

$$\frac{|S|}{\frac{1}{2}\left(s^{k-i+1}\right)^{1/2}} > 8\delta(\ell s)^{1/2} \tag{3.19}$$

is greater, than $1-\varepsilon$ uniformly for any fixed choice of the e_j's not in (3.15), so that for a random choice of the binary function $f : \mathcal{P}(T) \to \{-1, 1\}$ we have

$$P\left(\frac{|S|}{\frac{1}{2}\left(s^{k-i+1}\right)^{1/2}} > 8\delta(\ell s)^{1/2}\right) > 1 - \varepsilon.$$

Thus in order to prove (3.12), it suffices to show that the event (3.16) follows from the event (3.19).

Assume that (3.19) holds. For $t = 1, 2, \ldots, \ell$, denote the vertices of the subtrees T_t and T'_t by $P_t(r,j)$ (=the j-th vertex in the r's row) and $P'_t(r,j)$, respectively (so that $P_t(r,j) = P'_t(r,j)$ for $r = 1, 2, \ldots, k-i+1$, $j = 1, 2, \ldots, s^{r-1}$). Then by (3.14),(3.19) and Proposition 2, we have

$$\left|U(T_1, T_2, \ldots, T_\ell) - U(T'_1, T'_2, \ldots, T'_\ell)\right| = \left|\sum_{r=1}^{k-i+2}\sum_{j=1}^{s^{r-1}} f(P_1(r,j))f(P_2(r,j))\ldots\right.$$

$$\left. f(P_\ell(r,j)) - \sum_{r=1}^{k-i+1}\sum_{j=1}^{s^{r-1}} f(P'_1(r,j))f(P'_2(r,j))\ldots f(P'_\ell(r,j))\right|$$

$$= \left|\sum_{j=1}^{s^{k-i+1}} f(P_1(k-i+2,j))f(P_2(k-i+2,j))\ldots f(P_\ell(k-i+2,j))\right|$$

$$= \left|\sum_{j=1}^{s^{k-i+1}} f(P(k+1,j))f(P(k+1,j+s^{k-i+1}))\ldots\right.$$

$$\left. f(P_\ell(k+1, j+(\ell-1)s^{k-i+1}))\right| = \left|\sum_{j=1}^{s^{k-i+1}} e_{m+j}g_j\right| = |S|$$

$$> 4\delta\left(\ell s^{k-i+2}\right)^{1/2} > 4\delta\left(rs^{i-3}s^{k-i+2}\right)^{1/2} = 4\delta(rs^{k-1})^{1/2} = 2\delta\left(2r\frac{s^k}{s/2}\right)^{1/2}$$

$$> 2\delta\left(2r\frac{s^k-1}{s-1}\right)^{1/2} \geq 2\delta N^{1/2}.$$

It follows from this inequality that

$$\max\{U(T_1, T_2, \ldots, T_\ell), U(T'_1, T'_2, \ldots, T'_\ell)\} \geq \delta N^{1/2}$$

whence by Definitions 5 and 6,

$$\tilde{C}_\ell(f,T) = \max_j \overline{C}_{j,\ell}(f,T) \geq \max\{\overline{C}_{k-i+1,\ell}(f,T), \overline{C}_{k-i,\ell}(f,T)\}$$
$$\geq \max\{U(T_1,T_2,\ldots,T_\ell), U(T_1',T_2',\ldots,T_\ell')\} \geq \delta N^{1/2}$$

which proves (3.16) (assuming (3.19)) and thus also (3.12). □

Based on these facts, we may say that a binary function $f: \mathcal{P}(T) \to \{-1,1\}$ is a "good pseudorandom binary function", "it possesses strong pseudorandom properties" if $\overline{W}(f,T)$ and $\tilde{C}_\ell(f,T)$ (at least for small ℓ) are small in terms of $N = N(T)$; they must be $o(N)$, and ideally they are $O(N^{1/2+\varepsilon})$.

Note that if T is an r-almost s-uniform tree with $N = N(T)$, then any "good" pseudorandom binary sequence $E_N = (e_1, e_2, \ldots, e_N) \in \{-1,1\}^N$ induces a "good" pseudorandom binary function $f: \mathcal{P}(T) \to \{-1,1\}$. Namely, if we define f by $f(Q_n) = e_n$ (for $n = 1,2,\ldots,N$) and E_N possesses good pseudorandom properties, then by Definition 4 and Corollary 1 the binary function f also possesses good pseudorandom properties.

Referring to the statement of the last paragraph, the referee of this paper asked the following question: "What can be said about the converse?" The answer to this question is that the converse of this statement is not true: it may occur that the binary function $f: \mathcal{P}(T) \to \{-1,1\}$ possesses good pseudorandom properties but the associated binary sequence $E_N(f,T)$ does not; more precisely, it may occur that both $\overline{W}(f,T)(= W(E_N(f,T)))$ and $\tilde{C}_\ell(f,T)$ (for almost all "small" ℓ) are "small"$(= o\left(N^{1/2+\varepsilon}\right))$, however $C_2(E_N(f,T))$ is "large" ($\gg N$). This can be shown by the following example:

Example 1. Consider the uniform binary tree T of height k, so that it has $N = N(T) = 2^{k+1} - 1$ vertices, and its vertices in the last row are $Q_{2^k} = P(k+1,1)$, $Q_{2^k+1} = P(k+1,2), \ldots, Q_N = Q_{2^{k+1}-1} = P(k+1,2^k)$. Define the binary function $f: \mathcal{P}(T) \to \{-1,1\}$ so that $f(Q_1) = f(P(1,1)) = e_1$ $(\in \{-1,1\})$, $f(Q_2) = f(P(2,1)) = e_2$ $(\in \{-1,1\}), \ldots, f\left(Q_{2^k+2^{k-1}}\right) = f(P(k+1,2^{k-1}+1)) = e_{2^k+2^{k-1}}$ $(\in \{-1,1\})$ in a truly random way, and let

$$f\left(Q_{2^k+2^{k-1}+i}\right) = f(P(k+1,2^{k-1}+1+i)) = e_{2^k+2^{k-1}+i} = e_{2^k-1+i}$$
$$= f(P(k+1,i)) = f\left(Q_{2^k+i-1}\right).$$

Then we have

$$C_2(E_N(f,T)) \geq \left|\sum_{i=1}^{2^{k-1}-1} e_{2^k-1+i}e_{2^k+2^{k-1}+i}\right| = \left|\sum_{i=1}^{2^{k-1}-1} 1\right|$$
$$= 2^{k-1} - 1 \gg 2^{k+1} - 1 = N.$$

On the other hand, it is easy to see that with probability $> 1-\varepsilon$ we have $\overline{W}(f,T) \ll N^{1/2}(\log N)^c$ for some $c > 0$, and it could be shown with a little work that for every fixed ℓ with probability $> 1-\varepsilon$ we also have $\overline{C}_\ell(f,T) \ll N^{1/2}(\log N)^c$ for some $c = c(\ell) > 0$.

(Observe that for this f and e.g., $\ell = 2$ the estimate of the sum in Definition 5 can be reduced to $O(\log N)$ sums of form

$$\sum_{n=m}^{M} e_n e_{n+d}$$

where $1 \leq m < m + M \leq 2^k + 2^{k-1}$, and by the structure of the construction we have $d \neq 0$.)

Thus, e.g., we may construct a "good" pseudorandom binary function f on a given almost uniform tree T by using the Legendre symbol. Let $N = N(T)$, and let p denote the smallest odd prime with $N < p$ so that by Tchebycheff's theorem we have $p \leq 2N$. Then consider the Legendre symbol sequence $E_{p-1} = (e_1, e_2, \dots, e_{p-1})$ defined by

$$e_n = \left(\frac{n}{p}\right) \quad \text{for } n = 1, 2, \dots, p-1. \tag{3.20}$$

It was shown in [25] that for this binary sequence E_{p-1} we have

$$W(E_{p-1}) \ll p^{1/2} \log p \tag{3.21}$$

and

$$C_\ell(E_{p-1}) \ll \ell p^{1/2} \log p \tag{3.22}$$

(in [25] slightly different notation was used). Then consider the binary function

$$\lambda : \mathcal{P}(T) \to \{-1, 1\} \tag{3.23}$$

associated with the truncated binary sequence $E_N = (e_1, e_2, \dots, e_N)$. It follows from definition 4, Corollary 1, (3.21) and (3.22) that for this binary function we have

$$\overline{W}(\lambda, T) = W(E_N(\lambda, T)) \ll p^{1/2} \log p \ll N^{1/2} \log N$$

(recall that $p < N < 2p$) and

$$\tilde{C}_\ell(\lambda, T) \ll (\log N) C_\ell(E_N(\lambda, T)) \ll (\log N) \ell p^{1/2} \log p \ll \ell N^{1/2} (\log N)^2. \tag{3.24}$$

By Corollary 1, the correlation of order ℓ of a binary function f can be estimated from above in terms of the correlation of order ℓ of the associated sequence E_N, i.e., if $C_\ell(E_N(f,T))$ is small, then $\tilde{C}_\ell(f,T)$ also must be small. One might like to know whether the converse of this is also true, i.e., if $\tilde{C}_\ell(f,T)$ is small, then $C_\ell(E_N(f,T))$ also must be small? We will show that the answer to this question is negative. To simplify the discussion we will restrict ourselves to uniform binary trees and to the special case $\ell = 2$ but the proof could be extended to the general case.

We will start out from a slight modification of the Legendre symbol construction above. Consider a uniform binary tree T, write $N = N(T)$, and let p denote the smallest prime p with $4N < p$ (so that now p is slightly greater than above). Then we have

$$4N < p < 8N. \tag{3.25}$$

Define the Legendre symbol sequence (3.20) and the binary function λ in (3.23) as above. Note that (3.24) also holds for this modified construction (only the implicit constant changes).

Let T' denote the uniform binary tree obtained from T by adding one more row: Denote the height of T' by H so that $N = N(T) = 2^H - 1$ and, writing $N' = N(T')$, we have $N' = 2^{H+1} - 1$. Let $\lambda' : \mathcal{P}(T') \to \{-1, 1\}$ denote the extension of λ from T to T' defined by

$$\lambda'(P(i,j)) = \lambda(P(i,j)) \text{ for } 1 \le i \le H,\ 1 \le j \le 2^{i-1}$$

and

$$\lambda'(P(H+1, 2j-1)) = \lambda'(P(H+1, 2j)) = \lambda(P(H,j)) \text{ for } 1 \le j \le 2^{H-1}$$

(so that λ' assumes the same value on the vertices of the last row of T as on their children). Then the binary sequences $E_N = E_N(\lambda, T)$ and $E_{N'} = E_{N'}(\lambda', T')$ are

$$E_N = (e_1, e_2, \dots, e_N) = \left(\left(\frac{1}{p}\right)\left(\frac{2}{p}\right), \dots, \left(\frac{N}{p}\right)\right) \tag{3.26}$$

and

$$E_{N'} = (e_1, e_2, \dots, e_{N'}) = \left(\left(\frac{1}{p}\right)\left(\frac{2}{p}\right), \dots, \left(\frac{N}{p}\right), \left(\frac{2^{H-1}}{p}\right), \left(\frac{2^{H-1}}{p}\right),\right.$$
$$\left.\left(\frac{2^{H-1}+1}{p}\right), \left(\frac{2^{H-1}+1}{p}\right), \dots, \left(\frac{2^H-1}{p}\right), \left(\frac{2^H-1}{p}\right)\right). \tag{3.27}$$

Theorem 3. *Defining T', f' and N' in this way we have*

$$\tilde{C}_2(\lambda', T') \ll N^{1/2}(\log N)^2 \tag{3.28}$$

and

$$C_2(E_{N'}(\lambda', T')) \gg N'. \tag{3.29}$$

Proof of Theorem 3. In order to prove (3.28) we have to show that defining T_1, T_2 and $U(T_1, T_2)$ as in Definition 5 with T', λ' and 2 in place of T, f and ℓ, the inequality

$$\left|\sum_{i=1}^{k+1}\sum_{j=1}^{q(i)} \lambda'(P_1(i,j))\lambda'(P_2(i,j))\right| \ll N^{1/2}(\log N)^2 \tag{3.30}$$

holds. To prove this we have to distinguish three cases.

CASE 1. T_1, T_2 are subtrees of T, i.e., no vertex of the last row of T' is among the vertices of T_1 and T_2. Then by the construction we have

$$\left|\sum_{i=1}^{k+1}\sum_{j=1}^{q(i)} \lambda'(P_1(i,j))\lambda'(P_2(i,j))\right| = \left|\sum_{i=1}^{k+1}\sum_{j=1}^{q(i)} \lambda(P_1(i,j))\lambda(P_2(i,j))\right| \le \tilde{C}_2(\lambda, T)$$

whence (3.30) follows by (3.24).

CASE 2. Assume that one of T_1 and T_2, say T_1 is a subtree of T, but the last row of the other one, T_2 consists of consecutive elements of the last row of T'. Then we have

$$\left|\sum_{i=1}^{k+1}\sum_{j=1}^{q(i)}\lambda'(P_1(i,j))\lambda'(P_2(i,j))\right| \le \left|\sum_{i=1}^{k}\sum_{j=1}^{q(i)}\lambda'(P_1(i,j))\lambda'(P_2(i,j))\right| + \left|\sum_{j=1}^{q(k+1)}\lambda'(P_1(k+1,j))\lambda'(P_2(k+1,j))\right|. \tag{3.31}$$

Here the first double sum can be estimated as the one in Case 1:

$$\left|\sum_{i=1}^{k+1}\sum_{j=1}^{q(i)}\lambda'(P_1(i,j))\lambda'(P_2(i,j))\right| \le \tilde{C}_2(\lambda,T) \ll N^{1/2}(\log N)^2. \tag{3.32}$$

By (3.26), (3.27) and our assumption on T_1 and T_2, the last sum is of the form

$$\left|\sum_{j=1}^{q(k+1)}\left(\frac{a+j}{p}\right)\lambda'(P'(H+1,2b+j))\right| = \left|\sum_{u=1}^{\frac{q(k+1)}{2}}\left(\frac{a+2u}{p}\right)\left(\frac{d+u}{p}\right) + \sum_{u=1}^{\frac{q(k+1)}{2}}\left(\frac{a+2u-1}{p}\right)\left(\frac{d+u}{p}\right)\right| \le \left|\sum_{u=1}^{\frac{q(k+1)}{2}}\left(\frac{(a2^{-1}+u)(d+u)}{p}\right)\right| + \left|\sum_{u=1}^{\frac{q(k+1)}{2}}\left(\frac{((a-1)2^{-1}+u)(d+u)}{p}\right)\right| \tag{3.33}$$

where a, d are integers with

$$0 \le a \le N - q(k+1) = 2^H - 1 - q(k+1) \tag{3.34}$$
$$2^{H-1} - 1 \le d \le N' - 1 = 2^{H+1} - 1 \tag{3.35}$$

(and 2^{-1} denotes the multiplicative inverse of 2 modulo p).

Now we need the following well-known consequence of Weil's theorem [36]:

Lemma 1. *If p is a prime, χ is a non-principal character modulo p of order D, $g(x) \in \mathbb{F}_p[x]$ is not of the form $a(h(x))^d$ with $a \in \mathbb{F}_p$, $h(x) \in \mathbb{F}_p[x]$, $y \in \mathbb{Z}$, $z \in \mathbb{N}$ and $z \le p$, then we have*

$$\left|\sum_{n=y}^{y+z}\chi(g(n))\right| \le 10tp^{1/2}\log p$$

where t denotes the number of distinct zeros of $g(x)$.

(See e.g. [12], [25], [33] for different versions of this lemma.)

To use this lemma (with $\chi(n) = \left(\frac{n}{p}\right)$) for estimating the sums in (3.33), we have to show that the polynomials of u there are not perfect squares, i.e.,

$$a2^{-1} \not\equiv d \pmod p \text{ and } (a-1)2^{-1} \not\equiv d \pmod p$$

or, in equivalent form,

$$a \not\equiv 2d \pmod p \text{ and } a-1 \not\equiv 2d \pmod p. \tag{3.36}$$

By (3.25), (3.34) and (3.35) we have

$$-1 \le a-1 < a \le 2^H - 1 - q(k+1) \tag{3.37}$$

and

$$2^H - 2 = 2\left(2^{H-1}-1\right) \le 2d \le N' - 1 = 2N+1 \le 4N-1 < p-1. \tag{3.38}$$

The case $q(k+1) = 1$ is trivial (then the upper bound in (3.33) is $O(1)$). Thus we may assume that $q(k+1) > 1$. Then by (3.37) and (3.38) we have

$$-1 \le a-1 < a < 2d < p-1$$

and (3.36) follows from this. Thus the lemma can be applied, and then we obtain from (3.33) that

$$\left|\sum_{j=1}^{q(k+1)} \lambda'(P_1(k+1,j))\lambda'(P_2(k+1,j))\right| \ll p^{1/2}\log p \ll N^{1/2}\log N \tag{3.39}$$

(which also holds for $q(k+1) = 1$).

(3.30) follows from (3.32) and (3.39).

CASE 3. Assume that the last row of both T_1 and T_2 consists of consecutive elements of the last row of T'. Then as in (3.31) and (3.32) in Case 2, we obtain

$$\left|\sum_{i=1}^{k+1}\sum_{j=1}^{q(i)} \lambda'(P_1(i,j))\lambda'(P_2(i,j))\right| \ll N^{1/2}(\log N)^2 + \left|\sum_{j=1}^{q(k+1)} \lambda_1'(P_1(k+1,j))\lambda_2'(P_2(k+1,j))\right|. \tag{3.40}$$

The last term can be rewritten as

$$\left|\sum_{u=1}^{\frac{q(k+1)}{2}} \left(\frac{a+u}{p}\right)\left(\frac{b+u}{p}\right)\right| = \left|\sum_{u=1}^{\frac{q(k+1)}{2}} \left(\frac{(a+u)(b+u)}{p}\right)\right|$$

with some integers a, b such that

$$0 \le a < b < p.$$

Then again we may apply Lemma 1 to estimate this sum, and we obtain

$$\left|\sum_{j=1}^{q(k+1)} \lambda_1'(P_1(k+1,j))\lambda_2'(P_2(k+1,j))\right| \ll p^{1/2}\log p \ll N^{1/2}\log N. \tag{3.41}$$

(3.30) follows from (3.40) and (3.41), and this completes the proof of (3.28).

Now we will prove (3.29). $e_1, e_2, \ldots, e_{N'}$ are defined as in (3.27), then

$$\begin{aligned} C_2(E_{N'}(\lambda', T')) &\geq \left| \sum_{n=N+1}^{N'-1} e_n e_{n+1} \right| \\ &= \left| \sum_{\substack{N+1 \leq n \leq N'-1 \\ 2|n}} e_n e_{n+1} + \sum_{\substack{N+1 \leq n \leq N'-1 \\ 2\nmid n}} e_n e_{n+1} \right| \\ &= \left| \sum_{\substack{N+1 \leq n \leq N'-1 \\ 2|n}} 1 + \sum_{m=2^{H-1}}^{2^H-2} \left(\frac{m}{p}\right)\left(\frac{m+1}{p}\right) \right| \\ &= \left| \frac{N'-N}{2} + \sum_{m=2^{H-1}}^{2^H-2} \left(\frac{m(m+1)}{p}\right) \right|. \end{aligned} \tag{3.42}$$

Since $N = \frac{1}{2}N' + O(1)$ and, by (3.25) and Lemma 1,

$$\left| \sum_{m=2^{H-1}}^{2^H-2} \left(\frac{m(m+1)}{p}\right) \right| \ll p^{1/2} \log p \ll N^{1/2} \log N,$$

it follows from (3.42) that

$$C_2(E_{N'}(\lambda', T')) \gg N$$

which completes the proof of Theorem 3. □

Numerous other properties of the measures of pseudorandomness of binary sequences have been studied [3], [2], [5], [14], [15], [16], [28]. In particular, $\min_{E_N \in \{-1,1\}^N} W(E_N)$ and $\min_{E_N \in \{-1,1\}^N} C_k(E_N)$ have been estimated; an inequality between $W(E_N)$ and $C_2(E_N)$ (and later $C_k(E_N)$) has been proved; for fixed k and ℓ, the connection between $C_k(E_N)$ and $C_\ell(E_N)$ has been analyzed; an inequality between $C_2(E_N)$ and $C_3(E_N)$ has been proved. One might like to look for analogues of these results for binary functions. However, these problems seem to be much more difficult for binary functions than for binary sequences, and we have not been able to settle them. The constructions, resp. the crucial ideas of the proofs which were used in the case of binary sequences fail in the case of binary functions completely. Thus we just do not know what to expect in the latter case. Before formulating any conjectures in this directions, first one has to answer the following basic questions (that we have not been able to settle either): Is it true, that if r, s are fixed integers, $2k+1$ is a fixed odd integer, and $N = N(T) \longrightarrow \infty$, then $\min \tilde{C}_{2k+1}(f, T) = O(1)$ where the minimum is taken over all binary functions defined on r-almost s-uniform trees of N vertices? Is it true, that it follows from $\tilde{C}_2(f, T) = o(N)$ that $\overline{W}(f, T) = o(N)$?

4. The Normality Measure

So far we have introduced and studied the well-distribution and correlation measures of binary functions defined on r-almost s-uniform trees. In this section we will introduce and study the *normality measure of order* k of binary functions of this type. A simple way of defining the normality measure of order k of the binary function $f : \mathcal{P}(T) \to \{-1, 1\}$ would be to define it in the same manner as the well-distribution measure (Definition 4), i.e., by

$$\overline{N}_k(f, T) = N_k(E_N(f, T)).$$

However, this seems to be a rather artificial definition of not much use. Instead, we propose to use the following more natural definition:

Definition 7. The normality measure $\overline{N}_k(f, T)$ of order k ($k \in \mathbb{N}$, $k \geq 2$) of the binary function f over the r-almost s-uniform tree T is defined in the following way: Let τ_k denote the set of uniform binary subtrees of height k of T. If $G_{2^{k+1}-1} = (g_1, g_2, \dots, g_{2^{k+1}-1}) \in \{-1, 1\}^{2^{k+1}-1}$, then let $\phi(f, T, G_{2^{k+1}-1})$ denote the number of the subtrees $T' \in \tau_k$ such that the binary sequence $E_{2^{k+1}-1} = E_{2^{k+1}-1}(f, T')$ assigned to the binary function $f : \mathcal{P}(T') \to \{-1, 1\}$ (restricted to T' and defined after Proposition 2) is the given $2^{k+1} - 1$ tuple $G_{2^{k+1}-1}$:

$$\phi(f, T, G_{2^{k+1}-1}) = \left|\{T' : T' \in \tau_k,\ E_{2^{k+1}-1}(f, T') = G_{2^{k+1}-1}\}\right|.$$

Then define $\overline{N}_k(f, T)$ by

$$\overline{N}_k(f, T) = \max_{G_{2^{k+1}-1} \in \{-1,1\}^{2^{k+1}-1}} \left|\phi\left(f, T, G_{2^{k+1}-1}) - \frac{|\tau_k|}{2^{2^{k+1}-1}}\right)\right|.$$

(So that $\overline{N}_k(f, T)$ is defined as the maximal deviation between $\phi(f, T, G_{2^{k+1}-1})$ and its expected value for all the possible choices of $G_{2^{k+1}-1}$.)

As (3.8) shows, for fixed $k, N \to \infty$ and a truly random binary sequence $E_N \in \{-1, 1\}^N$ one has

$$C_\ell(E_N) = O(N^{1/2+\varepsilon}). \tag{4.43}$$

It was proved in [25] (see Proposition 1) that for all N, E_N and $k < N$ we have

$$N_k(E_N) \leq \max_{1 \leq t \leq k} |C_t(E_N)|. \tag{4.44}$$

Combining (4.43) and (4.44) we get that for fixed k, $N \to \infty$ and a truly random $E_N \in \{-1, 1\}^N$ we have

$$N_k(E_N) = O(N^{1/2+\varepsilon}).$$

One may expect that the measure $\overline{N}_k(f, T)$ possesses an analogous property: if k, r, s are fixed and we consider r-almost s-uniform trees T with $N(T) \to \infty$, then for a truly random binary function f over T one has

$$\overline{N}_k(f, T) = O\left(N(T)^{1/2+\varepsilon}\right). \tag{4.45}$$

This is probably true but, unfortunately, we have not been able to prove it. The difficulty is that the above argument cannot be adopted to binary functions over trees: namely, there is no inequality of type (4.44) in other words, it may occur that $\tilde{C}_t(f,T)$ is small for all $t \le k$, however, $\overline{N}_k(f,T)$ is large. Indeed, we will show that this is the case for the especially important Legendre symbol construction studied in (3.20) and (3.23) in Section 3. As (3.24) shows, for this construction $\tilde{C}_t(\lambda,T)$ is small for every fixed t, thus it remains to show that $\overline{N}_k(\lambda,T)$ is large (for fixed k). To simplify the discussion we will restrict ourselves to uniform binary trees.

Theorem 4. *Let $k, K \in \mathbb{N}$, $k \le K$, let T denote uniform binary tree of height K, and define the binary function $\lambda : \mathcal{P}(T) \to \{-1,1\}$ by* (3.20) *and* (3.23). *Then we have*

$$\overline{N}_k(\lambda,T) > \frac{2^{2^{k+1}-1-k}-1}{2^{2^{k+1}-1}}\left(2^{K-k+1}-1\right). \tag{4.46}$$

Note that for fixed k and $K \to \infty$ this implies that

$$\overline{N}_k(\lambda,T) \gg N(T)$$

(where the implicit constant factor depends on k).

Proof of Theorem 4. First we will compute $|\tau_k|$. Denote the height of T by K so that by Proposition 2 we have

$$N = N(T) = 2^{K+1}-1. \tag{4.47}$$

A subtree $T' \in \tau_k$ is uniquely determined by its root, and the root must belong to one of the first $K-k+1$ rows of T. Again by Proposition 2, the total number of vertices of these rows is 2^{K-k+1} so that

$$|\tau_k| = 2^{K-k+1}-1. \tag{4.48}$$

Let $\mathcal{G}$ denote the set of the $2^{k+1}-1$ tuples $G_{2^{k+1}-1}$ such that $\phi(\lambda,T,G_{2^{k+1}-1}) > 0$, i.e., there is at least one $T' \in \tau_k$ with

$$E_{2^{k+1}-1}(\lambda,T') = (e'_1, e'_2, \dots, e'_{2^{k+1}-1}) = G_{2^{k+1}-1}. \tag{4.49}$$

Now we will give an upper bound for $|\mathcal{G}|$. Assume that $G_{2^{k+1}-1} = (g_1, g_2, g_3, \dots, g_{2^{k+1}-1}) \in \mathcal{G}$, and there is a $T' \in \tau_k$ satisfying (4.49). We will show that

$$g_1, g_3, g_5, \dots, g_{2^{k+1}-1} \tag{4.50}$$

uniquely determine

$$g_2, g_4, g_6, \dots, g_{2^{k+1}-2}.$$

Indeed, assume that we are looking for $g_{2\ell}$ with $\ell \in \{1,2,\dots,2^k-1\}$. It follows from (4.44) that

$$e'_m = g_m \text{ for } m = 1,2,\dots,2^{k+1}-1 \tag{4.51}$$

and, in particular,

$$e'_{2\ell} = g_{2\ell}. \tag{4.52}$$

Now consider a uniform binary tree $\tilde{T}$. Denote its j-th vertex in the i-th row by $\tilde{P}(i,j)$. We will say that $\tilde{P}(i,j)$ is an even vertex if j is even, and it is said to be an odd vertex if j is odd. Note that if $\tilde{P}(i,j)$ is an odd vertex of $\tilde{T}$, and it is also a vertex of a binary uniform subtree $\tilde{T}'$ of $\tilde{T}$, then $\tilde{P}(i,j)$ is also an odd vertex in $\tilde{T}'$. It follows from Proposition 3 that if a binary function $f : \mathcal{P}(\tilde{T}) \to \{-1,1\}$ is given, and $\tilde{E_N} = \tilde{E_N}(f,\tilde{T}) = (\tilde{e}_1, \tilde{e}_2, \dots, \tilde{e}_N)$ (with $N = N(\tilde{T})$) is the binary sequence associated with it (in the sense described after Proposition 2), then we have

$$f\left(\tilde{P}(i,j)\right) = \tilde{e}_{2^{i-1}+j-1}. \tag{4.53}$$

It follows from this that the vertex $\tilde{P}(i,j)$ is odd if and only if the subscript of the associated $\tilde{e}_n$ is even:

$$f\left(\tilde{P}(i,j)\right) = \tilde{e}_{2q} \text{ (iff } \tilde{P}(i,j) \text{ is odd).} \tag{4.54}$$

Let us consider an odd vertex $P(i, 2u-1)$ of T which is not the root; this is the "left" child of the vertex $P(i-1,u)$. Then by (3.20) and (4.53) we have

$$\begin{aligned}\lambda(P(i,2u-1)) = e_{2^{i-1}+2u-2} &= \left(\frac{2^{i-1}+2u-2}{p}\right) = \left(\frac{2}{p}\right)\left(\frac{2^{i-2}+u-1}{p}\right) \\ &= \left(\frac{2}{p}\right) e_{2^{i-2}+u-1} = \left(\frac{2}{p}\right)\lambda(P(i-1,u)).\end{aligned} \tag{4.55}$$

Now we return to (4.52); then it follows from the discussion above that $e'_{2\ell}$ is associated with an odd vertex $P'(i,j)$ of T'. Let us consider the directed path starting from $P'(i,j)$ and ending with the root $P'(1,1)$ of T'. The root $P'(1,1)$ is an even vertex; let $P'(i,x,v)$ be the first even vertex along this path. Denote the subscript of the element of $E_{2^{k+1}-1}(\lambda, T')$ associated with it by $2t-1$ (it must be odd by (4.54)) so that, by (4.51),

$$\lambda(P'(i-x,v)) = e'_{2t-1} = g_{2t-1}. \tag{4.56}$$

Then using (4.55) repeatedly, we obtain from (4.56) that

$$g_{2\ell} = e'_{2\ell} = \lambda\left(P'(i,j)\right) = \left(\frac{2}{p}\right)^x \lambda\left(P'(i-x,v)\right) = \left(\frac{2}{p}\right)^x g_{2t-1}$$

so that, indeed, $g_{2\ell}$ is uniquely determined by g_{2t-1} (note that p is fixed, and x is determined by ℓ).

The sequence (4.50) can be selected in at most 2^k ways (since we have $g_i \in \{-1,1\}$ for all i) so that

$$|\mathcal{G}| \le 2^k. \tag{4.57}$$

It follows from (4.57) by the pigeon hole principle that there is at least one $G_{2^{k+1}-1}$ such that

$$\phi(\lambda, T, G_{2^{k+1}-1}) \ge \frac{|\mathcal{T}_k|}{|G|} \ge \frac{|\mathcal{T}_k|}{2^k}$$

whence, by (4.48),

$$\phi(\lambda, T, G_{2^{k+1}-1}) - \frac{|\tau_k|}{2^{2^{k+1}-1}} \geq \frac{|\tau_k|}{2^k} - \frac{|\tau_k|}{2^{2^{k+1}-1}} = \frac{2^{2^{k+1}-1-k}-1}{2^{2^{k+1}-1}}|\tau_k|$$
$$= \frac{2^{2^{k+1}-1-k}-1}{2^{2^{k+1}-1}}\left(2^{K-k+1}-1\right)$$

which proves (4.46). □

As we mentioned earlier, we conjecture that for truly random binary function f defined on a fixed r-almost s-uniform tree the normality measure of it is small (more precisely, (4.45) holds) but we have not been able to prove this. Thus one might like to present at least one example for a binary function with small normality measure. In case of binary sequences the Legendre symbol construction (3.20) is known to possess the best pseudorandom properties (and in some other situations also the Legendre symbol provides the best examples). However, as Theorem 4 shows here, somewhat unexpectedly, the most natural Legendre symbol construction defined by (3.20) and (3.23) fails, its normality measure is large. On the other hand, we will present another, slightly more complicated Legendre symbol construction where the normality measure (and the other pseudorandom measures as well) are small:

Theorem 5. *Let p be a prime such that $p > 3$ and 2 is primitive root mod p. Define the positive integer K by*

$$2^{K+1} \leq p < 2^{K+2}, \tag{4.58}$$

and let T denote the uniform binary tree of height K. Let c be a quadratic non-residue modulo p, and define the binary function $\rho : \mathcal{P}(T) \to \{-1, 1\}$ (where $\mathcal{P}(T) = \{Q_1, Q_2, \dots, Q_N\}$) by

$$\rho(Q_n) = \left(\frac{n^2 - c}{p}\right) \quad \text{for } n = 1, 2, \dots, N = N(T) = 2^{K+1} - 1 (\leq p - 1) \tag{4.59}$$

(note that $p \nmid n^2 - c$ since c is quadratic non-residue modulo p, thus $\left(\frac{n^2-c}{p}\right)$ is defined). Then we have

$$\overline{N}_k(\rho, T) \leq 20\left(2^{k-1} - 1\right)p^{1/2}\log p (\ll 2^k N^{1/2} \log N) \tag{4.60}$$

for every $k \in \mathbb{N}$ with $k \leq K$.

Note that it also follows from the results in [12], Definition 4 and Corollary 1 that $\overline{W}(\rho, T)$ and $\tilde{C}(\rho, T)$ are also small. Moreover, we remark that it could be shown that if (4.59) is replaced by

$$\rho(Q_n) = \left(\frac{n - c}{p}\right)$$

where $|c| = O(1)$ and $\ell \to +\infty$ slowly, then the normality measure of order ℓ is large.

Proof of Theorem 5. We will use the notations of Section 1 and Definition 7. For

$$G_{2^{k+1}-1} = (g_1, g_2, \dots, g_{2^{k+1}-1}) \in \{-1, 1\}^{2^{k+1}-1},$$

we have to estimate $\phi(\rho, T, G_{2^{k+1}-1})$, i.e., the number of the uniform binary subtrees $T' \in \tau_k$ such that

$$\begin{aligned} E_{2^{k+1}-1}(\rho, T') &= (e'_1, e'_2, \dots, e'_{2^{k+1}-1}) = (\rho(Q'_1), \rho(Q'_2), \dots, \rho(Q'_{2^{k+1}-1})) \\ &= G_{2^{k+1}-1} = (g_1, g_2, \dots, g_{2^{k+1}-1}) \end{aligned} \tag{4.61}$$

where $\mathcal{P}(T') = \{Q'_1, Q'_2, \dots, Q'_{2^{k+1}-1}\}$ is the vertex set of T'. To simplify the notation, write $2^{k+1} - 1 = H$. Then (4.61) holds if and only if

$$e'_t = \rho(Q'_t) = g_t \text{ for } 1 \le t \le H. \tag{4.62}$$

Now consider a subtree

$$T' \in \tau_k$$

whose root is the vertex Q_n of T so that

$$Q'_1 = Q_n, \tag{4.63}$$

and, by (4.48) here the possible values of n are

$$n \in \{1, 2, \dots, |\tau_k|\} = \{1, 2, \dots, 2^{K-k+1} - 1\}. \tag{4.64}$$

Since T and T' are uniform binary trees, it follows from (4.63) by Proposition 3 that the vertices of T' are

$$\begin{aligned} (Q'_1, Q'_2, Q'_3, Q'_4, \dots, Q'_H) &= (Q_n, Q_{2n}, Q_{2n+1}, Q_{4n}, \dots, Q_{2^k n + 2^k - 1}) \\ &= (Q_{p_1(n)}, Q_{p_2(n)}, Q_{p_3(n)}, Q_{p_4(n)}, \dots, Q_{p_H(n)},) \end{aligned}$$

where $p_i(x)$ denotes the i-th polynomial in the sequence $x, 2x, 2x+1, 4x, \dots, 2^k x + 2^{k-1}$. Then by (4.59), (4.62) can be rewritten as

$$e_{t'} = \rho(Q'_t) = \rho(Q_{p_t(n)}) = \left(\frac{p_t(n)^2 - c}{p}\right) = g_t \text{ for } 1 \le t \le H. \tag{4.65}$$

Clearly for fixed T', i.e., for a fixed n satisfying (4.64) we have

$$\frac{1}{2^H} \prod_{t=1}^{H} (e_{t'} g_t + 1) = \frac{1}{2^H} \prod_{t=1}^{H} \left(\left(\frac{p_t(n)^2 - c}{p}\right) g_t + 1 \right) = \begin{cases} 1 & \text{if (4.65) holds,} \\ 0 & \text{otherwise.} \end{cases}$$

Since n may run over the integers in (4.64), thus writing $M = |\mathcal{T}_k| = 2^{K-k+1} - 1$ we have

$$\left|\phi(\rho, T, G_{2^{k+1}-1}) - \frac{|\mathcal{T}_k|}{2^{2^{k+1}-1}}\right| = \left|\frac{1}{2^H}\sum_{n=1}^{M}\prod_{t=1}^{H}\left(\left(\frac{p_t(n)^2-c}{p}\right)g_t+1\right)-\frac{M}{2^H}\right|$$

$$= \left|\frac{1}{2^H}\sum_{\ell=1}^{H}\sum_{1\le t_1<\cdots<t_\ell\le H}\sum_{n=1}^{M-1}\left(\frac{p_{t_1}(n)^2-c}{p}\right)g_{t_1}\cdots\left(\frac{p_{t_\ell}(n)^2-c}{p}\right)g_{t_\ell}\right|$$

$$= \left|\frac{1}{2^H}\sum_{\ell=1}^{H}\sum_{1\le t_1<\cdots<t_\ell\le H} g_{t_1}\cdots g_{t_\ell}\sum_{n=1}^{M-1}\left(\frac{(p_{t_1}(n)^2-c_1)\dots(p_{t_\ell}(n)^2-c)}{p}\right)\right|$$

$$\le \frac{1}{2^H}\sum_{\ell=1}^{H}\sum_{1\le t_1<\cdots<t_\ell\le H}|g_{t_1}\cdots g_{t_\ell}|\left|\sum_{n=1}^{M-1}\left(\frac{(p_{t_1}(n)^2-c_1)\dots(p_{t_\ell}(n)^2-c)}{p}\right)\right|$$

$$= \frac{1}{2^H}\sum_{\ell=1}^{H}\sum_{1\le t_1<\cdots<t_\ell\le H}\left|\sum_{n=1}^{M-1}\left(\frac{(p_{t_1}(n)^2-c_1)\dots(p_{t_\ell}(n)^2-c)}{p}\right)\right|. \tag{4.66}$$

Now we will apply Lemma 1 for estimating the innermost sum. To be able to use this lemma with $\chi(n) = \left(\frac{n}{p}\right)$, we have to show that here a typical polynomial

$$f(n) = (p_{t_1}(n)^2 - c_1)\dots(p_{t_\ell}(n)^2 - c) \text{ with } 1 \le t_1 < \cdots < t_\ell \le H(< p) \tag{4.67}$$

is not of the form $a(g(n))^2$ with $a \in \mathbb{F}_p$, $g(x) \in \mathbb{F}_p[x]$.

The polynomial $n^2 - c$ is irreducible over $\mathbb{F}_p$ since c is a quadratic non-residue modulo p. Thus for $A \neq 0$ a polynomial of form $(An + B)^2 - c$ is also irreducible. It follows that the polynomials $p_i(n)^2 - c$ are irreducible. Then writing $p_i(n)^2 - c$ as constant times a monic polynomial:

$$p_i(n)^2 - c = C_i q_i(n)$$

(where the coefficient of the highest degree term of $q_i(n)$ is 1), the polynomials $q_i(n)$ are irreducible. Now we will show:

Lemma 2. *For $1 \le i < j \le H$ the polynomials $q_i(n), q_j(n)$ are different.*

Proof of Lemma 2. Let

$$p_i(n) = 2^u n + v \text{ (with } 1 \le u \le k < p,\ 0 \le v < 2^u),$$
$$p_j(n) = 2^y n + z \text{ (with } 1 \le y \le k < p,\ 0 \le z < 2^y),$$

so that

$$p_i(n)^2 - c = 2^{2u}\left(n^2 + 2^{1-u}vn + 2^{-2u}(v^2 - c)\right) = 2^{2u}q_i(n)$$

and

$$p_j(n)^2 - c = 2^{2y}\left(n^2 + 2^{1-y}zn + 2^{-2y}(z^2 - c)\right) = 2^{2y}q_j(n)$$

If $u = y$ then, by $i \neq j$, we must have $v \neq z$, so that clearly the coefficients of n are different in $q_i(n)$ and $q_j(n)$, thus $q_i(n)$ and $q_j(n)$ are different.

If $u \neq y$, then we may assume that $u < y$. Write $y - u = s\ (> 0)$ so that

$$2^s < 2^y \leq 2^k \leq 2^K \leq \frac{1}{2}p. \tag{4.68}$$

Then q_j can be rewritten as

$$q_j(n) = n^2 + 2^{1-u-s}zn + 2^{-2u-2s}(z^2 - c).$$

Assume that contrary to the conclusion of the lemma we have $q_i(n) = q_j(n)$. Then comparing the coefficients of n in $q_i(n)$ and $q_j(n)$ we get

$$2^{1-u}v \equiv 2^{1-u-s}z \pmod p, \tag{4.69}$$

$$2^{-2u}(v^2 - c) \equiv 2^{-2u-2s}(z^2 - c) \pmod p. \tag{4.70}$$

It follows from (4.69) that

$$z \equiv 2^s v \pmod p. \tag{4.71}$$

By (4.70) and (4.71) we have

$$2^{2s}(v^2 - c) \equiv z^2 - c \equiv 2^{2s}v^2 - c \pmod p$$

whence

$$2^{2s}c \equiv c \pmod p,$$

$$2^{2s} \equiv 1 \pmod p.$$

Since 2 is a primitive root modulo p, it follows that $p - 1 \mid 2s$ so that by $s > 0$, we have

$$p - 1 \leq 2s$$

$$\frac{p-1}{2} \leq s$$

which contradicts (4.68), and this completes the proof of the lemma. □

The polynomial $f(n)$ in (4.67) can be written as

$$f(n) = C_1 C_2 \dots C_\ell q_1(n) q_2(n) \dots q_\ell(n)$$

where $C_1 C_2 \dots C_\ell$ is constant and, by Lemma 2, $q_1(n), q_2(n), \dots, q_\ell(n)$ are different irreducible polynomials. Then clearly, $f(n)$ cannot be of the form $a(g(n))^2$ so that, indeed, we may apply Lemma 1 to estimate the innermost sum in (4.66). Then we get from (4.66) that

$$\left|\phi(\rho, T, G_{2^{k+1}-1}) - \frac{|\tau_k|}{2^{2^{k+1}-1}}\right| \leq \frac{1}{2^H} \sum_{\ell=1}^{H} \sum_{1 \leq t_1 < \dots < t_\ell \leq H} 10(2\ell)p^{1/2} \log p$$
$$< 20Hp^{1/2} \log p = 20(2^{k-1} - 1)p^{1/2} \log p$$

which proves (4.60). □

Acknowledgment

We would like to thank the (anonymous) referee for a couple of good questions and for an important reference.

Research partially supported by Hungarian National Foundation for Scientific Research, Grants No. K67676, K72731 and PD72264, French-Hungarian exchange program F-06/48, and the János Bolyai Research Fellowship.

References

[1] R. Ahlswede, C. Mauduit and A. Sárközy, Large families of pseudorandom sequences of k symbols and their complexity, I, II, in: General Theory of Information Transfer and Combinatorics (eds., R. Ahlswede et al.), *Lecture Notes in Comput. Sci.*, Vol. 4123, Springer, Berlin, 2006, pp. 293–307 and 308–325.

[2] N. Alon, Y. Kohayakawa, C. Mauduit, C. G. Moreira and V. Rödl, Measures of pseudorandomness for finite sequences: minimal values, *Combin. Probab. Comput.* 15 (2006), 1–29.

[3] N. Alon, Y. Kohayakawa, C. Mauduit, C. G. Moreira and V. Rödl, Measures of pseudorandomness for finite sequences: typical values, *Proc. London Math. Soc.* 95 (2007), 778–812.

[4] G. Bérczi, On finite pseudorandom binary sequences of k symbols, *Period. Math. Hungar.* 47 (2003), 29–44.

[5] J. Cassaigne, C. Mauduit and A. Sárközy, On finite pseudorandom binary sequences VII: The measures of pseudorandomness, *Acta Arith.* 103 (2002), 97–118.

[6] C. Dartyge, E. Mosaki and A. Sárközy, On large families of subsets of the set of the integers not exceeding N, *Ramanujan J.* 18 (2009), 209–229.

[7] C. Dartyge and A. Sárközy, Large families of pseudorandom subsets formed by power residues, *Unif. Distrib. Theory* 2 (2007), 73–88.

[8] C. Dartyge and A. Sárközy, On pseudorandom subsets of the set of integers not exceeding N, *Period. Math. Hungar.* 54 (2007), 183–200.

[9] C. Dartyge and A. Sárközy, On pseudo-random subsets of $\mathbb{Z}_n$, *Monatsh. Math.,* to appear.

[10] C. Dartyge, A. Sárközy and M. Szalay, *On the pseudo-randomness of subsets related to primitive roots*, Combinatorica, submitted.

[11] H. Furstenberg and B. Weiss, Markov processes and Ramsey theory for trees, Special issue on Ramsey theory, *Combin. Probab. Comput.* 12 (2003), 547–563.

[12] L. Goubin, C. Mauduit and A. Sárközy, Construction of large families of pseudorandom binary sequences, *J. Number Theory* 106 (2004), 56–69.

[13] E. Grant, J. Shalit and T. Stoll, Bounds for the discrete correlation of infinite sequences on k symbols and generalized Rudin-Shapiro sequences, *Acta Arith.,* to appear.

[14] K. Gyarmati, An inequality between the measures of pseudorandomness, *Ann. Univ. Sci. Budapest. Eötvös Sect. Math.* 46 (2003), 157–166.

[15] K. Gyarmati, On a fast version of a pseudorandom generator, in: General Theory of Information Transfer and Combinatorics (eds., R. Ahlswede et al.), *Lecture Notes in Comput. Sci.,* Vol. 4123, Springer, Berlin, 2006, pp. 326–342.

[16] K. Gyarmati, On the correlation of binary sequences, *Studia Sci. Math. Hungar.* 42 (2005), 59–75.

[17] K. Gyarmati, C. Mauduit and A. Sárközy, Constructions of pseudorandom binary lattices, *Unif. Distrib. Theory,* submitted.

[18] K. Gyarmati, A. Sárközy and C. L. Stewart, On Legendre symbol lattices, *Unif. Distrib. Theory,* to appear.

[19] P. Hubert, C. Mauduit and A. Sárközy, On pseudorandom binary lattices, *Acta Arith.* 125 (2006), 51–62.

[20] H. Liu, A large family of pseudorandom binary lattices, *Proc. Amer. Math. Soc.* 137 (2009), 793–803.

[21] H. Liu, T. Zhan and X. Wang, Large families of elliptic curve pseudorandom binary sequences, *Acta Arith.,* to appear.

[22] H. Liu, T. Zhan and X. Wang, On the correlation of pseudorandom binary sequences with composite moduli, *Publ. Math. Debrecen* 74 (2009), 195-214.

[23] R. Marzouk and A. Winterhof, On the pseudorandomness of binary and quaternary sequences linked by the Gray mapping, *Period. Math. Hungar.*, to appear.

[24] C. Mauduit and A. Sárközy, Construction of pseudorandom binary lattices by using the multiplicative inverse, *Monatsh. Math.* 153 (2008), 217–231.

[25] C. Mauduit and A. Sárközy, On finite pseudorandom binary sequences I: Measure of pseudorandomness, the Legendre symbol, *Acta Arith.* 82 (1997), 365–377.

[26] C. Mauduit and A. Sárközy, On finite pseudorandom binary sequences of k symbols, *Indag. Math.* 13 (2002), 89-101.

[27] C. Mauduit and A. Sárközy, On large families of pseudorandom binary lattices, *Unif. Distrib. Theory* 2 (2007), 23–37.

[28] C. Mauduit and A. Sárközy, On the measures of pseudorandomness of binary sequences, *Discrete Math.* 271 (2003), 195–207.

[29] L. Mérai, Construction of pseudorandom binary lattices based on multiplicative characters, *Period. Math. Hungar.*, to appear.

[30] L. Mérai, On finite pseudorandom lattices of k symbols, *Monatsh. Math.,* submitted.

[31] H. Niederreiter, J. Rivat and A. Sárközy, Pseudorandom sequences of binary vectors, *Acta Arith.* 133 (2008), 109–125.

[32] B. R. Preiss, *Data Structures and Algorithms with Object Oriented Design Patterns in Java*, Wiley, 1999.

[33] W. M. Schmidt, Equations over Finite Fields: An Elementary Approach, *Lecture Notes in Math.*, Vol. 536, Springer, Berlin, 1976.

[34] J.-P. Serre, Trees, *Springer Monographs in Math.,* 2nd ed., Springer, Berlin, 2003.

[35] J. A. Storer, *An Introduction to Data Structures and Algorithms*, Springer, Berlin, 2002.

[36] A. Weil, Sur les courbes algébriques et les variétés qui s'en déduisent, *Act. Sci. Ind.*, Vol. 1041, Hermann, Paris, 1948.

In: Frontiers of Combinatorics and Number Theory, Volume 2 ISBN: 978-1-61942-918-5
Editor: Zhi-Wei Sun, pp. 27-39

Chapter 2

COMPOSITES THAT REMAIN COMPOSITE AFTER CHANGING A DIGIT

***Michael Filaseta*[1]*, Mark Kozek*[2]*, Charles Nicol*[1] *and John Selfridge*[3]**
[1]Mathematics Department, University of South Carolina
Columbia, SC 29208-0001
[2] Department of Mathematics, Whittier College
Whittier, CA 90608-0634
[3] Department of Mathematical Sciences, Northern Illinois University
DeKalb, IL 60115-2888

Abstract

We show that there are infinitely many composite numbers N, relatively prime to 10, with the property that if any one of its digits is replaced by an arbitrary digit, then the resulting number is composite. We also show that there are infinitely many composite numbers N, relatively prime to 10, having the property that if an arbitrary digit is inserted once into the decimal representation of N, then the resulting number is composite.

2010 Mathematics Subject Classification: 11B25 (11A07, 11A63).

Keywords: covering, covering system

1. Introduction

We begin with a simple example. Let m be a positive integer. Let

$$N = (11 \cdot 13 \cdot 17 \cdot 19 \cdot 10)\, m + 15.$$

The right-most digit of N is 5. Thus, N is composite, and if we replace any digit of N, other than the right-most digit, with a different digit, then the new number obtained will be divisible by 5 and, hence, be composite. On the other hand, if we replace the right-most digit of N by a different digit, then we obtain one of $N \pm j$ where $j \in \{1, 2, 3, 4, 5\}$ and each of these is easily seen to be composite. Thus, N has the property that if we replace any one of its digits with an arbitrary digit, the number we obtain is composite.

Constructing examples of N having the above property is simplified by the fact that changing digits (arbitrarily many) of a natural number N, other than the right-most digit, results in a number divisible by $\gcd(N, 10)$. Thus, if we allow $\gcd(N, 10) \neq 1$, then we can easily construct examples of N with the property that replacing any one of its digits with an arbitrary digit results in a composite number.

Our goal is to address here the more difficult issue of constructing such N but with the restriction $\gcd(N, 10) = 1$. For example, the number $N = 212159$ has this property. In other words, every number in the set,

$$\{d12159, 2d2159, 21d159, 212d59, 2121d9, 21215d : d \in \{0, 1, 2, \ldots, 9\}\}$$

is composite. In fact, it can be checked that 212159 is the smallest composite natural number, coprime to 10, such that if we replace any one of the digits of its decimal expansion with a $d \in \{0, \ldots, 9\}$, then the number created by this replacement is composite. We prove the following:

Theorem 1. *There are infinitely many composite natural numbers N, coprime to 10, with the property that if we replace any digit in the decimal expansion of N with $d \in \{0, \ldots, 9\}$, then the number created by this replacement is composite.*

After establishing Theorem 1, we will examine a similar problem in which the *insertion* of any one digit into a composite natural number N results in a number that is composite. As in the previous paragraph, we are interested in N that are coprime to 10. For example, the number $N = 25011$ has this property. In other words, every number in the set,

$$\{d25011, 2d5011, 25d011, 250d11, 2501d1, 25011d : d \in \{0, 1, 2, \ldots, 9\}\}$$

is composite. The number 25011 is the smallest composite natural number, coprime to 10, such that if you insert any digit $d \in \{0, \ldots, 9\}$ anywhere in its decimal expansion, then the number created by this insertion is composite. Analogous to Theorem 1, we establish the following:

Theorem 2. *There are infinitely many composite natural numbers N, coprime to 10, with the property that if you insert any digit $d \in \{0, \ldots, 9\}$ anywhere in the decimal expansion of N, then the number created by this insertion is composite.*

Another problem that could be considered along similar lines is whether there are infinitely many composite numbers such that when you remove any one digit, then the number remains composite. This problem is easier, however, as can be seen by considering a number consisting of a string of 3's. Further, a string of 1's of the form $(10^{15k+10} - 1)/9$, where k is an arbitrary positive integer, provides a further slightly less obvious example.

Our approach to establishing the above results is based on determining appropriate covering systems, or coverings, of the integers. We define coverings in the next section, but mention here the references [2]-[11] as a sampling of the many works connected to this subject.

We end the introduction with some open questions related to this paper. We were able to obtain analogous results to Theorem 1 and Theorem 2 for all bases $b < 12$, but do analogous results hold for all bases? Do there exist infinitely many composite numbers

N satisfying the conditions of both Theorem 1 and Theorem 2? Are there infinitely many *primes* p that are composite for every replacement (or insertion) of a digit? Does there exist a k_0 such that for every positive integer $k \geq k_0$ there is a composite number N with exactly k digits having the property of Theorem 1 (or Theorem 2)? Do there exist infinitely many composite numbers N that remain composite when any two digits (not necessarily consecutive) are changed (or inserted)? Based on heuristics, we conjecture that the answers to all of the above questions except the last is in the affirmative.

Acknowledgment: The first two authors note that Theorem 1 was established some years ago by the latter two authors. As it never appeared in print, the first two authors reconstructed a proof and included Theorem 2 which, as we will see, follows rather easily from the argument for Theorem 1.

2. Proof of Theorem 1

We give a construction of an infinite sequence of composite natural numbers N satisfying the conditions in the theorem. We write the decimal expansion of N as

$$N = d_{n-1}d_{n-2}\ldots d_1d_0, \quad d_i \in \{0,\ldots,9\}, \quad n \geq 1, \quad d_{n-1} \neq 0.$$

We also express N in the form

$$N = \frac{10^n - 1}{9} + M,$$

where M is a fixed natural number to be determined and n is a large natural number. Observe that our last expression for N is equivalent to

$$N = \underbrace{11\ldots11}_{n\text{-many 1's}} + M$$

where the number $11\ldots11$ is a string of n digits that are 1. To establish the theorem, we will find an infinite arithmetic progression of natural numbers n, a fixed natural number M, and a finite set of primes $\mathcal{P}$, such that when we replace any digit in the decimal expansion of N with $x \in \{0,\ldots,9\}$, then the number created by this replacement is divisible by at least one of the primes in $\mathcal{P}$. We suppose throughout that n is large enough to imply that the left-most digit of N is 1.

Let $N^{(k)}(x)$ be the number that is obtained by replacing the digit d_k of the decimal expansion of N with $x \in \{0,\ldots,9\}$. For example, if $N = 212159 = d_5d_4d_3d_2d_1d_0$, then $N^{(2)}(3) = 212359$. Since M is fixed, there exists a non-negative integer $K \leq n-1$ such that $d_k = 1$ for all $k \geq K$. We first consider these large values of k. For these k, we have that $N^{(k)}(x)$ takes the form

$$N^{(k)}(x) = \frac{10^n - 1}{9} + M + (x-1)\cdot 10^k.$$

Observe that if

$$\begin{aligned} n &\equiv 0 \pmod 3 \\ M &\equiv 1 \pmod 3 \\ x &\equiv 0 \pmod 3, \end{aligned}$$

then $N^{(k)}(x)$ is divisible by 3 since, in this case,

$$N^{(k)}(x) = \frac{10^n - 1}{9} + M + (x-1) \cdot 10^k$$
$$\equiv 0 + 1 + (-1) \pmod 3.$$

Thus, the conditions $n \equiv 0 \pmod 3$ and $M \equiv 1 \pmod 3$ imply that for large k if we replace d_k with $x \in \{0, 3, 6, 9\}$, then we obtain a composite number $N^{(k)}(x)$.

Similarly, we observe that if

$$\begin{aligned} n &\equiv 0 \pmod 6 \\ M &\equiv 0 \pmod 7 \\ x &\equiv 1 \pmod 7, \end{aligned}$$

then $N^{(k)}(x)$ is divisible by 7. Thus, for large k we know that our original N is necessarily composite and that if we replace d_k by $x = 8$, then we obtain the composite number $N^{(k)}(x)$.

Combining the above, we see that if

$$n \equiv 0 \pmod 6 \quad \text{and} \quad M \equiv 7 \pmod{21}$$

and if, for an arbitrary $k \geq K$, we replace d_k in N with any digit in $\{0, 1, 3, 6, 8, 9\}$, then the number $N^{(k)}(x)$ obtained is divisible by either 3 or 7.

To address the remaining cases of $N^{(k)}(x)$ for $x \in \{2, 4, 5, 7\}$ when k is large, we will use the following.

Definition. A finite system of congruences $x \equiv a_i \pmod{m_i}$, $1 \leq i \leq t$, is called a *covering of the integers* if each integer satisfies at least one congruence in the system.

As an example, we note

$$x \equiv 0 \pmod 2$$
$$x \equiv 1 \pmod 3$$
$$x \equiv 3 \pmod 4$$
$$x \equiv 5 \pmod 6$$
$$x \equiv 9 \pmod{12}$$

is a covering of the integers. In the way of notation, for a prime p and an integer a not divisible by p, we use $\mathrm{ord}_p(a)$ to denote the order of a modulo p, that is the least positive integer k such that $a^k \equiv 1 \pmod p$.

Lemma 1. *Let N and M be natural numbers such that*

$$N = \frac{10^n - 1}{9} + M,$$

where N has the decimal expansion

$$N = d_{n-1}d_{n-2}\ldots d_1 d_0, \quad d_i \in \{0, \ldots, 9\}, \quad n \geq 1, \quad d_{n-1} \neq 0.$$

Let K be a non-negative integer $\leq n-1$ such that $d_k = 1$ for $k \in \{K, K+1, \dots, n-1\}$. For a fixed $x \in \{0, \dots 9\}$, suppose we have distinct primes $p_1, \dots, p_t$, each > 5, for which

(i) *there exists a covering of the integers*

$$k \equiv b_i \pmod{c_i}, \quad 1 \leq i \leq t,$$

where $c_i = \text{ord}_{p_i}(10)$,

(ii) $n \equiv 0 \pmod{\text{lcm}(c_1, \dots, c_t)}$,

(iii) *M is a solution to the system of congruences*

$$M \equiv -(x-1) \cdot 10^{b_i} \pmod{p_i}, \quad 1 \leq i \leq t.$$

Then, for each $k \in \{K, K+1, \dots, n-1\}$,

$$N^{(k)}(x) = \frac{10^n - 1}{9} + M + (x-1) \cdot 10^k$$

is divisible by at least one of the primes p_i where $1 \leq i \leq t$.

Proof. Suppose the conditions in the lemma hold. Let $k \in \{K, K+1, \dots, n-1\}$. By (i), there is an $i \in \{1, \dots, t\}$ such that $k \equiv b_i \pmod{c_i}$. Since $c_i = \text{ord}_{p_i}(10)$ and $n \equiv 0 \pmod{\text{lcm}(c_1, \dots, c_t)}$, we have $p_i \mid (10^n - 1)/9$. Hence,

$$N^{(k)}(x) = \frac{10^n - 1}{9} + M + (x-1) \cdot 10^k \equiv M + (x-1) \cdot 10^{b_i} \pmod{p_i}.$$

From (iii), we deduce $N^{(k)}(x) \equiv 0 \pmod{p_i}$, completing the proof. □

Set

$$\mathcal{P}_0 = \mathcal{P}_3 = \mathcal{P}_6 = \mathcal{P}_9 = \{3\}, \quad \mathcal{P}_1 = \mathcal{P}_8 = \{7\}.$$

For $x \in \{2, 4, 5, 7\}$, let $\mathcal{P}_x$ denote a set of primes, if it exists, as in Lemma 1 so that $N^{(k)}(x)$ is divisible by some prime from $\mathcal{P}_x$ for each $k \in \{K, K+1, \dots, n-1\}$ provided (ii) and (iii) hold for some covering as in (i). For primes $p \notin \{2, 5\}$, we define $c(p) = \text{ord}_p(10)$. For $x \in \{2, 4, 5, 7\}$, we write the covering system in (i) as

$$k \equiv b(x, p) \pmod{c(p)}, \quad p \in \mathcal{P}_x. \tag{1}$$

To establish Theorem 1, we will take n in the definition of N so that it is divisible by the least common multiple of the numbers in the set

$$\bigcup_{x=0}^{9} \{c(p) : p \in \mathcal{P}_x\}.$$

We will also take M so that the various congruences

$$\begin{aligned} &M \equiv 7 \pmod{21} \\ &M \equiv -(x-1) \cdot 10^{b(x,p)} \pmod{p}, \quad \text{where } x \in \{2, 4, 5, 7\} \text{ and } p \in \mathcal{P}_x, \end{aligned} \tag{2}$$

simultaneously hold. It is clear that we can find infinitely many n as above, but we need our choice of coverings as in (1) and prime sets $\mathcal{P}_x$ to be such that (2) has a solution. We justify that this is possible with explicit sets $\mathcal{P}_x$ and coverings as in (1).

To obtain a set of primes $\mathcal{P}_x$ and a covering as in (1), we make use of the tables of factorizations of $10^c - 1$ in [1] to determine, for a given positive integer c, the set of primes p such that $c = \text{ord}_p(10)$. For the purposes of finding a covering as in (1), each such p corresponds to a congruence of the form $k \equiv b \pmod{c}$ that we can use, where we have some freedom on how to choose b. Observe that, given c, there may be more than one prime p for which $c = \text{ord}_p(10)$. Choosing b differently for different p allows us to use two or more different congruences with the same modulus in a covering. On the other hand, given c, it is also possible that there are no primes p for which $c = \text{ord}_p(10)$. In this case, we cannot use the modulus c in (1).

To clarify, one can check that

$$\begin{aligned} 2 &= \text{ord}_{11}(10) \\ 4 &= \text{ord}_{101}(10) \\ 8 &= \text{ord}_{73}(10) \\ 8 &= \text{ord}_{137}(10). \end{aligned}$$

Thus, for a covering of the integers, we may use 2 and 4 each once as the modulus of a congruence of a covering, and we may use 8 twice as the modulus of a congruence. One checks that

$$\begin{aligned} k &\equiv 0 \pmod{2} \\ k &\equiv 1 \pmod{4} \\ k &\equiv 3 \pmod{8} \\ k &\equiv 7 \pmod{8} \end{aligned}$$

is a covering of the integers. To apply Lemma 1, we take

$$\mathcal{P}_2 = \{11, 73, 101, 137\},$$

n divisible by 8, and M so that

$$\begin{aligned} M &\equiv -(2-1)\cdot 10^0 \pmod{11} \\ M &\equiv -(2-1)\cdot 10^1 \pmod{101} \\ M &\equiv -(2-1)\cdot 10^3 \pmod{73} \\ M &\equiv -(2-1)\cdot 10^7 \pmod{137}. \end{aligned}$$

Note that these congruences on M are equivalent to a single congruence modulo $11 \cdot 73 \cdot 101 \cdot 137$ by the Chinese Remainder Theorem; in particular, such M exist. We deduce from Lemma 1 that, under these conditions, the number $N^{(k)}(2)$ is composite for $K \le k \le n-1$.

We write the congruence on M just alluded to above as

$$M \equiv B(2) \pmod{11 \cdot 73 \cdot 101 \cdot 137}.$$

Combining this with our previous results, we have that if

$$n \equiv 0 \pmod{24}$$

and M is a solution of the system of congruences

$$\begin{aligned} M &\equiv 7 \pmod{3 \cdot 7} \\ M &\equiv B(2) \pmod{11 \cdot 73 \cdot 101 \cdot 137}, \end{aligned}$$

which we know exists by the Chinese Remainder Theorem, then $N^{(k)}(x)$ will be divisible by at least one prime in $\{3, 7, 11, 73, 101, 137\}$ for each $x \in \{0, 1, 2, 3, 6, 8, 9\}$ and $K \le k \le n-1$.

Observe that to assure that M exists by using the Chinese Remainder Theorem above, we chose $\mathcal{P}_2$ so that it had an empty intersection with the previous sets

$$\mathcal{P}_0, \mathcal{P}_1, \mathcal{P}_3, \mathcal{P}_6, \mathcal{P}_8, \mathcal{P}_9$$

already constructed. This will be our strategy then for establishing that the congruences in (2) simultaneously hold. Thus, we follow the same approach for $N^{(k)}(x)$ when $x \in \{4, 5, 7\}$ as we did for $N^{(k)}(2)$, but, in each case, the covering of the integers will be more complicated because we do not wish to repeat using the same primes previously chosen. We list the coverings and the primes used for $x \in \{4, 5, 7\}$.

We begin with $x = 4$. Table 1 contains the needed information for us. To conserve space, we note here that

$$p_7 = 440334654777631, \quad p_{14} = 3199044596370769.$$

Table 1. Covering used in Lemma 1 (i) for $N^{(k)}(4)$

row	congruence	prime p_i
1	$k \equiv 0 \pmod{3}$	37
2	$k \equiv 1 \pmod{6}$	13
3	$k \equiv 2 \pmod{9}$	333667
4	$k \equiv 5 \pmod{18}$	19
5	$k \equiv 14 \pmod{18}$	52579
6	$k \equiv 8 \pmod{27}$	757
7	$k \equiv 17 \pmod{27}$	p_7

row	congruence	prime p_i
8	$k \equiv 26 \pmod{54}$	70541929
9	$k \equiv 53 \pmod{54}$	14175966169
10	$k \equiv 4 \pmod{12}$	9901
11	$k \equiv 10 \pmod{24}$	99990001
12	$k \equiv 22 \pmod{72}$	3169
13	$k \equiv 46 \pmod{72}$	98641
14	$k \equiv 70 \pmod{72}$	p_{14}

We observe that we are taking $\mathcal{P}_4$ to be the set of 14 primes appearing in the columns with the heading "prime p_i". The congruences in (1) appear in the columns with heading "congruence". Thus, if p appears in the column with heading "prime p_i", then the congruence in this same row takes the form $k \equiv b(4, p) \pmod{c(p)}$. In each row, one can check directly that $\text{ord}_p(10) = c(p)$.

We also need to justify that the 14 congruences displayed in Table 1 form a covering. The least common multiple of the moduli appearing in these congruences is 216. One checks directly (by hand or computer) that the numbers $0, 1, 2, \ldots, 215$ each satisfy at least one of the congruences in the table. Now, suppose that m is an arbitrary integer. Let $m' \in \{0, 1, \ldots, 215\}$ such that $m \equiv m' \pmod{216}$. We deduce then, since each modulus in the congruences divides 216, that m is a solution of whatever congruence m' satisfies. Hence, the verification that the congruences in Table 1 form a covering is complete.

For $x \in \{5, 7\}$, we use the same approach as above. We give the tables explicitly but do not elaborate on the details. We simply note that in each case one should check that the congruence $k \equiv b(x, p) \pmod{c(p)}$ and p appearing on a row are such that $\mathrm{ord}_p(10) = c(p)$. For each table, one should also check that the congruences form a covering, and this can be done by simply checking if the non-negative integers up to one less that the least common multiple of the moduli each satisfy at least one congruence from the table.

For $x = 5$, we make use of the information in Table 2. To conserve space, we set

$$p_{10} = 5964848081, \quad p_{17} = 418550283013311072 1.$$

Table 2. Covering used in Lemma 1 (i) for $N^{(k)}(5)$

row	congruence	prime p_i
1	$k \equiv 0 \pmod{5}$	41
2	$k \equiv 1 \pmod{5}$	271
3	$k \equiv 2 \pmod{10}$	9091
4	$k \equiv 3 \pmod{20}$	3541
5	$k \equiv 13 \pmod{20}$	27961
6	$k \equiv 7 \pmod{30}$	211
7	$k \equiv 17 \pmod{30}$	241
8	$k \equiv 27 \pmod{30}$	2161
9	$k \equiv 8 \pmod{40}$	1676321
10	$k \equiv 28 \pmod{40}$	p_{10}

row	congruence	prime p_i
11	$k \equiv 18 \pmod{60}$	61
12	$k \equiv 38 \pmod{60}$	4188901
13	$k \equiv 58 \pmod{60}$	39526741
14	$k \equiv 4 \pmod{15}$	31
15	$k \equiv 9 \pmod{15}$	2906161
16	$k \equiv 14 \pmod{45}$	238681
17	$k \equiv 29 \pmod{45}$	p_{17}
18	$k \equiv 44 \pmod{90}$	29611
19	$k \equiv 89 \pmod{90}$	3762091

For $x = 7$, we use the information in Table 3. We note here that

$$p_{11} = 102598800232111471, \; p_{13} = 265212793249617641, \; p_{14} = 30703738801,$$

$$p_{15} = 625437743071, \; p_{16} = 5780205030878619196540944 1,$$

$$p_{21} = 4458192223320340849, \; p_{27} = 127522001020150503761, \; p_{31} = 60368344121,$$

$$p_{32} = 8486544838794975628 21, \; p_{34} = 7376575589640313840 1,$$

$$p_{35} = 11189053009, \; p_{36} = 603812429055411913, \; p_{37} = 1480294234007505065 53.$$

For $K \le k \le n-1$, our argument is complete upon noting that the sets $\{3, 7\}$, $\mathcal{P}_2$, $\mathcal{P}_4$, $\mathcal{P}_5$ and $\mathcal{P}_7$ are pairwise disjoint. This allows us then to take n so that it is divisible by 6 and each modulus appearing in (1) independent of the value of $x \in \{2, 4, 5, 7\}$ and also assures that there is a simultaneous solution to the congruences appearing in (2).

Table 3. Covering used in Lemma 1 (i) for $N^{(k)}(7)$

row	congruence	prime p_i
1	$k \equiv 0 \pmod{7}$	239
2	$k \equiv 1 \pmod{7}$	4649
3	$k \equiv 2 \pmod{21}$	43
4	$k \equiv 9 \pmod{21}$	1933
5	$k \equiv 16 \pmod{21}$	10838689
6	$k \equiv 3 \pmod{14}$	909091
7	$k \equiv 10 \pmod{28}$	29
8	$k \equiv 24 \pmod{28}$	281
9	$k \equiv 4 \pmod{35}$	71
10	$k \equiv 11 \pmod{35}$	123551
11	$k \equiv 18 \pmod{35}$	p_{11}
12	$k \equiv 25 \pmod{70}$	4147571
13	$k \equiv 60 \pmod{70}$	p_{13}
14	$k \equiv 32 \pmod{105}$	p_{14}
15	$k \equiv 67 \pmod{105}$	p_{15}
16	$k \equiv 102 \pmod{105}$	p_{16}
17	$k \equiv 5 \pmod{42}$	127
18	$k \equiv 26 \pmod{42}$	2689
19	$k \equiv 12 \pmod{42}$	459691

row	congruence	prime p_i
20	$k \equiv 33 \pmod{84}$	226549
21	$k \equiv 75 \pmod{84}$	p_{21}
22	$k \equiv 19 \pmod{63}$	10837
23	$k \equiv 40 \pmod{63}$	23311
24	$k \equiv 61 \pmod{63}$	45613
25	$k \equiv 6 \pmod{28}$	121499449
26	$k \equiv 13 \pmod{56}$	7841
27	$k \equiv 41 \pmod{56}$	p_{27}
28	$k \equiv 20 \pmod{140}$	421
29	$k \equiv 48 \pmod{140}$	3471301
30	$k \equiv 76 \pmod{140}$	13489841
31	$k \equiv 104 \pmod{140}$	p_{31}
32	$k \equiv 132 \pmod{140}$	p_{32}
33	$k \equiv 27 \pmod{112}$	113
34	$k \equiv 83 \pmod{112}$	p_{34}
35	$k \equiv 55 \pmod{168}$	p_{35}
36	$k \equiv 111 \pmod{168}$	p_{36}
37	$k \equiv 167 \pmod{168}$	p_{37}

We fix C to be the least common multiple of the moduli appearing in the congruences in (1) for $x \in \{2, 4, 5, 7\}$. Note that 6 divides C. Thus, taking

$$\mathcal{P} = \{3, 7\} \cup \mathcal{P}_2 \cup \mathcal{P}_4 \cup \mathcal{P}_5 \cup \mathcal{P}_7,$$

we see that if $n \equiv 0 \pmod{C}$, $K \le k \le n-1$ and $x \in \{0, 1, \ldots, 9\}$, then $N^{(k)}(x)$ is divisible by at least one prime from $\mathcal{P}$.

It is worth noting here that the primes 2 and 5 are not in $\mathcal{P}$. This is done deliberately as Theorem 1 requires that N be coprime to 10. Combining the congruences appearing in (2) with the congruence $M \equiv 0 \pmod{10}$, for example, allows us to deduce that we can in fact take N coprime to 10.

For the remainder of this section, we fix $M \equiv 0 \pmod{10}$ so that N is coprime to 10 and if $n \equiv 0 \pmod{C}$, $K \le k \le n-1$ and $x \in \{0, 1, \ldots, 9\}$, then $N^{(k)}(x)$ is divisible by at least one prime from $\mathcal{P}$. Fix also some $n_0 \equiv 0 \pmod{C}$. We suppose that n_0 satisfies $10^{n_0-2} > M$. Then for $n \ge n_0$, the number

$$N = \frac{10^n - 1}{9} + M$$

has left-most digit 1. More importantly, as M is fixed, the number K is well-defined and does not vary as n varies over the integers $\ge n_0$.

There are a finitely many ways, independent of $n \geq n_0$, to have $k \in \{0, 1, \ldots, K-1\}$ and $x \in \{0, 1, \ldots, 9\}$. Fix

$$N_0 = \frac{10^{n_0} - 1}{9} + M.$$

Observe that N_0 is coprime to 10. Analogous to our previous notation, we refer to $N_0^{(k)}(x)$ as the number obtained by replacing the digit $x \in \{0, 1, \ldots, 9\}$ for the digit of N_0 appearing in the $(k+1)$st position on the right. For each $k \in \{0, 1, \ldots, K-1\}$ and $x \in \{0, 1, \ldots, 9\}$, we consider the least prime $q = q(k, x)$ dividing $N_0^{(k)}(x)$. If $q > 5$, $n \equiv n_0 \pmod{c(q)}$ and $n \geq n_0$, then q divides $N^{(k)}(x)$ since

$$N^{(k)}(x) - N_0^{(k)}(x) = \frac{10^n - 10^{n_0}}{9} = \frac{10^{n_0}}{9}\left(10^{n-n_0} - 1\right).$$

If $q \in \{2, 5\}$, then N_0 being coprime to 10 implies $k = 0$. It follows in this case that $N^{(k)}(x) = N^{(0)}(x)$ is also divisible by q. The definition of n_0 implies that if $q = 3$ and $n \equiv n_0 \pmod 3$, then both n and n_0 are divisible by 3 so that $10^{n-n_0} - 1$ is divisible by 27. Thus, in this case, we also have that $N^{(k)}(x)$ is divisible by q.

Set C' to be the least common multiple of C and the numbers $c(q)$ where q varies over the primes $q(k, x) > 5$ with $0 \leq k \leq K-1$ and $x \in \{0, 1, \ldots, 9\}$. Then we deduce that if

$$N = \frac{10^n - 1}{9} + M,$$

where

$$n \equiv n_0 \pmod{C'},$$

then $N^{(k)}(x)$ is divisible by the prime $q(k, x)$ provided $0 \leq k \leq K-1$ and $x \in \{0, 1, \ldots, 9\}$. As such n are necessarily divisible by C, we further obtain for such n that $N^{(k)}(x)$ is divisible by at least one of the primes in $\mathcal{P}$ for $K \leq k \leq n-1$ and $x \in \{0, 1, \ldots, 9\}$. Since the number N tends to infinity with n and the number of primes $q(k, x)$ as well as the set $\mathcal{P}$ are finite, the theorem follows.

3. Proof of Theorem 2

For a number $N' = d_r d_{r-1} \ldots d_0$ written in base 10 with $d_r \neq 0$ and digit $x \in \{0, 1, \ldots, 9\}$, we denote by $\widehat{N'}^{(k)}(x)$ the number obtained by inserting the digit x between d_k and d_{k-1}. For example,

$$\widehat{N'}^{(2)}(x) = d_r d_{r-1} \ldots d_3 d_2 x d_1 d_0.$$

We also define $\widehat{N'}^{(0)}(x)$ as the number obtained by inserting in N' the digit x to the right of d_0 and $\widehat{N'}^{(r+1)}(x)$ as the number obtained by inserting in N' the digit x to the left of d_r. Thus,

$$\widehat{N'}^{(0)}(x) = d_r d_{r-1} \ldots d_1 d_0 x \quad \text{and} \quad \widehat{N'}^{(r+1)}(x) = x d_r d_{r-1} \ldots d_1 d_0.$$

We fix $\mathcal{P}$ to be the set of primes from the proof of Theorem 1 and C, M and K as determined there so that if $n \equiv 0 \pmod{C}$, then the number

$$N = \frac{10^n - 1}{9} + M$$

has the property that, for $K \le k \le n-1$, replacing the digit 1 that appears as the $(k+1)$st digit from the right with $x \in \{0, 1, \ldots, 9\}$ results in a number $N^{(k)}(x)$ that is divisible by a prime in $\mathcal{P}$. Setting

$$N' = \frac{10^{n-1} - 1}{9} + M,$$

we see that

$$\widehat{N'}^{(k)}(x) = N^{(k)}(x) \quad \text{for } K \le k \le n-1.$$

Observe that, in the notation of the previous paragraph, $r = n-2$. Also, N' has the property that if $n \equiv 0 \pmod{C}$, then the insertion $\widehat{N'}^{(k)}(x)$, for $x \in \{0, 1, \ldots, 9\}$ and $K \le k \le n-1$, is divisible by some prime from the set $\mathcal{P}$. Note that the exponent $n-1$ appearing in the definition of N' above is -1 modulo C.

The rest of our argument is analogous to what we did at the end of the previous section. We fix M so that

$$N' = \frac{10^n - 1}{9} + M$$

is coprime to 10 and if $n \equiv -1 \pmod{C}$, $K \le k \le n$ and $x \in \{0, 1, \ldots, 9\}$, then $\widehat{N'}^{(k)}(x)$ is divisible by at least one prime from $\mathcal{P}$. Fix also some $n_0 \equiv -1 \pmod{C}$. We suppose that n_0 satisfies $10^{n_0-2} > M$. Then for $n \ge n_0$, the number N' has left-most digit 1. As M is fixed, the number K is well-defined and does not vary as n varies over the integers $\ge n_0$.

There are a finitely many ways, independent of $n \ge n_0$, to have $k \in \{0, 1, \ldots, K-1\}$ and $x \in \{0, 1, \ldots, 9\}$. Fix

$$N'_0 = \frac{10^{n_0} - 1}{9} + M.$$

Observe that N'_0 is coprime to 10. We refer to $\widehat{N'_0}^{(k)}(x)$ as the number obtained by inserting the digit $x \in \{0, 1, \ldots, 9\}$ to the left of the kth digit of N_0. For each $k \in \{0, 1, \ldots, K-1\}$ and $x \in \{0, 1, \ldots, 9\}$, we consider the least prime $q = q(k, x)$ dividing $\widehat{N'_0}^{(k)}(x)$. Recall the definition $c(p) = \mathrm{ord}_p(10)$, for p a prime, used in the previous section. Observe that if $q > 5$, $n \equiv n_0 \pmod{c(q)}$ and $n \ge n_0$, then q divides $\widehat{N'}^{(k)}(x)$ since

$$\widehat{N'}^{(k)}(x) - \widehat{N'_0}^{(k)}(x) = \frac{10^{n+1} - 10^{n_0+1}}{9} = \frac{10^{n_0+1}}{9}\left(10^{n-n_0} - 1\right).$$

If $q \in \{2, 5\}$, then N'_0 being coprime to 10 implies $k = 0$. It follows in this case that $\widehat{N'}^{(k)}(x) = \widehat{N'}^{(0)}(x)$ is also divisible by q. If $q = 3$ and $n \equiv n_0 \pmod{3}$, then $n - n_0$ is divisible by 3 so that $10^{n-n_0} - 1$ is divisible by 27. Thus, in this case, we also have that $\widehat{N'}^{(k)}(x)$ is divisible by q.

Set C'' to be the least common multiple of C and the numbers $c(q)$ where q varies over the primes $q(k,x) > 5$ with $0 \le k \le K-1$ and $x \in \{0,1,\ldots,9\}$. Then we deduce that if

$$N' = \frac{10^n - 1}{9} + M,$$

where

$$n \equiv n_0 \pmod{C''},$$

then $\widehat{N'}^{(k)}(x)$ is divisible by the prime $q(k,x)$ provided $0 \le k \le K-1$ and $x \in \{0,1,\ldots,9\}$. As such n are necessarily -1 modulo C, we further obtain for such n that $\widehat{N'}^{(k)}(x)$ is divisible by at least one of the primes in $\mathcal{P}$ for $K \le k \le n$ and $x \in \{0,1,\ldots,9\}$. Since the number N' tends to infinity with n and the number of primes $q(k,x)$ as well as the set $\mathcal{P}$ are finite, the theorem follows.

Reference

[1] J. Brillhart, D. H. Lehmer, J. L. Selfridge, B. Tuckerman, and S. S. Wagstaff, Jr., Factorizations of $b^n \pm 1$, $b = 2,3,5,6,7,10,11,12$ Up to High Powers, 3rd edition, *Contemporary Mathematics*, Vol. 22, American Math. Soc., Providence, 2002 (available online).

[2] F. Cohen and J. L. Selfridge, Not every number is the sum or difference of two prime powers, Collection of articles dedicated to Derrick Henry Lehmer on the occasion of his seventieth birthday. *Math. Comp.* 29 (1975), 79–81.

[3] P. Erdős, On integers of the form $2^k + p$ and some related problems, *Summa Brasil. Math.* 2 (1950), 113–123.

[4] M. Filaseta, C. Finch and M. Kozek, On powers associated with Sierpinski numbers, Riesel numbers and Polignac's conjecture, *J. Number Theory* 128 (2008), 1916–1940.

[5] M. Filaseta, K. Ford, S. Konyagin, C. Pomerance and G. Yu, Sieving by large integers and covering systems of congruences, *J. Amer. Math. Soc.* 20 (2007), 495–517.

[6] R. K. Guy, Unsolved Problems in Number Theory (3rd ed.), Sections A19, B21, F13 and F14, *Problem Books in Mathematics,* Springer-Verlag, New York, 2004.

[7] J. A. Haight, Covering systems of congruences, a negative result, *Mathematika* 26 (1979), 53–61.

[8] Š. Porubský and J. Schönheim, Covering systems of Paul Erdős: Past, present and future, in *Paul Erdős and his mathematics,* vol. I, János Bolyai Math. Soc., Budapest, 2002, pp. 581–627.

[9] P. Nielsen, A covering system whose smallest modulus is 40, *J. Number Theory* 129 (2009), 640–666.

[10] Z. W. Sun, On integers not of the form $\pm p^a \pm q^b$, *Proc. Amer. Math. Soc.* 128 (2000), 997–1002.

[11] K. J. Wu and Z. W. Sun, Covers of the integers with odd moduli and their applications to the forms $x^m - 2^n$ and $x^2 - F_{3n}/2$, *Math. Comp.* 78 (2009), 1853–1866.

In: Frontiers of Combinatorics and Number Theory, Volume 2 ISBN: 978-1-61942-918-5
Editor: Zhi-Wei Sun, pp. 41-58

Chapter 3

CONSTRUCTION OF FAMILIES OF IMAGINARY PRIME-CYCLOTOMIC INTEGERS WITH PRESCRIBED PROPERTIES*

F. Thaine†
Department of Mathematics and Statistics - CICMA, Concordia University
1455, de Maisonneuve Blvd. W., Montreal, Quebec, H3G 1M8, Canada

Abstract

Let $m > 2$ be an even integer. We develop a method to construct families of prime numbers $q = mf + 1$ with f odd, and of complex nonreal cyclotomic integers $\beta_q \in \mathbb{Z}[\zeta_q]$ of degree m, with some prescribed properties, which are expressed as linear combinations, over $\mathbb{Z}$, of Gaussian periods. We apply that method in the construction of families of numbers β_q, as above, such that $\beta_q\overline{\beta_q} \in \mathbb{Z}$. Some of those families bring us families of cyclic difference sets. We also apply the method in the construction of families of cyclic polynomials of degree m whose roots are complex nonreal q-cyclotomic integers with absolute norm q. This complements results of a previous article where we show how to construct families of cyclic polynomials whose roots are real q-cyclotomic units. We give several examples for small values of m, and a MAPLE program that can be used to perform the calculations and to search for more examples.

2000 Mathematics Subject Classification: Primary 11R18, 11T22; Secondary 05B10, 12E10.

1. Introduction

Let $m > 2$ be an even integer and, for each prime number $q \equiv 1 \bmod m$, let ζ_q be a primitive q-th root of 1, $f = f_q = (q-1)/m$, $s = s_q$ a primitive root modulo q and K_q the only subfield of degree m of $\mathbb{Q}[\zeta_q]$. In this article we develop a method to find families of prime numbers $q \equiv 1 \bmod m$, and complex nonreal cyclotomic integers $\beta_q \in \mathbb{Z}[\zeta_q]$

*This work was supported in part by a grant from NSERC.
†E-mail address: ftha@alcor.concordia.ca

of degree m, with some prescribed properties. Those elements β_q are expressed as linear combinations, over $\mathbb{Z}$, of certain matrices of rational integers which represent the Gaussian periods of degree m in $\mathbb{Z}[\zeta_q]$. The method is an enhancement of one that first appeared in [T2], and complements the (somehow simpler) method used in [T3] to construct families of cyclic polynomials whose roots are real q-cyclotomic units. We show two applications: the construction of families of complex nonreal cyclotomic numbers $\beta_q \in \mathbb{Z}[\zeta_q]$ of degree m, such that $\beta_q\overline{\beta_q} \in \mathbb{Z}$, which, following an idea of Emma Lehmer, includes the construction of families of cyclic difference sets, and the construction of families of cyclic polynomials of degree m whose roots are complex nonreal q-cyclotomic integers of norm q in $K_q/\mathbb{Q}$.

For $0 \le i \le m-1$, define the Gaussian periods of degree m in $\mathbb{Q}[\zeta_q]$ by

$$\eta_i = \sum_{j=0}^{f-1} \zeta_q^{s^{i+mj}}. \tag{1.1}$$

We have that $K_q = \mathbb{Q}[\eta_0]$. The set $\{\eta_0, \eta_1, \ldots, \eta_{m-1}\}$ is a normal integral basis of $K_q/\mathbb{Q}$ and we have $\sum_{i=0}^{m-1} \eta_i = -1$. The Galois group of $K_q/\mathbb{Q}$ is generated by the automorphism τ such that $\tau(\eta_i) = \eta_{i+1}$ (indices modulo m). We will assume that f is odd, then the numbers η_i are complex nonreal. We have that $\tau^{m/2}$ is complex conjugation; that is $\overline{\eta_i} = \eta_{i+m/2}$ for all i (indices modulo m).

1.1. Difference Sets

Since our more interesting examples are related to cyclic difference sets, here we recall some basic facts about those sets. Let v, k and λ be positive integers. A set $D = \{d_1, \ldots, d_k\}$ of integers modulo v is a v, k, λ-cyclic difference set if for any integer $a \not\equiv 0 \bmod v$ there are exactly λ pairs (d_i, d_j), with $d_i, d_j \in D$, satisfying the congruence $d_i - d_j \equiv a \bmod v$. In this article difference set means cyclic difference set (for more general definitions and an introduction to the subject see [B]).

Difference sets have applications in several areas like the construction of error correcting codes, finite geometries and the design of experiments, and have been studied extensively. For references see [BEW] Chapter 5.

Given a v, k, λ-difference set, it follows from the definition that $k(k-1) = \lambda(v-1)$. Call $n = k - \lambda$. To avoid the trivial cases we assume that $n \ge 2$.

If $D = \{d_1, \ldots, d_k\}$ is a difference set and $b, l \in \mathbb{Z}$, with $\gcd(l, v) = 1$, then the sets $D + b = \{d_1 + b, \ldots, d_k + b\}$ and $lD = \{ld_1, \ldots, ld_k\}$ modulo v are also difference sets, with the same parameters v, k, λ. If $D' = lD + b$, for some b, l as above, then D and D' are said to be equivalent difference sets. If D is a v, k, λ-difference set and D' is the set of integers modulo v such that $D \dot\cup D' = \{0, 1, \ldots, v-1\}$ (disjoint union), then D' is a $v, v-k, v-2k+\lambda$-difference set called the complement of D.

Let ζ_v be a primitive v-th root of 1. If $D = \{d_1, \ldots, d_k\}$ is a v, k, λ-difference set, the polynomial $\theta(x) = x^{d_1} + \cdots + x^{d_k}$ is called the Hall polynomial of D. We have $\theta(1) = k$ and

$$\theta(\zeta_v^i)\theta(\zeta_v^{-i}) = n \ \text{ for } 1 \le i \le v-1. \tag{1.2}$$

Conversely, if $\theta(x) = x^{d_1} + \cdots + x^{d_k}$, with $0 \le d_1, \ldots, d_k \le v-1$ distinct integers, is a polynomial satisfying (1.2), then $D = \{d_1, \ldots, d_k\}$ is a v, k, λ-difference set, with

$\lambda = k - n$. (For the proofs of these facts see the Introduction of [B].)

Assume now that $v = q$ is a prime number and that $q = mf + 1$, with f odd. We will use the theory of cyclotomic fields to construct families of Hall polynomials $\theta(x)$ such that the elements $\theta(\zeta_q^i)$ are either sums of distinct Gaussian periods of degree m of $\mathbb{Q}[\zeta_q]$ or 1 plus sums of distinct Gaussian periods of degree m. This includes the construction of the m-th power residue difference sets and the modified m-th power residue difference sets considered by Emma Lehmer in [L]. Lehmer's interesting results in [L] were the original motivation for this work.

Let $r \geq 1$ and $0 \leq i_1 < \cdots < i_r \leq m-1$ be integers. To a Hall polynomial $\theta(x)$ such that

$$\theta(1) = rf, \quad \theta(\zeta_q) = \eta_{i_1} + \eta_{i_2} + \cdots + \eta_{i_r} \quad \text{and} \quad \theta(\zeta_q)\overline{\theta(\zeta_q)} = n, \tag{1.3}$$

with $n \in \mathbb{Z}$, $n \geq 2$, corresponds a q, k, λ-difference set

$$D = \{s^{i_a+mj} : 1 \leq a \leq r, 0 \leq j \leq f-1\} = \bigcup_{a-1}^{r} \{s^{i_a+mj} : 0 \leq j \leq f-1\} \mod q, \tag{1.4}$$

with $k = rf$ and $\lambda = k - n$. To a Hall polynomial $\theta(x)$ such that

$$\theta(1) = rf + 1, \quad \theta(\zeta_q) = 1 + \eta_{i_1} + \eta_{i_2} + \cdots + \eta_{i_r} \quad \text{and} \quad \theta(\zeta_q)\overline{\theta(\zeta_q)} = n, \tag{1.5}$$

with $n \in \mathbb{Z}$, $n \geq 2$, corresponds a q, k, λ-difference set

$$D_0 = \{0\} \cup \{s^{i_a+mj} : 1 \leq a \leq r, 0 \leq j \leq f-1\} = \{0\} \cup \bigcup_{a=1}^{r} \{s^{i_a+mj} : 0 \leq j \leq f-1\} \mod q, \tag{1.6}$$

with $k = rf + 1$ and $\lambda = k - n$.

When $r = 1$ and $i_1 = 0$, the difference set D in (1.4) is called an m-th power residue difference set and the difference set D_0 in (1.6) is called a modified m-th power residue difference set.

Since (1.3) or (1.5) are necessary and sufficient conditions for D or D_0 (in (1.4) and (1.6)) respectively to be difference sets, we can state our objective as follows: Find integers $\varepsilon, \varepsilon_0, \ldots, \varepsilon_{m-1} \in \{0, 1\}$, where at least one of the ε_i is nonzero, such that

$$\left(\varepsilon + \sum_{i=0}^{m-1} \varepsilon_i \eta_i\right)\left(\varepsilon + \sum_{i=0}^{m-1} \varepsilon_i \eta_{i+m/2}\right) = n, \tag{1.7}$$

for some integer $n \geq 2$. The cases $\varepsilon_0 = 1$, $\varepsilon = \varepsilon_1 = \cdots = \varepsilon_{m-1} = 0$ and $\varepsilon = \varepsilon_0 = 1$, $\varepsilon_1 = \cdots = \varepsilon_{m-1} = 0$ correspond to the m-th power residue difference sets and to the modified m-th power residue difference sets respectively.

We define the degree of a cyclic difference set, with $v = q$ and Hall polynomial $\theta(x)$, as the degree of the extension $\mathbb{Q}[\theta(\zeta_q)]/\mathbb{Q}$. When studying examples, in Section 3, for a given value of m, of cyclic difference sets corresponding to elements of the form $\varepsilon +$

$\sum_{i=0}^{m-1} \varepsilon_i \eta_i$, we are not going to consider those of degree smaller than m. In particular notice that, denoting by $\eta_i^{(m)}$ the Gaussian periods defined in (1.1), if $l \mid m$ and $\theta(\zeta_q) = \sum_{k=0}^{m/l-1} \eta_{i+kl}^{(m)} = \eta_i^{(l)}$, or $\theta(\zeta_q) = 1 + \sum_{k=0}^{m/l-1} \eta_{i+kl}^{(m)} = 1 + \eta_i^{(l)}$, then the corresponding difference sets have degree l and will not be considered if $l < m$. For example, if $q \equiv 3 \mod 4$, the difference sets with $\theta(\zeta_q) = \eta_0 + \eta_2 + \cdots + \eta_{m-2} = (-1 \pm \sqrt{-q})/2$ or $\theta(\zeta_q) = 1 + \eta_0 + \eta_2 + \cdots + \eta_{m-2} = (1 \pm \sqrt{-q})/2$ will not appear as examples, since $m > 2$.

The theory above motivated us to consider the more general problem of constructing families of primes q and elements $\beta = \beta_q = \varepsilon + \sum_{i=0}^{m-1} \varepsilon_i \eta_i \in K_q$, with $\varepsilon, \varepsilon_i \in \mathbb{Z}$ (not necessarily $\in \{0, 1\}$), such that $\beta\overline{\beta} \in \mathbb{Z}$ (that is, such that (1.7) holds).

1.2. Objective and Contents of the Paper

In the next section, using ideas of Gauss, Kummer and Dickson, we show how to express the periods η_i as matrices of cyclotomic numbers and how to construct such matrices. In Section 3, we apply those results in the construction of the mentioned families of difference sets and (more generally) of families of cyclotomic integers $\beta = \varepsilon + \sum_{i=0}^{m-1} \varepsilon_i \eta_i$, of degree m, with $\varepsilon, \varepsilon_i \in \mathbb{Z}$, such that $\beta\overline{\beta} \in \mathbb{Z}$, and give examples of such families for small values of m. More families, and for larger values of m, can surely be obtained by using better computer resources. In Section 4, we show how to construct families of complex nonreal cyclotomic integers of norm q in K_q and their characteristic polynomials, and show several examples. In our examples, those cyclotomic integers (which are associates of the elements $\rho_q = \prod_{j=0}^{f-1}(1 - \zeta_q^{s^{mj}}) \in K_q$) are expressed in the form $\sum_{i=0}^{m-1} a_i \eta_i$ with $a_i \in \mathbb{Z}$. In Section 5, we give a MAPLE program that can be used to check our results of Sections 3 and 4 and to search for new ones.

I thank Javier Thaine for helping me to improve some programs used in the calculation of the examples.

2. The Multiplication Matrix of the Gaussian Periods when f is Odd

Recall that $q = mf + 1$ is a prime with $m > 2$ even and f odd. Let $c_{i,j}$, $0 \le i, j \le m - 1$, be the integers such that

$$\eta_0 \eta_i = \sum_{j=0}^{m-1} c_{i,j} \eta_j \tag{2.8}$$

and let $C = [c_{i,j}]_{0 \le i,j < m}$. We call C the multiplication matrix of the η_i.

For $i, j \in \mathbb{Z}$, we define

$$\delta_{i,j} = \begin{cases} 1 & \text{if } i \equiv j \bmod m \\ 0 & \text{if } i \not\equiv j \bmod m \end{cases} \qquad \text{and} \qquad \delta_i = \delta_{m/2,i}. \tag{2.9}$$

We have that $c_{i,j} = (i, j) - f\delta_i$, where the (i, j) are the cyclotomic numbers of order m. The number (i, j) is defined as the number of ordered pairs of integers $\langle k, l \rangle$, $0 \le k, l \le f - 1$, such that $1 + s^{km+i} \equiv s^{lm+j} \mod q$ (see, for example, [D] formulas (4), (5) and

(6)). The numbers (i, j) can be efficiently calculated using the following formulas. For $0 \le i, j \le m-1$,

$$(i,j) \equiv -\frac{1}{m^2}\sum_{k=0}^{m}\sum_{l=0}^{m-1}\binom{fk}{fl} s^{f(li-kj)} \mod q \quad \text{and} \quad 0 \le (i,j) < \sqrt{q}+(q-1)/m^2 < q$$

(see, for example, [T2] page 259).

Let K be the $m\times m$ matrix $[\delta_{i+1,j}]_{i,j}$; that is $K = \begin{bmatrix} 0 & 1 & 0 & \dots & 0 \\ 0 & 0 & 1 & \dots & 0 \\ \vdots & \vdots & \vdots & \ddots & \vdots \\ 0 & 0 & 0 & \dots & 1 \\ 1 & 0 & 0 & \dots & 0 \end{bmatrix}$. We have that $K^m = I$, the identity matrix. Let $P(x) = (x-\eta_0)(x-\eta_1)\dots(x-\eta_{m-1}) \in \mathbb{Z}[x]$, the irreducible polynomial of the η_i. We have that $P(x)$ is the characteristic polynomial of C. Also, for $i = 0, \dots, m-1$,

$$C(K^{-i}CK^i) = (K^{-i}CK^i)C, \tag{2.10}$$

and the field $\mathbb{Q}[C]$ is a decomposition field for $P(x)$. In $\mathbb{Q}[C]$ we have

$$P(x) = (x-C)(x-K^{-1}CK)\dots(x-K^{-(m-1)}CK^{(m-1)}),$$

so we can identify the Gaussian periods η_i with the matrices $K^{-i}CK^i$. (For the proofs of these facts see, for example, [T2].) For example, since $\sum_{i=0}^{m-1}\eta_i = -1$ we have that $\sum_{i=0}^{m-1} K^{-i}CK^i = -I$.

The numbers $c_{i,j}$ and (i,j), $i, j \in \mathbb{Z}$, indices modulo m, also satisfy the properties

$$c_{i,j} = c_{-i,j-i}, \quad \sum_{k=0}^{m-1} c_{i,k} = f - q\delta_i, \quad \sum_{k=0}^{m-1} c_{k,j} = -\delta_{0,j} \quad \text{and} \quad (i,j) = (j+m/2, i+m/2)$$

(see [D] Part I, [S] or [T2] Proposition 8).

The coefficients $c_{i,j}$ can be expressed in terms of Jacobi sums. Let ζ_m be a primitive m-th root of 1. For $a, b \in \mathbb{Z}$, define the Jacobi sums $J_{a,b}$ by

$$J_{a,b} = -\sum_{k=2}^{q-1} \zeta_m^{a\,\mathrm{ind}_s(k)+b\,\mathrm{ind}_s(1-k)},$$

where $\mathrm{ind}_s(k)$ is the least nonnegative integer such that $s^{\mathrm{ind}_s(k)} \equiv k \mod q$. For $0 \le i, j \le m-1$ we have

$$c_{i,j} = -f\delta_i - \frac{1}{m^2}\sum_{a=0}^{m-1}\sum_{b=0}^{m-1}(-1)^a\zeta_m^{-ia-jb}J_{a,b} =$$
$$\frac{q+1}{m^2} - \frac{q\delta_i}{m} - \frac{\delta_{0,j}}{m} - \frac{\delta_{i,j}}{m} - \frac{1}{m^2}\sum_{\substack{1\le a,b\le m-1\\ a+b\ne m}}(-1)^a\zeta_m^{-ia-jb}J_{a,b} \tag{2.11}$$

(see, for example, [BEW] Theorem 2.5.1 or [T2], formulas (22) and (23)). Now, the nice thing is that, when the primes of $\mathbb{Q}[\zeta_m]$ over q are principal [1], we can use Stickelberger

[1] We assume that the prime ideals over q are principal, generated, say, by conjugates of an element $\alpha \in \mathbb{Z}[\zeta_m]$ since, in order to construct our families, we will write α as a polynomial in some parameters. This can be generalized to nonprincipal ideals by, for example, using a trick shown in Section 2 of [T1].

Theorem to construct the principal ideals $(J_{a,b})$ of $\mathbb{Z}[\zeta_m]$ and then find the actual elements $J_{a,b}$ by multiplying generators of the ideals by suitable powers of ζ_m. For c relatively prime with m let σ_c be the automorphism of $\mathbb{Q}[\zeta_m]$ such that $\sigma_c(\zeta_m) = \zeta_m^c$. We have that q splits completely in $\mathbb{Z}[\zeta_m]$. Suppose that $\alpha \in \mathbb{Z}[\zeta_m]$ is such that $q = \prod_{\substack{1\le c<m \\ (c,m)=1}} \sigma_c(\alpha)$. Let s be a primitive root modulo q such that $s^f \equiv \zeta_m \bmod \alpha$. For $a, b \in \mathbb{Z}$ such that $m \nmid a, b, a+b$, we have that $(J_{a,b}) = (\gamma_{a,b})$, where

$$\gamma_{a,b} = \prod_{\substack{1\le c<m \\ (c,m)=1}} \sigma_c^{-1}(\overline{\alpha})^{\left[\frac{(a+b)c}{m}\right]-\left[\frac{ac}{m}\right]-\left[\frac{bc}{m}\right]} \tag{2.12}$$

(see, for example, [W] Lemma 6.2 (d) and page 98, or [T2] formula (21)). Here the bar denotes complex conjugation σ_{-1}, $[x]$ the integral part of a real number x, and (c, m) the greatest common divisor of the integers c and m. Hence we have that, when $m \nmid a, b, a+b$, $J_{a,b} = \epsilon_{a,b}\gamma_{a,b}$, for some units $\epsilon_{a,b}$ of $\mathbb{Z}[\zeta_m]$. The $\epsilon_{a,b}$ are in fact roots of unity, as can be shown by using Dedekind's Unit Theorem (see [T2], the observation preceding Proposition 11).

For $x \in \mathbb{Z}$ denote by $|x|_m$ the least nonnegative residue of x modulo m. The following result arises from [T2], Theorem 2 and the observation preceding Proposition 11.

Proposition 2.1. *For $1 \le a, b \le m-1$ such that $a+b \ne m$ and $\gamma_{a,b}$ as in (2.12), we have $J_{a,b} = \epsilon_{a,b}\gamma_{a,b}$, where $\epsilon_{a,b} = \zeta_m^{l_{a+b}-l_a-l_b}$ and the numbers l_i, $i \in \mathbb{Z} - m\mathbb{Z}$, $0 \le l_i \le m-1$, satisfy the following conditions:*
i) $l_1, l_2, \ldots, l_{m-1}$ are integers such that, for all a and c with $1 \le a, c \le m-1$ and $\gcd(c, m) = 1$, $l_{|ca|_m} \equiv cl_a + al_c - [ca/m](l_{m-1} + m/2) \bmod m$, and
ii) for $1 \le a \le m-1$ and $k \in \mathbb{Z}$, $l_{a+km} = l_a + k(l_{m-1} + m/2)$.
For numbers l_i satisfying conditions (i) and (ii), the elements

$$d_{i,j} = \frac{q+1}{m^2} - \frac{q\delta_i}{m} - \frac{\delta_{0,j}}{m} - \frac{\delta_{i,j}}{m} - \frac{1}{m^2} \sum_{\substack{1\le a,b\le m-1 \\ a+b\ne m}} (-1)^a \zeta_m^{-ia-jb} \zeta_m^{l_{a+b}-l_a-l_b} \gamma_{a,b}, \tag{2.13}$$

$0 \le i, j \le m-1$, are always in $\mathbb{Q}$. We have $d_{i,j} = c_{i,j}$ (and $\zeta_m^{l_{a+b}-l_a-l_b}\gamma_{a,b} = J_{a,b}$) precisely when all $d_{i,j} \in \mathbb{Z}$.

To calculate the examples of Sections 3 and 4, we use Proposition 2.1 for finding the matrices $C = [c_{i,j}]$. The construction of C requires the knowledge of solutions $[l_1, \ldots, l_{m-1}]$ for the congruences in Proposition 2.1 (i). It follows from induction that integers l_i, satisfying those congruences must also satisfy

$$l_{|c^k a|_m} \equiv c^k l_a + kc^{k-1} a l_c - \left[\frac{c^k a}{m}\right]\left(l_{m-1} + \frac{m}{2}\right) \mod m,$$

where $1 \le a, c \le m-1$, $\gcd(c, m) = 1$ and $k \ge 1$. We solved those congruences (mod m) for $m = 4, \ldots, 14$.

For $m = 4$: $l_1 \equiv 0$, $l_2 \equiv x$, $l_3 \equiv 2l_2 + 2$, $0 \le x \le 3$.

For $m = 6$: $l_1 \equiv 0$, $l_2 \equiv x$, $l_3 \equiv y$, $l_4 \equiv 5l_2 + 2l_3 + 3$, $l_5 \equiv 2l_3$, $0 \leq x, y \leq 5$.

For $m = 8$: $l_1 \equiv 0$, $l_2 \equiv 2x$, $l_3 \equiv 2y$, $l_4 \equiv l_3 + 4z$, $l_5 \equiv 5l_3 + 4$, $l_6 \equiv 3l_2 + 2l_3$, $l_7 \equiv 6l_3 + 4$, $0 \leq x, y \leq 3$, $0 \leq z \leq 1$.

For $m = 10$: $l_1 \equiv 0$, $l_2 \equiv x$, $l_3 \equiv 2l_5 + 5$, $l_4 \equiv 7l_2 + 8l_5 + 5$, $l_5 \equiv y$, $l_6 \equiv 3l_2 + 4l_5$, $l_7 \equiv 5$, $l_8 \equiv 9l_2 + 2l_5 + 5$, $l_9 \equiv 2l_5$, $0 \leq x, y \leq 9$.

For $m = 12$: $l_1 \equiv 0$, $l_2 \equiv 2x$, $l_3 \equiv y$, $l_4 \equiv 2y + 6z$, $l_5 \equiv 6l_2 + 8l_3$, $l_6 \equiv l_4$, $l_7 \equiv 8l_3 + 6$, $l_8 \equiv 6l_2 + 4l_3 + 5l_4$, $l_9 \equiv 6l_2 + 3l_3 + 6$, $l_{10} \equiv 5l_2 + 4l_3$, $l_{11} \equiv 6l_2 + 4l_3 + 6$, $0 \leq x \leq 5$, $0 \leq y \leq 11$, $0 \leq z \leq 1$.

For $m = 14$: $l_1 \equiv 0$, $l_2 \equiv x$, $l_3 \equiv 10l_7 + 7$, $l_4 \equiv 9l_2 + 6l_7 + 7$, $l_5 \equiv 12l_7$, $l_6 \equiv 3l_2 + 6l_7$, $l_7 \equiv y$, $l_8 \equiv 11l_2 + 10l_7 + 7$, $l_9 \equiv 4l_7$, $l_{10} \equiv 5l_2 + 10l_7$, $l_{11} \equiv 6l_7 + 7$, $l_{12} \equiv 13l_2 + 2l_7 + 7$, $l_{13} \equiv 2l_7$, $0 \leq x, y \leq 13$.

This gives, for example, the following lists $[l_1, \ldots, l_{m-1}]$:

$m = 4$. $[0, 0, 2], [0, 1, 0], [0, 2, 2], [0, 3, 0]$.

$m = 6$. $[0, 0, 0, 3, 0], [0, 0, 1, 5, 2], [0, 0, 2, 1, 4], [0, 0, 3, 3, 0], [0, 0, 4, 5, 2], [0, 0, 5, 1, 4],$
$[0, 1, 0, 2, 0], [0, 1, 1, 4, 2], [0, 1, 2, 0, 4], [0, 1, 3, 2, 0], [0, 1, 4, 4, 2], [0, 1, 5, 0, 4],$
$[0, 2, 0, 1, 0], [0, 2, 1, 3, 2], [0, 2, 2, 5, 4], [0, 2, 3, 1, 0], [0, 2, 4, 3, 2], [0, 2, 5, 5, 4],$
$[0, 3, 0, 0, 0], [0, 3, 1, 2, 2], [0, 3, 2, 4, 4], [0, 3, 3, 0, 0], [0, 3, 4, 2, 2], [0, 3, 5, 4, 4],$
$[0, 4, 0, 5, 0], [0, 4, 1, 1, 2], [0, 4, 2, 3, 4], [0, 4, 3, 5, 0], [0, 4, 4, 1, 2], [0, 4, 5, 3, 4],$
$[0, 5, 0, 4, 0], [0, 5, 1, 0, 2], [0, 5, 2, 2, 4], [0, 5, 3, 4, 0], [0, 5, 4, 0, 2], [0, 5, 5, 2, 4]$.

$m = 8$. $[0, 0, 0, 0, 4, 0, 4], [0, 0, 0, 4, 4, 0, 4], [0, 0, 2, 2, 6, 4, 0], [0, 0, 2, 6, 6, 4, 0],$
$[0, 0, 4, 4, 0, 0, 4], [0, 0, 4, 0, 0, 0, 4], [0, 0, 6, 6, 2, 4, 0], [0, 0, 6, 2, 2, 4, 0], [0, 2, 0, 0, 4, 6, 4],$
$[0, 2, 0, 4, 4, 6, 4], [0, 2, 2, 2, 6, 2, 0], [0, 2, 2, 6, 6, 2, 0], [0, 2, 4, 4, 0, 6, 4], [0, 2, 4, 0, 0, 6, 4],$
$[0, 2, 6, 6, 2, 2, 0], [0, 2, 6, 2, 2, 2, 0], [0, 4, 0, 0, 4, 4, 4], [0, 4, 0, 4, 4, 4, 4], [0, 4, 2, 2, 6, 0, 0],$
$[0, 4, 2, 6, 6, 0, 0], [0, 4, 4, 4, 0, 4, 4], [0, 4, 4, 0, 0, 4, 4], [0, 4, 6, 6, 2, 0, 0], [0, 4, 6, 2, 2, 0, 0],$
$[0, 6, 0, 0, 4, 2, 4], [0, 6, 0, 4, 4, 2, 4], [0, 6, 2, 2, 6, 6, 0], [0, 6, 2, 6, 6, 6, 0], [0, 6, 4, 4, 0, 2, 4],$
$[0, 6, 4, 0, 0, 2, 4], [0, 6, 6, 6, 2, 6, 0], [0, 6, 6, 2, 2, 6, 0]$.

3. Construction of Families of q-Cyclotomic Integers β with $\beta\overline{\beta} \in \mathbb{Z}$ and Cyclic Difference Sets

In this section we show how to construct families of cyclotomic integers $\beta \in K_q$, of degree m, with $\beta\overline{\beta} \in \mathbb{Z}$. That includes the construction of some families of cyclic difference sets.

Let $\nu = \varphi(m)$. Let $\alpha = t_0 + t_1\zeta_m + \cdots + t_{\nu-1}\zeta_m^{\nu-1}$, where the t_i are parameters which are supposed to take integer values, but which we often regard as indeterminates. Let $q = N(\alpha)$, where N is the norm from $\mathbb{Q}(t_0, \ldots, t_{\nu-1})[\zeta_m]$ to $\mathbb{Q}(t_0, \ldots, t_{\nu-1})$. Suppose that q is irreducible in $\mathbb{Z}[t_0, \ldots, t_{\nu-1}]$. Let $l_1, l_2, \ldots, l_{m-1}$ be integers satisfying condition (i) of Proposition 2.1. For $1 \leq a \leq m-1$ and $k \in \mathbb{Z}$, define $l_{a+km} = l_a + k(l_{m-1} + m/2)$. For $0 \leq i, j \leq m-1$, define the elements $d_{i,j}$ by formulas (2.12) and (2.13). By Proposition 2.1 we have that $d_{i,j} \in \mathbb{Q}$, for all integer values of the parameters t_i. When the parameters take integer values such that $d_{i,j} \in \mathbb{Z}$ and q is a prime number, we have that $d_{i,j} = c_{i,j}$ are as in (2.8). Hence, if $C = [c_{i,j}]_{0 \leq i,j < m}$, by the results shown in Section 2, we can identify, in formula (1.7), the Gaussian periods η_i with the matrices $K^{-i}CK^i$ to obtain the following equation in the unknowns $\varepsilon, \varepsilon_i \in \{0,1\}$.

$$\left(\varepsilon I + \sum_{i=0}^{m-1} \varepsilon_i K^{-i}CK^i\right)\left(\varepsilon I + \sum_{i=0}^{m-1} \varepsilon_i K^{-(i+m/2)}CK^{i+m/2}\right) = nI, \tag{3.14}$$

with $n \geq 2$ integer. Any solution of this equation such that at least one ε_i is nonzero corresponds to a cyclic difference set of the form D (if $\varepsilon = 0$) or D_0 (if $\varepsilon = 1$) described in (1.4) and (1.6), where s is a primitive root modulo q such that $s^f \equiv \zeta_m \bmod \alpha$, and $\{i_a : 1 \leq a \leq r\} = \{i : \varepsilon_i = 1\}$.

More generally, any solution of equation (3.14) with $\varepsilon, \varepsilon_i \in \mathbb{Z}$ corresponds to an element $\beta = \varepsilon + \sum_{i=0}^{m-1} \varepsilon_i \eta_i$ such that $\beta\overline{\beta} = n$.

Examples

First notice that, for m even, f odd, $q = mf + 1$ a prime and η_i as in (1.1), the elements $\gamma_0 = \sum_{i=0}^{m/2-1} \eta_{2i}$ and $\gamma_1 = \sum_{i=0}^{m/2-1} \eta_{2i+1}$ are quadratic Gaussian periods with $\gamma_0 + \gamma_1 = -1$, and $\gamma_0 - \gamma_1$ is a quadratic Gauss sum equal to $\pm\sqrt{(-1)^{m/2}q}$. When $m \equiv 0 \bmod 4$, γ_0 and γ_1 are real numbers. In this case we have $(\gamma_0 - \gamma_1)(\overline{\gamma_0 - \gamma_1}) = q$. When $m \equiv 2 \bmod 4$, γ_0 is nonreal and $\overline{\gamma_0} = \gamma_1$. In this case we have $(a + b\gamma_0 + c\gamma_1)(\overline{a + b\gamma_0 + c\gamma_1}) \in \mathbb{Z}$ for any $a, b, c \in \mathbb{Z}$. These cases will not be listed in our examples below.

$\underline{m = 4.}$ We have

$$C = \begin{bmatrix} \mathbf{a} & \mathbf{b} & \mathbf{c} & \mathbf{d} \\ \mathbf{e} & \mathbf{e} & \mathbf{d} & \mathbf{b} \\ \mathbf{r} & \mathbf{s} & \mathbf{r} & \mathbf{s} \\ \mathbf{e} & \mathbf{d} & \mathbf{b} & \mathbf{e} \end{bmatrix}, \tag{3.15}$$

with $\mathbf{r} = -\mathbf{a} - 2\mathbf{e} - 1$ and $\mathbf{s} = -\mathbf{b} - \mathbf{d} - \mathbf{e}$.

Taking $[l_1, l_2, l_3] = [0, 3, 0]$ and $\alpha = 4t + 2 + \zeta_4$, we get $q = 16t^2 + 16t + 5$, C as in (3.15) with $\mathbf{a} = t^2 + t$, $\mathbf{b} = t^2$, $\mathbf{c} = t^2 + t$, $\mathbf{d} = t^2 + 2t + 1$, $\mathbf{e} = t^2 + t$, and we have $C(K^{-2}CK^2) = \text{diagonal}\,(3t^2 + 3t + 1)$. So $\eta_0\overline{\eta_0} = n = 3t^2 + 3t + 1$.

Taking $[l_1, l_2, l_3] = [0, 1, 0]$ and $\alpha = 4t + 2 + 3\zeta_4$, we get $q = 16t^2 + 16t + 13$, C as in (3.15) with $\mathbf{a} = t^2 + t$, $\mathbf{b} = t^2 + 2t + 1$, $\mathbf{c} = t^2 + t + 2$, $\mathbf{d} = t^2$, $\mathbf{e} = t^2 + t + 1$, and we have $(I + C)(I + K^{-2}CK^2) = \text{diagonal}\,(3t^2 + 3t + 3)$. So $(1 + \eta_0)(\overline{1 + \eta_0}) = n = 3t^2 + 3t + 3$.

These examples correspond, respectively, to Chowla's family of 4th power residue difference sets and Hall's family of modified 4th power residue difference sets mentioned in [B] pages 123, 124.

Consider now the more general case of cyclotomic integers β with $\beta\overline{\beta} \in \mathbb{Z}$. We obtained the following families.

Taking $[l_1, l_2, l_3] = [0, 0, 2]$ and $\alpha = 4t + 1 + 2\zeta_4$, we get
$q = 16t^2 + 8t + 5$, C as in (3.15) with $\mathbf{a} = t^2 + t$, $\mathbf{b} = t^2 + t + 1$, $\mathbf{c} = t^2 - t$, $\mathbf{d} = t^2 + t$, $\mathbf{e} = t^2$. We have, for example, that equation (3.14) is satisfied by $[\varepsilon, \varepsilon_1, \varepsilon_2, \varepsilon_3, \varepsilon_4] = [0, 1, 1, -1, 3]$ and $n = 32t^2 + 16t + 11$. So we have
$(\eta_0 + \eta_1 - \eta_2 + 3\eta_3)(\overline{\eta_0 + \eta_1 - \eta_2 + 3\eta_3}) = 32t^2 + 16t + 11.$

Similarly we get
$(2+\eta_0-5\eta_1-3\eta_2-\eta_3)(\overline{2+\eta_0-5\eta_1-3\eta_2-\eta_3}) = 80t^2+40t+41$.

Taking $[l_1,l_2,l_3]=[0,0,2]$ and $\alpha = 4t+1+6\zeta_4$, we get
$q=16t^2+8t+37$, C as in (3.15) with $\mathbf{a}=t^2+t+2$, $\mathbf{b}=t^2+t+4$, $\mathbf{c}=t^2-t+2$, $\mathbf{d}=t^2+t+1$, $\mathbf{e}=t^2+2$, and we have, for example,
$(\eta_0+5\eta_1+3\eta_2+3\eta_3)(\overline{\eta_0+5\eta_1+3\eta_2+3\eta_3}) = 32t^2+16t+83$,
$(\eta_0-3\eta_1+3\eta_2-5\eta_3)(\overline{\eta_0-3\eta_1+3\eta_2-5\eta_3}) = 160t^2+80t+371$,
$(4+\eta_0-\eta_1+5\eta_2-5\eta_3)(\overline{4+\eta_0-\eta_1+5\eta_2-5\eta_3}) = 208t^2+104t+497$.

Taking $[l_1,l_2,l_3]=[0,3,0]$ and $\alpha = 4t+2+\zeta_4$, we get
$q=16t^2+16t+5$, C as in (3.15) with $\mathbf{a}=t^2+t$, $\mathbf{b}=t^2$, $\mathbf{c}=t^2+t$, $\mathbf{d}=t^2+2t+1$, $\mathbf{e}=t^2+t$, and we have, for example,
$\eta_0\overline{\eta}_0 = 3t^2+3t+1$ (corresponding to the power residue difference sets mentioned above),
$(\eta_0+2\eta_1+4\eta_2+2\eta_3)(\overline{\eta_0+2\eta_1+4\eta 2+2\eta_3}) = 19t^2+19t+11$,
$(5+\eta_0-\eta_1-4\eta_2-\eta_3)(\overline{5+\eta_0-\eta_1-4\eta_2-\eta_3}) = 51t^2+51t+55$,
$(\eta_0-2\eta_1+4\eta_2-2\eta_3)(\overline{\eta_0-2\eta_1+4\eta_2-2\eta_3}) = 99t^2+99t+31$,
$(4+\eta_0+3\eta_1+\eta_2-5\eta_3)(\overline{4+\eta_0+3\eta_1+\eta_2-5\eta_3}) = 144t^2+144t+61$.

Taking $[l_1,l_2,l_3]=[0,1,0]$ and $\alpha = 4t+2+3\zeta_4$, we get
$q=16t^2+16t+13$, C as in (3.15) with $\mathbf{a}=t^2+t$, $\mathbf{b}=t^2+2t+1$, $\mathbf{c}=t^2+t+2$, $\mathbf{d}=t^2$, $\mathbf{e}=t^2+t+1$, and we have, for example,
$(1+\eta_0)(\overline{1+\eta_0}) = 3t^2+3t+3$ (corresponding to the modified power residue difference sets mentioned above),
$(\eta_0+\eta_2-\eta_3)(\overline{\eta_0+\eta_2-\eta_3}) = 11t^2+11t+9$,
$(-4+\eta_0+3\eta_1+4\eta_2+3\eta_3)(\overline{-4+\eta_0+3\eta_1+4\eta_2+3\eta_3}) = 19t^2+19t+61$,
$(4+\eta_0-3\eta_1-4\eta_2-3\eta_3)(\overline{4+\eta_0-3\eta_1-4\eta_2-3\eta_3}) = 59t^2+59t+87$.

Taking $[l_1,l_2,l_3]=[0,3,0]$ and $\alpha = 4t+2+5\zeta_4$, we get
$q=16t^2+16t+29$, C as in (3.15) with $\mathbf{a}=t^2+t+2$, $\mathbf{b}=t^2+2$, $\mathbf{c}=t^2+t$, $\mathbf{d}=t^2+2t+3$, $\mathbf{e}=t^2+t+1$, and we have, for example,
$(-1+\eta_0)(\overline{-1+\eta_0}) = 3t^2+3t+7$,
$(\eta_0-\eta_1+2\eta_2-\eta_3)(\overline{\eta_0-\eta_1+2\eta_2-\eta_3}) = 27t^2+27t+49$,
$(\eta_0+2\eta_1+\eta_2+5\eta_3)(\overline{\eta_0+2\eta_1+\eta_2+5\eta_3}) = 43t^2+43t+83$,
$(\eta_0-2\eta_1+\eta_2-5\eta_3)(\overline{\eta_0-2\eta_1+\eta_2-5\eta_3}) = 99t^2+99t+181$.

Taking $[l_1,l_2,l_3]=[0,1,0]$ and $\alpha = 4t+2+7\zeta_4$, we get
$q=16t^2+16t+53$, C as in (3.15) with $\mathbf{a}=t^2+t+2$, $\mathbf{b}=t^2+2t+3$, $\mathbf{c}=t^2+t+6$, $\mathbf{d}=t^2+2$, $\mathbf{e}=t^2+t+4$, and we have, for example,
$(2+\eta_0)(\overline{2+\eta_0}) = 3t^2+3t+13$,
$(\eta_0-2\eta_1+2\eta_2-2\eta_3)(\overline{\eta_0-2\eta_1+2\eta_2-2\eta_3}) = 51t^2+51t+169$,
$(\eta_0-4\eta_1-2\eta_2-4\eta_3)(\overline{\eta_0-4\eta_1-2\eta_2-4\eta_3}) = 67t^2+67t+227$,
$(\eta_0+4\eta_1-2\eta_2+4\eta_3)(\overline{\eta_0+4\eta_1-2\eta_2+4\eta_3}) = 99t^2+99t+331$,
$(3+\eta_0-5\eta_1-4\eta_2-5\eta_3)(\overline{3+\eta_0-5\eta_1-4\eta_2-5\eta_3}) = 99t^2+99t+367$.

$\underline{m = 6.}$ We have

$$C = \begin{bmatrix} \mathbf{a} & \mathbf{b} & \mathbf{c} & \mathbf{d} & \mathbf{e} & \mathbf{f} \\ \mathbf{g} & \mathbf{h} & \mathbf{i} & \mathbf{e} & \mathbf{c} & \mathbf{i} \\ \mathbf{h} & \mathbf{j} & \mathbf{g} & \mathbf{f} & \mathbf{i} & \mathbf{b} \\ \mathbf{r} & \mathbf{s} & \mathbf{t} & \mathbf{r} & \mathbf{s} & \mathbf{t} \\ \mathbf{g} & \mathbf{f} & \mathbf{i} & \mathbf{b} & \mathbf{h} & \mathbf{j} \\ \mathbf{h} & \mathbf{i} & \mathbf{e} & \mathbf{c} & \mathbf{i} & \mathbf{g} \end{bmatrix}, \tag{3.16}$$

with $\mathbf{r} = -\mathbf{a} - 2\mathbf{g} - 2\mathbf{h} - 1$, $\mathbf{s} = -\mathbf{b} - \mathbf{f} - \mathbf{h} - \mathbf{i} - \mathbf{j}$ and $\mathbf{t} = -\mathbf{c} - \mathbf{e} - \mathbf{g} - 2\mathbf{i}$.

Taking $[l_1, l_2, l_3, l_4, l_5] = [0, 0, 3, 3, 0]$ and $\alpha = 6t - 1 + 6\zeta_6$, we get $q = 36t^2 + 24t + 31$, C as in (3.16) with $\mathbf{a} = t^2 + 2t + 1$, $\mathbf{b} = \mathbf{c} = t^2 + t + 2$, $\mathbf{d} = t^2 - 2t$, $\mathbf{e} = \mathbf{f} = t^2 + t$, $\mathbf{g} = t^2 + 1$, $\mathbf{h} = t^2$, $\mathbf{i} = \mathbf{j} = t^2 + t + 1$, and we have $(C + K^{-3}CK^3 + K^{-4}CK^4)\big(K^{-3}(C + K^{-3}CK^3 + K^{-4}CK^4)K^3\big) = \text{diagonal}\,(9t^2 + 6t + 8)$. So $(\eta_0 + \eta_3 + \eta_4)(\overline{\eta_0 + \eta_3 + \eta_4}) = n = 9t^2 + 6t + 8$.

This example corresponds to Hall's family of cyclic difference sets mentioned in [B], Theorem 5.23.

Taking $[l_1, l_2, l_3, l_4, l_5] = [0, 0, 1, 5, 2]$ and $\alpha = -12\zeta_6 t - 3 + \zeta_6$, we get $q = 144t^2 + 12t + 7$, C as in (3.16) with $\mathbf{a} = 4t^2$, $\mathbf{b} = \mathbf{e} = 4t^2 - 2t$, $\mathbf{c} = \mathbf{j} = 4t^2 + t + 1$, $\mathbf{d} = 4t^2 + 4t$, $\mathbf{f} = \mathbf{i} = 4t^2 + t$, $\mathbf{g} = 4t^2 + 2t$, $\mathbf{h} = 4t^2 - t$, and we have, for $[\varepsilon, \varepsilon_0, \varepsilon_1, \varepsilon_2, \varepsilon_3, \varepsilon_4, \varepsilon_5] = [0, 1, 1, 3, -3, 1, 3]$, $\big(\varepsilon I + \sum_{i=0}^{5} \varepsilon_i K^{-i}CK^i\big)\big(K^{-3}\big(\varepsilon I + \sum_{i=0}^{5} \varepsilon_i K^{-i}CK^i\big)K^3\big) = \text{diagonal}\,(576t^2 + 48t + 29)$. So $(\varepsilon + \sum_{i=0}^{5} \varepsilon_i\eta_i)(\overline{\varepsilon + \sum_{i=0}^{5} \varepsilon_i\eta_i}) = n = 576t^2 + 48t + 29$.

Taking $[l_1, l_2, l_3, l_4, l_5] = [0, 0, 1, 5, 2]$ and $\alpha = -12\zeta_6 t + 3 - 5\zeta_6$, we get $q = 144t^2 + 84t + 19$, C as in (3.16) with $\mathbf{a} = 4t^2 + 2t$, $\mathbf{b} = \mathbf{e} = 4t^2$, $\mathbf{c} = \mathbf{j} = 4t^2 + 3t$, $\mathbf{d} = 4t^2 + 6t + 2$, $\mathbf{f} = \mathbf{i} = 4t^2 + 3t + 1$, $\mathbf{g} = 4t^2 + 4t + 1$, $\mathbf{h} = 4t^2 + t$, and we have, for $[\varepsilon, \varepsilon_0, \varepsilon_1, \varepsilon_2, \varepsilon_3, \varepsilon_4, \varepsilon_5] = [2, 1, 1, 3, -3, 1, 3]$, $(\varepsilon + \sum_{i=0}^{5} \varepsilon_i\eta_i)(\overline{\varepsilon + \sum_{i=0}^{5} \varepsilon_i\eta_i}) = n = 576t^2 + 336t + 77$.

Taking $[l_1, l_2, l_3, l_4, l_5] = [0, 4, 5, 3, 4]$ and $\alpha = (6 + 6\zeta_6)t + 1 + 5\zeta_6$, we get $q = 108t^2 + 108t + 31$, C as in (3.16) with $\mathbf{a} = \mathbf{i} = \mathbf{j} = 3t^2 + 3t + 1$, $\mathbf{b} = \mathbf{c} = 3t^2 + t$, $\mathbf{d} = 3t^2 + 3t$, $\mathbf{e} = \mathbf{f} = 3t^2 + 5t + 2$, $\mathbf{g} = 3t^2 + 2t$, $\mathbf{h} = 3t^2 + 4t + 1$, and we have, for $[\varepsilon, \varepsilon_0, \varepsilon_1, \varepsilon_2, \varepsilon_3, \varepsilon_4, \varepsilon_5] = [0, 1, 0, 1, 0, 1, 3]$, $(\varepsilon + \sum_{i=0}^{5} \varepsilon_i\eta_i)(\overline{\varepsilon + \sum_{i=0}^{5} \varepsilon_i\eta_i}) = n = 108t^2 + 108t + 32$.

Taking $[l_1, l_2, l_3, l_4, l_5] = [0, 1, 5, 0, 4]$ and $\alpha = (6 + 6\zeta_6)t - 1 + 7\zeta_6$, we get $q = 108t^2 + 108t + 43$, C as in (3.16) with $\mathbf{a} = 3t^2 + 3t$, $\mathbf{b} = \mathbf{c} = 3t^2 + 5t + 2$, $\mathbf{d} = 3t^2 + 3t + 3$, $\mathbf{e} = \mathbf{f} = 3t^2 + t$, $\mathbf{g} = 3t^2 + 4t + 2$, $\mathbf{h} = 3t^2 + 2t + 1$, $\mathbf{i} = \mathbf{j} = 3t^2 + 3t + 1$, and we have, for $[\varepsilon, \varepsilon_0, \varepsilon_1, \varepsilon_2, \varepsilon_3, \varepsilon_4, \varepsilon_5] = [3, 1, 0, 1, 0, 1, 3]$, $(\varepsilon + \sum_{i=0}^{5} \varepsilon_i\eta_i)(\overline{\varepsilon + \sum_{i=0}^{5} \varepsilon_i\eta_i}) = n = 108t^2 + 108t + 47$.

$\underline{m = 8.}$ Write

$$C = \begin{bmatrix} \mathbf{a} & \mathbf{b} & \mathbf{c} & \mathbf{d} & \mathbf{e} & \mathbf{b} & \mathbf{f} & \mathbf{d} \\ \mathbf{g} & \mathbf{g} & \mathbf{h} & \mathbf{i} & \mathbf{b} & \mathbf{d} & \mathbf{i} & \mathbf{j} \\ \mathbf{k} & \mathbf{i} & \mathbf{k} & \mathbf{j} & \mathbf{f} & \mathbf{i} & \mathbf{c} & \mathbf{h} \\ \mathbf{g} & \mathbf{i} & \mathbf{i} & \mathbf{g} & \mathbf{d} & \mathbf{j} & \mathbf{h} & \mathbf{b} \\ \mathbf{r} & \mathbf{s} & \mathbf{t} & \mathbf{s} & \mathbf{u} & \mathbf{s} & \mathbf{t} & \mathbf{s} \\ \mathbf{g} & \mathbf{d} & \mathbf{j} & \mathbf{h} & \mathbf{b} & \mathbf{g} & \mathbf{i} & \mathbf{i} \\ \mathbf{k} & \mathbf{j} & \mathbf{f} & \mathbf{i} & \mathbf{c} & \mathbf{h} & \mathbf{k} & \mathbf{i} \\ \mathbf{g} & \mathbf{h} & \mathbf{i} & \mathbf{b} & \mathbf{d} & \mathbf{i} & \mathbf{j} & \mathbf{g} \end{bmatrix}, \tag{3.17}$$

with $\mathbf{r} = -\mathbf{a} - 4\mathbf{g} - 2\mathbf{k} - 1$, $\mathbf{s} = -\mathbf{b} - \mathbf{g} - 2\mathbf{i} - \mathbf{d} - \mathbf{j} - \mathbf{h}$, $\mathbf{t} = -\mathbf{c} - \mathbf{h} - \mathbf{k} - 2\mathbf{i} - \mathbf{j} - \mathbf{f}$, and $\mathbf{u} = -\mathbf{e} - 2\mathbf{b} - \mathbf{f} - 2\mathbf{d} - \mathbf{c}$.

Taking $[l_1, l_2, l_3, l_4, l_5, l_6, l_7] = [0, 4, 4, 4, 0, 4, 4]$ and
$\alpha = 1 + (2u + 1)\zeta_8 + 4w\zeta_8^2 + (2u + 1)\zeta_8^3$, we get
$q = 256w^4 - 256u^2w^2 - 256uw^2 - 32w^2 + 64u^4 + 128u^3 + 112u^2 + 48u + 9$ and C as in (3.17) with
$\mathbf{a} = (q - 32w^2 + 16u + 16u^2 - 9)/64$,
$\mathbf{b} = (q + 96w^2 - 48u^2 + 64w - 48u - 9)/64$,
$\mathbf{c} = \mathbf{a} - w$, $\mathbf{d} = \mathbf{b} - 2w$,
$\mathbf{e} = (q + 55 + 144u - 288w^2 + 144u^2)/64$,
$\mathbf{f} = \mathbf{a} + w$, $\mathbf{g} = \mathbf{a}$,
$\mathbf{h} = (q + 16u - 160w^2 + 80u^2 - 9)/64$,
$\mathbf{i} = \mathbf{k} = \mathbf{b} - w$,
$\mathbf{j} = (q + 55 + 144u + 80u^2 - 160w^2)/64$.

If the parameters u, w are chosen so that $2w^2 = u^2 + u$, then $q = 64w^2 + 9$ and $C(K^{-4}CK^4) = \text{diagonal}\,(7w^2 + 1)$. So $\eta_0\overline{\eta_0} = n = 7w^2 + 1$.

This corresponds to one of Emma Lehmer's results ([L] Section 6) mentioned in [B] page 124 (example iv).

For $u = 1$, $w = 1$, we get $q = 73$,

$$C = \begin{bmatrix} 1 & 2 & 0 & 0 & 2 & 2 & 2 & 0 \\ 1 & 1 & 0 & 1 & 2 & 0 & 1 & 3 \\ 1 & 1 & 1 & 3 & 2 & 1 & 0 & 0 \\ 1 & 1 & 1 & 1 & 0 & 3 & 0 & 2 \\ -8 & -8 & -8 & -8 & -8 & -8 & -8 & -8 \\ 1 & 0 & 3 & 0 & 2 & 1 & 1 & 1 \\ 1 & 3 & 2 & 1 & 0 & 0 & 1 & 1 \\ 1 & 0 & 1 & 2 & 0 & 1 & 3 & 1 \end{bmatrix}$$

and $C(K^{-4}CK^4) = \text{diagonal}\,(8)$. So $\eta_0\overline{\eta_0} = n = 8$.

For $u = 239^2$, $w = 239 \times 169$, we get $q = 104411704393$, $C(K^{-4}CK^4) = \text{diagonal}\,(11420030168)$. So $\eta_0\overline{\eta_0} = n = 11420030168$. (Note: there is a misprint in [B] page 124.)

Taking $[l_1, l_2, l_3, l_4, l_5, l_6, l_7] = [0, 0, 4, 0, 0, 0, 4]$ and
$\alpha = 3 + \big(2 + (2/7)(3u - 2w)\big)\zeta_8 - \big(2 + (4/7)(u - 3w)\big)\zeta_8^2 + \big(4 + (2/7)(3u - 2w)\big)\zeta_8^3$,
we get $q = 64((2w^2 - u^2)/7)^2 - 128u(2w^2 - u^2)/7 - 32w^2 + 112u^2 + 336u + 441$ and C as in (3.17) with
$\mathbf{a} = (q - 57 - 16u + 16(2w^2 - u^2)/7)/64$,
$\mathbf{b} = (q + 71 + 48u - 64w - 48(2w^2 - u^2)/7)/64$,
$\mathbf{c} = \mathbf{a} + 2 + w$, $\mathbf{d} = \mathbf{b} + 2w$,
$\mathbf{e} = (q - 377 - 144u + 144(2w^2 - u^2)/7)/64$,
$\mathbf{f} = \mathbf{a} + 2 - w$, $\mathbf{g} = \mathbf{a} + 1$,
$\mathbf{h} = (q + 71 - 16u + 80(2w^2 - u^2)/7)/64$,
$\mathbf{i} = \mathbf{b} + w$, $\mathbf{k} = \mathbf{b} - 1 + w$,
$\mathbf{j} = (q - 377 - 144u + 80(2w^2 - u^2)/7)/64$.

If the parameters u, w are chosen so that $2w^2 = u^2 + 7u$, then $q = 64w^2 + 441$ and $(I + C)(I + K^{-4}CK^4) = \text{diagonal}\,(7w^2 + 49)$. So $(1 + \eta_0)(\overline{1 + \eta_0}) = n = 7w^2 + 49$.

(Observe that if $2w^2 = u^2 + 7u$, then $7|(u+3w)(u-3w)$. In order for α to be an algebraic integer we must choose u and w integers such that $7|u-3w$.)

This corresponds to one of Emma Lehmer's results ([L] Section 7) mentioned in [B] page 124 (example v).

For $u = 5^2$, $w = 5 \times 4$, we get $q = 26041$, $(I+C)(I+K^{-4}CK^4) = \text{diagonal}\,(2849)$. So $(1+\eta_0)(\overline{1+\eta_0}) = n = 2849$.
For $u = 379^2$, $w = 379 \times (-268)$, we get $q = 660279756217$, $(I+C)(I+K^{-4}CK^4) = \text{diagonal}\,(72218098337)$. So $(1+\eta_0)(\overline{1+\eta_0}) = n = 72218098337$.

Taking $[l_1, l_2, l_3, l_4, l_5, l_6, l_7] = [0, 4, 4, 4, 0, 4, 4]$ and
$\alpha = 5 + 4\zeta_8 + (2u+1)\zeta_8 + 4w\zeta_8^2 + (2u+1)\zeta_8^3$, we get
$q = 64u^4 - 256u^2w^2 + 256w^4 + 384u^3 - 768uw^2 + 1392u^2 - 1280uw + 480w^2 + 2448u - 1920w + 1801$ and C as in (3.17) with
$\mathbf{a} = (q + 55 + 16u^2 + 48u - 32w^2)/64$,
$\mathbf{b} = (q - 265 - 48u^2 - 272u + 320w + 96w^2)/64$,
$\mathbf{c} = (q - 9 + 16u^2 + 176u - 320w - 32w^2)/64$,
$\mathbf{d} = (q + 119 - 48u^2 - 16u - 320w + 96w^2)/64$,
$\mathbf{e} = (q + 631 + 144u^2 + 432u - 288w^2)/64$,
$\mathbf{f} = (q - 393 + 16u^2 - 80u + 320w - 32w^2)/64$,
$\mathbf{g} = (q - 73 + 48u - 32w^2 + 16u^2)/64$,
$\mathbf{h} = (q - 265 - 80u - 160w^2 + 80u^2 + 256w)/64$,
$\mathbf{i} = (q - 73 - 144u + 96w^2 - 48u^2)/64$,
$\mathbf{j} = (q + 695 + 560u - 160w^2 + 80u^2 - 256w)/64$,
$\mathbf{k} = (q + 55 - 48u^2 - 144u + 96w^2)/64$.

If the parameters u, w are chosen so that $2w^2 = u^2 + 3u - 2$, then $q = 2112w^2 - 1280uw - 1920w + 3689$ and $(-2I+C)(K^{-4}(-2I+C)K^4) = \text{diagonal}\,(231w^2 - 140uw - 210w + 408)$. So $(-2+\eta_0)(\overline{-2+\eta_0}) = n = 231w^2 - 140uw - 210w + 408$.

For $u = 1$, $w = -1$, we get $q = 9001$ and $(-2+\eta_0)(\overline{-2+\eta_0}) = 989$.
For $u = 625$, $w = 443$, we get $q = 59231017$ and $(-2+\eta_0)(\overline{-2+\eta_0}) = 6478397$.
For $u = 2257$, $w = -1597$, we get $q = 10003202857$ and $(-2+\eta_0)(\overline{-2+\eta_0}) = 1094100317$.

Taking $[l_1, l_2, l_3, l_4, l_5, l_6, l_7] = [0, 4, 4, 4, 0, 4, 4]$ and
$\alpha = -3 - 4z + 4\zeta_8^3 + (2u+1)\zeta_8 + 4w\zeta_8^2 + (2u+1)\zeta_8^3$, we get
$q = 64u^4 - 256u^2w^2 + 256w^4 + 128u^3 - 256uw^2 + 752u^2 - 1536uw + 1248w^2 + 688u - 768w + 697$ and C as in (3.17) with
$\mathbf{a} = (q - 57 + 16u - 32w^2 + 16u^2)/64$,
$\mathbf{b} = (q + 71 + 208u - 192w + 96w^2 - 48u^2)/64$,
$\mathbf{c} = (q + 199 - 240u + 192w - 32w^2 + 16u^2)/64$,
$\mathbf{d} = (q - 185 - 304u + 192w + 96w^2 - 48u^2)/64$,
$\mathbf{e} = (q - 377 + 144u - 288w^2 + 144u^2)/64$,
$\mathbf{f} = (q + 455 + 272u - 192w - 32w^2 + 16u^2)/64$,
$\mathbf{g} = (q + 135 + 16u + 16u^2 - 32w^2)/64$,
$\mathbf{h} = (q + 71 + 272u - 160w^2 - 512w + 80u^2)/64$,

$\mathbf{i} = (q - 57 - 48u + 96w^2 - 48u^2)/64,$
$\mathbf{j} = (q - 121 - 112u - 160w^2 + 512w + 80u^2)/64,$
$\mathbf{k} = (q - 249 - 48u^2 + 96w^2 - 48u)/64.$

If the parameters u, w are chosen so that $2w^2 = u^2 + u + 6$, then $q = 2624w^2 - 1536uw - 768w - 1127$ and $(3I + C)(K^{-4}(3I + C)K^4) = \text{diagonal}\,(287w^2 - 168uw - 84w - 115)$. So $(3 + \eta_0)\overline{(3 + \eta_0)} = n = 287w^2 - 168uw - 84w - 115$.

For $u = 3$, $w = 3$, we get $q = 6361$ and $(3 + \eta_0)\overline{(3 + \eta_0)} = 704$.
For $u = 428$, $w = -303$, we get $q = 440333017$ and $(3 + \eta_0)\overline{(3 + \eta_0)} = 48161432$.
For $u = 5949432723$, $w = 4206884223$, we get $q = 7995332969457225644761$ and $(3 + \eta_0)\overline{(3 + \eta_0)} = 874489543534384054904$.

$\underline{m = 10.}$ Let $[l_1, l_2, l_3, l_4, l_5, l_6, l_7, l_8, l_9] = [0, 3, 5, 6, 0, 9, 5, 2, 0]$. Taking $\alpha = 2\zeta_{10} + \zeta_{10}^2$ we get $q = 31$ and

$$C = \begin{bmatrix} 0 & 0 & 0 & 0 & 0 & 2 & 0 & 0 & 1 & 0 \\ 1 & 0 & 1 & 1 & 0 & 0 & 0 & 0 & 0 & 0 \\ 0 & 0 & 0 & 1 & 0 & 0 & 0 & 0 & 1 & 1 \\ 0 & 0 & 0 & 0 & 0 & 1 & 0 & 1 & 0 & 1 \\ 0 & 0 & 0 & 0 & 1 & 0 & 0 & 1 & 1 & 0 \\ -3 & -2 & -3 & -3 & -3 & -3 & -2 & -3 & -3 & -3 \\ 1 & 0 & 0 & 1 & 1 & 0 & 0 & 0 & 0 & 0 \\ 0 & 0 & 1 & 0 & 1 & 0 & 1 & 0 & 0 & 0 \\ 0 & 1 & 0 & 0 & 0 & 0 & 1 & 1 & 0 & 0 \\ 0 & 1 & 1 & 0 & 0 & 0 & 0 & 0 & 0 & 1 \end{bmatrix},$$

and we have $(C + K^{-1}CK)K^{-5}(C + K^{-1}CK)K^5 = 5I$. So $(\eta_0 + \eta_1)\overline{(\eta_0 + \eta_1)} = n = 5$. This corresponds to Hayashi's cyclic difference set mentioned in [B], Theorem 5.24. Here we can take $s = 11$, a primitive root modulo 31 such that $s^3 \equiv \zeta_{10} \bmod \alpha$.

We do not know families of cyclotomic integers β with $\beta\overline{\beta} \in \mathbb{Z}$ for $m = 10$. Particular examples are not difficult to find, like that following.

Taking $\alpha = -\zeta_{10} + \zeta_{10}^2 + \zeta_{10}^3$ we get $q = 11$. In this case $\eta_0 = \zeta_{11}$, so $\eta_0\overline{\eta_0} = 1$. We also have, for example,
$(-1 + 2\eta_0 + 2\eta_1 + 2\eta_2 - 2\eta_3 - 2\eta_4)\overline{(-1 + 2\eta_0 + 2\eta_1 + 2\eta_2 - 2\eta_3 - 2\eta_4)} = 23$.

Taking $\alpha = -1 - 2\zeta_{10} - 3\zeta_{10}^2 + \zeta_{10}^4$ we get $q = 191$, and we have
$(3 + 2\eta_0 - 2\eta_2 - \eta_3 + \eta_4)\overline{(3 + 2\eta_0 - 2\eta_2 - \eta_3 + \eta_4)} = 200$.

Taking $\alpha = 2 + 3\zeta_{10} - 2\zeta_{10}^2 + \zeta_{10}^3$ we get $q = 311$, and we have
$(-1 + 2\eta_0 + 2\eta_1 + 2\eta_2 - 2\eta_3 - 2\eta_4)\overline{(-1 + 2\eta_0 + 2\eta_1 + 2\eta_2 - 2\eta_3 - 2\eta_4)} = 611$.

4. Construction of Families of Cyclic Polynomials of Degree m Whose Roots are Complex Nonreal Cyclotomic Integers of Norm q in $K_q/\mathbb{Q}$

Recall that, given a prime $q \equiv 1 \mod m$, $K_q = \mathbb{Q}[\eta_0]$ is the subfield of degree m of $\mathbb{Q}[\zeta_q]$ and τ is the generator of $\text{Gal}(K_q/\mathbb{Q})$ such that $\tau(\eta_i) = \eta_{i+1}$ (indices mod m). Denote by $\mathcal{N}$ the norm from K_q to $\mathbb{Q}$. Let $\beta = \sum_{i=0}^{m-1} a_i\eta_i \in K_q$ with $a_i \in \mathbb{Z}$. We saw, in Section 2, that β can be represented by the matrix $M = \sum_{i=0}^{m-1} a_i K^{-i} C K^i$. As is shown, for example, in [T2] Proposition 1, the characteristic polynomial of β (with respect to $K_q/\mathbb{Q}$) is equal to $P(x) = \det(xI - M)$, the characteristic polynomial of M. In this section we search for elements β, as above, such that $\mathcal{N}(\beta) = \pm q$ or, equivalently, for matrices M, as above, whose characteristic polynomials $P(x)$ are monic irreducible (of degree m), with coefficients in $\mathbb{Z}$ and constant terms $= \pm q$. Note that the polynomials $P(x)$ are cyclic, that is, irreducible with cyclic Galois groups. When $f = (q-1)/m$ is odd, as is the case that we consider here, those elements β are complex nonreal. We considered in [T3] the case f even, as a step in the construction of families of units. When f is odd and greater than 1, the units in K_q are real; in particular the elements $\tau^j(\beta)/\beta$ (which are units since β divides q, and q ramifies totally in $K_q/\mathbb{Q}$) belong to the real subfield of K_q of degree $m/2$. Also notice that, when f is odd, $\mathcal{N}(\beta) > 0$ (since complex conjugation is an automorphism of K_q) and the constant term of $P(x)$ is $(-1)^m\mathcal{N}(\beta) = \mathcal{N}(\beta)$.

Our methods to find the elements β of norm q and their characteristic polynomials $P(x)$ in the case f odd are essentially the same as in the case f even, except in the construction of the matrices C. The construction of C is a bit more complicated in the case f odd and was already done in Sections 2 and 3. The method we used to find most of the examples is the following: construct a matrix C on several parameters $t_1, t_2, \ldots, t_{\varphi(m)}$ and search for values of the t_i, depending on one parameter t, and integers a_i such that $M = \sum_{i=0}^{m-1} a_i K^{-i} C K^i$ has determinant q.

A variation of the program of Section 5 can be used to check the examples. Just replace line 50 (N0:=evalm(...)) and the following lines by:

```
q:=q; irreduc(q);
det(M);
P:=collect(charpoly(M,x),x);
galois(P,x);
```

Examples

<u>$m = 4$.</u> Taking $[l_1, l_2, l_3] = [0, 0, 2]$ and $\alpha = 1+2\zeta_4+(3+4\zeta_4)t$, we get $q = 25t^2+22t+5$, and for $M = 2C + K^{-1}CK - 2K^{-2}CK^2 - K^{-3}CK^3$ we get $P(x) = \det(xI - M) = x^4 + 5qx^2 + q$. So $P(x)$ is the characteristic polynomial of $\beta = 2\eta_0 + \eta_1 - 2\eta_2 - \eta_3$ in the subfield of degree 4 of $\mathbb{Q}[\zeta_q]$, when q is prime.

Taking $[l_1, l_2, l_3] = [0, 1, 0]$ and $\alpha = -2+3\zeta_4+(-3+4\zeta_4)t$, we get $q = 25t^2+36t+13$, and for $M = C - 3K^{-1}CK - K^{-2}CK^2 + 3K^{-3}CK^3$ we get $P(x) = \det(xI - M) = x^4 + 10qx^2 + q$. So $P(x)$ is the characteristic polynomial of $\beta = \eta_0 - 3\eta_1 - \eta_2 + 3\eta_3$ in the subfield of degree 4 of $\mathbb{Q}[\zeta_q]$, when q is prime.

Taking $[l_1, l_2, l_3] = [0, 2, 2]$ and $\alpha = -1 + 2\zeta_4 - t$, we get $q = t^2 + 2t + 5$, and for $M = C - K^{-2}CK^2$ we get $P(x) = \det(xI - M) = x^4 + qx^2 + q$. So $P(x)$ is the characteristic polynomial of $\beta = \eta_0 - \eta_2$ in the subfield of degree 4 of $\mathbb{Q}[\zeta_q]$, when q is prime.

Taking $[l_1, l_2, l_3] = [0, 0, 2]$ and $\alpha = 1 + 2\zeta_4 + \zeta_4 t$, we get $q = t^2 + 4t + 5$, and for $M = C - K^{-1}CK - K^{-2}CK^2 + K^{-3}CK^3$ we get $P(x) = \det(xI - M) = x^4 + 2qx^2 + q$. So $P(x)$ is the characteristic polynomial of $\beta = \eta_0 - \eta_1 - \eta_2 + \eta_3$ in the subfield of degree 4 of $\mathbb{Q}[\zeta_q]$, when q is prime.

Taking $[l_1, l_2, l_3] = [0, 0, 2]$ and $\alpha = u + \zeta_4 v$, we get $q = u^2 + v^2$, and for $M = 2C - 3K^{-1}CK - 2K^{-2}CK^2 + 3K^{-3}CK^3$ we get $P(x) = \det(xI - M) = x^4 + 13qx^2 + \frac{1}{4}(12u - 5v)^2 q$. If we take $u = 5t+6$ and $v = 12t+14$ then we get $q = 169t^2+396t+232$

and, if q is prime, then the characteristic polynomial of $\beta = 2\eta_0 - 3\eta_1 - 2\eta_2 + 3\eta_3$ is $P(x) = x^4 + 13qx^2 + q$.

Note that the sets $\{q = q(t) : t \in \mathbb{Z}$ and q is prime$\}$ of the above examples are all distinct.

$\underline{m = 6.}$ Taking $[l_1, l_2, l_3, l_4, l_5] = [0, 0, 1, 5, 2]$ and $\alpha = -3 + \zeta_6 + (-2 + \zeta_6)t$, we get $q = 3t^2 + 9t + 7$, and for $M = C + 2K^{-1}CK - 2K^{-2}CK^2 - K^{-3}CK^3 - 2K^{-4}CK^4 + 2K^{-5}CK^5$ we get $P(x) = \det(xI - M) = x^6 + 9qx^4 + 6qx^2 + q$. So $P(x)$ is the characteristic polynomial of $\beta = \eta_0 + 2\eta_1 - 2\eta_2 - \eta_3 - 2\eta_4 + 2\eta_5$ in the subfield of degree 6 of $\mathbb{Q}[\zeta_q]$, when q is prime.

Taking $[l_1, l_2, l_3, l_4, l_5] = [0, 1, 3, 2, 0]$ and $\alpha = -2 + 3\zeta_6 - t$, we get $q = t^2 + t + 7$. For $M_1 = C - K^{-1}CK - K^{-3}CK^3 + K^{-4}CK^4$, we get $P_1(x) = \det(xI - M_1) = x^6 + 2qx^4 + q(q-4)x^2 + q$, and for $M_2 = C - K^{-3}CK^3$ we get $P_2(x) = \det(xI - M_2) = x^6 + qx^4 + 2qx^2 + q$. So $P_1(x)$ is the characteristic polynomial of $\beta_1 = \eta_0 - \eta_1 - \eta_3 + \eta_4$ and $P_2(x)$ is the characteristic polynomial of $\beta_2 = \eta_0 - \eta_3$ in the subfield of degree 6 of $\mathbb{Q}[\zeta_q]$, when q is prime.

Taking $[l_1, l_2, l_3, l_4, l_5] = [0, 0, 0, 3, 0]$ and $\alpha = u + \zeta_6 v$, we get $q = u^2 + uv + v^2$, and for $M = C - K^{-1}CK - 2K^{-2}CK^2 - K^{-3}CK^3 + K^{-4}CK^4 + 2K^{-5}CK^5$ we get $P(x) = \det(xI - M) = x^6 + 6qx^4 + 9q^2x^2 + (2u^2 + 2uv - v^2)^2 q$. So, for the prime numbers $q = u^2 + uv + v^2$ such that $2u^2 + 2uv - v^2 = \pm 1$, the characteristic polynomial of $\beta = \eta_0 - \eta_1 - 2\eta_2 - \eta_3 + \eta_4 + 2\eta_5$ is $P(x) = x^6 + 6qx^4 + 9q^2x^2 + q$.

Taking $[l_1, l_2, l_3, l_4, l_5] = [0, 0, 1, 5, 2]$ and $\alpha = 3u + \zeta_6 v$, we get $q = 9u^2 + 3uv + v^2$, and for $M = C + K^{-1}CK + 2K^{-2}CK^2 - K^{-3}CK^3 - K^{-4}CK^4 - 2K^{-5}CK^5$ we get $P(x) = \det(xI - M) = x^6 + 6qx^4 + (5u^2 + 3uv + 9v^2)qx^2 + (u^2 - 2v^2)^2 q$. So, for the prime numbers $q = 9u^2 + 3uv + v^2$ such that $u^2 - 2v^2 = \pm 1$, the characteristic polynomial of $\beta = \eta_0 + \eta_1 + 2\eta_2 - \eta_3 - \eta_4 - 2\eta_5$ is $P(x) = x^6 + 6qx^4 + (5u^2 + 3uv + 9v^2)qx^2 + q$.

Note that the sets $\{q = q(t) : t \in \mathbb{Z}$ and q is prime$\}$ of the above examples are all distinct.

$\underline{m = 8.}$ We could not find families as above for $m = 8$. (Families of cyclotomic polynomials of degree 8 with real roots of absolute norm q in $\mathbb{Z}[\zeta_q]$ can be obtained from Example 4.6 in [T3].) But let us see how the method works in particular cases.

Taking $[l_1, l_2, \ldots, l_7] = [0, 4, 6, 2, 2, 0, 0]$ and $\alpha = 1 - 2\zeta_8 - \zeta_8^2 + \zeta_8^3$, we get $q = 41$ and

$$C = \begin{bmatrix} 0 & 2 & 1 & 2 & 0 & 0 & 0 & 0 \\ 1 & 0 & 1 & 0 & 0 & 2 & 0 & 1 \\ 1 & 0 & 1 & 1 & 0 & 0 & 1 & 1 \\ 0 & 0 & 0 & 1 & 0 & 1 & 1 & 2 \\ -5 & -4 & -4 & -5 & -5 & -4 & -4 & -5 \\ 1 & 0 & 1 & 1 & 2 & 0 & 0 & 0 \\ 1 & 1 & 0 & 0 & 1 & 1 & 1 & 0 \\ 0 & 1 & 0 & 0 & 2 & 0 & 1 & 1 \end{bmatrix}, \text{ and for } M =$$

$$C - K^{-1}CK - K^{-3}CK^3 - K^{-4}CK^4 + K^{-5}CK^5 + K^{-7}CK^7 =$$

$$\begin{bmatrix} 5 & 7 & 5 & 7 & 9 & 1 & 4 & 3 \\ -7 & -5 & -4 & -6 & -9 & -2 & -4 & -4 \\ 0 & -1 & 0 & 0 & -1 & 1 & 1 & 0 \\ -7 & -4 & -5 & -5 & -7 & -4 & -5 & -4 \\ -9 & -1 & -4 & -3 & -5 & -7 & -5 & -7 \\ 9 & 2 & 4 & 4 & 7 & 5 & 4 & 6 \\ 1 & -1 & -1 & 0 & 0 & 1 & 0 & 0 \\ 7 & 4 & 5 & 4 & 7 & 4 & 5 & 5 \end{bmatrix}$$

we get $P(x) = \det(xI - M) = x^8 + 3qx^6 + 12qx^4 + 7qx^2 + q$. So $P(x)$ is the characteristic polynomial of $\beta = \eta_0 - \eta_1 - \eta_3 - \eta_4 + \eta_5 + \eta_7$ in the subfield of degree 8 of $\mathbb{Q}[\zeta_{41}]$. Here we can take $s = 6$.

Taking $[l_1, l_2, \ldots, l_7] = [0, 4, 6, 2, 2, 0, 0]$ and $\alpha = -1 + 2\zeta_8 - 3\zeta_8^2 + 3\zeta_8^3$, we get $q = 137$, and for $M = C - K^{-1}CK - K^{-2}CK^2 - K^{-3}CK^3 - K^{-4}CK^4 + K^{-5}CK^5 + K^{-6}CK^6 + K^{-7}CK^7$ we get $P(x) = \det(xI - M) = x^8 + 4qx^6 + 590qx^4 + 25844qx^2 + q$. So $P(x)$ is the characteristic polynomial of $\beta = \eta_0 - \eta_1 - \eta_2 - \eta_3 - \eta_4 + \eta_5 + \eta_6 + \eta_7$ in the subfield of degree 8 of $\mathbb{Q}[\zeta_{137}]$. Here we can take $s = 132$.

$\underline{m = 10.}$ Taking $[l_1, l_2, \ldots, l_9] = [0, 0, 5, 5, 0, 0, 5, 5, 0]$ and $\alpha = 2 - \zeta_{10} + 2\zeta_{10}^2 + t$, we get $q = t^4 + 5t^3 + 15t^2 + 25t + 25$, and for $M = C - K^{-5}MK^5$ we get $P(x) = \det(xI - M) = x^{10} + qx^8 + (2t^2 + 6t + 8)qx^6 + q(q + t^3 + 4t^2 + 5t - 2)x^4 + q(q - 2t^2 - 6t - 10)x^2 + q$. So $P(x)$ is the characteristic polynomial of $\beta = \eta_0 - \eta_5$ in the subfield of degree 10 of $\mathbb{Q}[\zeta_q]$, when q is prime. Let $N = -K^{-2}MK^2M^{-1}$. We have that the characteristic polynomial of N is the square of Emma Lehmer's polynomial $x^5 + t^2x^4 - (2t^3 + 6t^2 + 10t + 10)x^3 + (t^4 + 5t^3 + 11t^2 + 15t + 5)x^2 + (t^3 + 4t^2 + 10t + 10)x + 1$ (see, for example, [T3] Example 4.3).

5. A MAPLE Program

The following program can be used to check the results of Section 3. A similar program is shown at the end of [T1]. In [T1] and [T3] we search for families of cyclic polynomials and consider primes $q = mf + 1$ with f even. Here we deal with f odd, and our aim is to look for families of q-cyclotomic integers β with $\beta\overline{\beta} \in \mathbb{Z}$, and cyclic difference sets. The program can be modified to check the examples of Section 4 and to search for more examples, as is explained in that section.

To start, give the values of m, $\alpha = t_0 + t_1 * z + \cdots + t_{\varphi(m)-1} * z^{\varphi(m)-1}$ where the t_i are, say, linear polynomials in a parameter t (here z stands for ζ_m), L=$[l_1, \ldots, l_{m-1}]$, a list of suitable numbers as described in Section 2, and Epsilon=$[\varepsilon, \varepsilon_0, \varepsilon_1, \ldots, \varepsilon_{m-1}]$ a list of integers. The program gives the values of the corresponding q, C, M=$\varepsilon I + \sum_{i=0}^{m-1} \varepsilon_i K^{-i}CK^i$, which plays the role of $\beta = \varepsilon + \sum_{i=0}^{m-1} \varepsilon_i\eta_i$, and N=M$(K^{-m/2}MK^{m/2})$, which plays the role of $\beta\overline{\beta}$. The values of the t_i must be such that q is irreducible in $\mathbb{Z}[t]$ and that the entries of C (which are automatically in $\mathbb{Q}[t]$) belong to $\mathbb{Z}[t]$, and the values of ε, ε_i must be such that N is a scalar matrix N=nI (with $n \in \mathbb{Z}[t]$, $n \neq 0, 1$). For the values of t such that q is a prime number, we have then that $\beta\overline{\beta} = n$.

The program can be modified to search for new families, by taking (many) different values of α and making ε and ε_i run through (small) intervals of integers, and after choosing, to test, some entries of M and some nondiagonal entries of N, discarding the cases where those entries of M are not in $\mathbb{Z}[t]$ and those entries of N are nonzero. (After that, of course, you must check if all entries of M are in $\mathbb{Z}[t]$ and that N is a scalar matrix.)

```
with(numtheory): with(linalg):
m:=6; alpha:=z->-12*z*t-3+z; L:=[0, 0, 1, 5, 2]; Epsilon:=[0,1,1,3,-3,1,3];
l[0]:=0:
for i from 1 to m-1 do
l[i]:=L[i]: od:
R:=cyclotomic(m,z):
for i from 0 to m-1 do
l[i+m]:=l[i]+l[m-1]+m/2; od:
```

```
for i0 from 0 to m-1 do
T[i0]:=modp(i0∧ (phi(m)-1),m); od:
for i1 from 0 to m-1 do
if igcd(i1,m)=1 then u[i1]:=1;
else u[i1]:=0; fi; od;
q:=sort(collect(rem(expand(product(alpha(z∧ c)∧ u[c], c=0..m-1)),R,z),t),t):
f:=(q-1)/m:
U:=array(1..m-1,1..m-1,1..m):
for i2 from 1 to m-1 do
for j2 from 1 to m-1 do
for k2 from 1 to m do
U[i2,j2,k2]:=(floor((i2+j2)*(k2-1)/m)-
floor(i2*(k2-1)/m)-floor(j2*(k2-1)/m))*u[k2-1];
od:od:od:
B:=array(1..m-1,1..m-1):
for i3 from 1 to m-1 do
for j3 from 1 to m-1 do
B[i3,j3]:=expand(z∧ (modp(l[i3+j3]-l[i3]-l[j3],m))*
product(alpha(z∧ (m-T[k3-1]))∧ U[i3,j3,k3],k3=1..m),z); od:od:
H:=array(1..m-1,1..m-1):
for i4 from 1 to m-1 do
for j4 from 1 to m-1 do
H[i4,j4]:=sort(collect(rem(B[i4,j4],R,z),z)); od:od:
Id:=array(identity,1..m,1..m):
C:=array(1..m,1..m):
for i5 from 1 to m do
for j5 from 1 to m do
C[i5,j5]:=sort(rem((1/m∧ 2)*(q+1-m*q*Id[1+m/2,i5]-
m*Id[1,j5]-m*Id[i5,j5]-sum(sum((-1)∧ a*
z∧ ((m-i5+1)*a+(m-j5+1)*b)*H[a,b],a=1..m-1),
b=1..m-1)+sum((-1)∧ l5*z∧ ((m-i5+j5)*l5)*
H[l5,m-l5],l5=1..m-1)),R,z),t);
od:od:
K:=array(1..m,1..m):
for i6 from 1 to m-1 do
for j6 from 2 to m do
K[i6,j6]:=Id[i6+1,j6];
K[i6,1]:=0; K[m,j6]:=0;
K[m,1]:=1; od: od:
evalm(K);
M:=evalm(Epsilon[1]*Id+Epsilon[2]*C+
sum(Epsilon[a]*K∧(-(a-2))&*C&*K∧ (a-2),a=3..m+1)):
N0:=evalm(M&*K∧(-m/2)&*M&*K∧(m/2)):
N:=array(1..m,1..m):
for i from 1 to m do
```

```
for j from 1 to m do
N[i,j]:=normal(N0[i,j]); od;od;
q:=q; C:=evalm(C); N:=evalm(N); n:=N[1,1];
```

References

[B] L. D. Baumert, *Cyclic Difference Sets*, Lecture Notes in Mathematics 182, Springer-Verlag, Berlin, 1971.

[BEW] B. Berndt, R. Evans and K. Williams, *Gauss and Jacobi sums*, John Wiley & sons Inc., New York-Toronto, 1998.

[D] L. E. Dickson, Cyclotomy, higher congruences and Waring's problem, *Amer. J. Math.* 57 (1935), 391–424.

[L] E. Lehmer, On residue difference sets, *Canad. J. Math.* 5 (1953), 425–432.

[S] T. Storer, *Cyclotomy and Difference Sets*, Lectures in Advanced Mathematics, vol. 2, Markham Publishing Company, Chicago, 1967.

[T1] F. Thaine, Jacobi sums and new families of irreducible polynomials of Gaussian periods, *Math. Comp.* 70 (2001), no. 236, 1617–1640.

[T2] F. Thaine, Cyclic polynomials and the multiplication matrices of their roots, *J. Pure Appl. Algebra* 188 (2004), 247–286.

[T3] F. Thaine, On the construction of families of cyclic polynomials whose roots are units, *Experiment. Math.* 17 (2008), no. 3, 315–331.

[W] L. C. Washington, Introduction to Cyclotomic Fields, *Graduate Texts in Mathematics,* vol. 83, 2nd ed., Springer, New York, 1996.

In: Frontiers of Combinatorics and Number Theory, Volume 2 ISBN: 978-1-61942-918-5
Editor: Zhi-Wei Sun, pp. 59-77

Chapter 4

ANALYSIS OF KASAMI-WELCH FUNCTIONS IN ODD DIMENSION USING STICKELBERGER'S THEOREM

***Philippe Langevin*[1], *Gregor Leander*[2], *Gary McGuire*[3] *and Eugen Zălinescu*[4]**
[1]GRIM-USTV, University of Toulon, France
[2]DTU Mathematics, Technical University of Denmark, Denmark
[3]School of Mathematical Sciences
University College Dublin, Ireland
[4]MSR-INRIA Joint Center, Orsay, France.

Abstract

In this article we apply some number theoretical techniques to derive results on Boolean functions. We apply Stickelberger's theorem on 2-adic valuations of Gauss sums to the Kasami-Welch functions $\mathrm{tr}_L(x^{4^k-2^k+1})$ on $\mathbb{F}_{2^n}$, where n is odd and $(k,n)=1$. We obtain information on the Fourier spectrum, including a characterization of the support of the Fourier transform. One interesting feature is that the behaviour is different for different values of k. We also apply the Gross-Koblitz formula to the Gold functions $\mathrm{tr}_L(x^{2^k+1})$.

Keywords: Kasami-Welch; Fourier transform; Walsh-Hadamard transform; finite field; Stickelberger; Gross-Koblitz.

AMS Classification: 05C38, 11T24, 94B15.

1. Introduction

Let $L=\mathbb{F}_q$, the finite field with $q=2^n$ elements. Let tr_L denote the trace map from L to $\mathbb{F}_2$. The *Fourier transform* of any real-valued function F defined on L is the function $\widehat{F}$ defined by

$$\widehat{F}(a)=\sum_{x\in L}F(x)(-1)^{\mathrm{tr}_L(ax)}$$

for $a\in L$. The *Fourier spectrum* of F is the set of values of $\widehat{F}$, that is the set

$$\{\widehat{F}(a) : a\in L\}.$$

To a Boolean function $f : L \longrightarrow \mathbb{F}_2$ we associate the real-valued function $F = (-1)^f$. When we say the Fourier spectrum of f, we mean the Fourier spectrum of the associated function $F = (-1)^f$. For a general f it is difficult to find the Fourier spectrum.

When the Fourier spectrum of a Boolean function is known, further information can be obtained using Stickelberger's theorem on 2-adic valuations of Gauss sums, but often the information is difficult to extract. Studying Boolean functions using Stickelberger's theorem leads to the study of a set which we call the J-set. We will give the definition in section 2. It is this J-set that is often hard to find explicitly, but when found it reaps rich rewards. Finding the J-set leads to finer information about the Fourier spectrum, and its support. For example, in section 5 we easily determine the degree of the characteristic function of the support using our results.

A Boolean function is said to be *bent* if its Fourier spectrum consists of the two values $\pm 2^{n/2}$, and thus n is necessarily even. An analysis of the J-set of Kasami-Welch bent functions, functions of the form $\mathrm{tr}_L(x^{4^k - 2^k + 1})$, was carried out in Langevin-Leander [7]. In this article we undertake an analysis of the J-set of $\mathrm{tr}_L(x^{4^k - 2^k + 1})$ for n odd. This is different from the n even case, and the complete result depends on paths in a certain graph (see sections 3 and 4).

One motivation for this article is a question of Hans Dobbertin, who asked for a trace description of the support of the Fourier transform of the Kasami-Welch functions. Dillon gives a description in [2], answering Dobbertin's question, although the description is not a trace description. In particular, he proves the No-Chung-Yun conjecture for n odd, which concerns one particular k, when $3k \equiv \pm 1 \pmod{n}$, and does give a trace description in that case. In the Dillon-Dobbertin [1] paper the n even case is proved. Our description is by its nature a trace description, which coincides with that of [2] in the $3k \equiv \pm 1 \pmod{n}$ case (see section 6).

We were also motivated by a desire to explain the data in Langevin-Veron [8]. As part of that research they searched on a computer for examples of the simplest possible J-set. It was observed that a single Kasami exponent has a J-set of this simple form. We are able to prove this result (see section 6).

Viewing the Kasami-Welch exponent as the $t = 3$ case of the family $(2^{tk}+1)/(2^k+1)$, we also consider in detail the next case, when $t = 5$. We are able to determine the J-set in terms of paths in two graphs (see section 7). The family of exponents $(2^{tk}+1)/(2^k+1)$ is of great interest because most known cases of a Fourier spectrum with five or fewer values belong to this family. The $t = 5$ case is known by results of Kasami to have 5-valued spectrum.

Finally, we discuss the Gross-Koblitz formula in our context in section 8, and apply this to the Gold functions in section 9. This allows us to determine the sign of the Fourier coefficient at 1, showing yet again how finer information can be deduced using these techniques. Besides Stickelberger's theorem and Gross-Koblitz formula, all our analyses make heavy use of the fundamental formula in [4] on modular multiplication.

2. Stickelberger's Theorem and the J-Set

Let n be odd. Let $\mu(y) = (-1)^{\mathrm{tr}_L(y)}$, so μ is the canonical additive character of L. We let $L^\times = L\backslash\{0\}$.

Let $\mathbf{Q}_2$ denote the field of 2-adic (or dyadic) numbers, and let $\overline{\mathbf{Q}_2}$ denote its algebraic closure. As usual $\mathbf{Z}_2$ denotes the ring of 2-adic integers. We consider characters of $L^\times$ taking values in $\overline{\mathbf{Q}_2}$. In fact, the character values will lie in the unramified algebraic extension of $\mathbf{Q}_2$ of degree n.

Let $\widehat{L^\times}$ denote the group of characters of $L^\times$ taking values in $\overline{\mathbf{Q}_2}$. Elements of $\widehat{L^\times}$ are sometimes called multiplicative characters of L. The group $\widehat{L^\times}$ is a cyclic group of order $2^n - 1$, generated by the *Teichmüller character*. In order to define the Teichmüller character, let i be any primitive $(2^n - 1)$-th root of unity in $\overline{\mathbf{Q}_2}$. The ring of integers of the extension $\mathbf{Q}_2(i)$ is $\mathbf{Z}_2[i]$, and this local ring has a unique maximal ideal generated by 2. The quotient $\mathbf{Z}_2[i]/(2)$ is a finite field of order 2^n, and is therefore isomorphic to L. We use this representation of L for the definition.

The Teichmüller character $\omega : L \longrightarrow \mathbf{Q}_2(i)$ is defined by the relation

$$\omega(i^j \mod 2) = i^j, \quad j = 0, 1, \ldots, 2^n - 2. \tag{1}$$

Using the convention $\omega(0) = 0$, the above relation is equivalent to saying that

$$\forall a \in L, \quad \omega(a) \mod 2 = a. \tag{2}$$

Next we recall Gauss sums. The Gauss sum associated to $\chi \in \widehat{L^\times}$ is

$$\tau(\chi) = - \sum_{x \in L^\times} \chi(x)\mu(x). \tag{3}$$

There are good justifications to introduce the minus sign in the definition of Gauss sums. The reader who is not familiar with the above definitions will find the basic material (characters and Gauss sums) in the book of Lidl and Niederreiter [9], and in the course of Koblitz [5] for the p-adic approach.

The following theorem can be found in [7].

Theorem 1 *Continuing the above notation, if $f(x) = \mu(x^d)$ then*

$$\widehat{f}(a) \equiv - \sum_{j=1}^{2^n-1} \tau(\overline{\omega}^j)\, \tau(\omega^{jd})\, \overline{\omega}^{jd}(a) \pmod{2^n}. \tag{4}$$

For any integer m, let $\mathrm{wt}_2(m)$ denote the 2-weight of m, namely $\sum_i m_i$ where $m \mod (2^n - 1) = \sum_i m_i 2^i$ is the base 2 expansion of $m \mod (2^n - 1)$.

A well known result of Stickelberger shows that for all integers $0 \leq j < 2^n - 1$, the following congruence holds:

$$\tau(\overline{\omega}^j) \equiv 2^{\mathrm{wt}_2(j)} \pmod{2^{1+\mathrm{wt}_2(j)}} \tag{5}$$

and so this gives us the 2-adic valuation of the Gauss sums in equation (4). It follows that the 2-adic valuation of each term in the sum (4) for $\widehat{f}(a)$ is $\mathrm{wt}_2(j)+\mathrm{wt}_2(-jd)$, and the overall 2-adic valuation of $\widehat{f}(a)$ is not less than the minimum of the numbers $\mathrm{wt}_2(j)+\mathrm{wt}_2(-jd)$ over all $j=1,2,\dots,2^n-2$.

Definition 1 *Continuing the above notation, if $f(x)=\mu(x^d)$ let*

$$M_d = \min_{j\in\{1,2,\dots,2^n-2\}} [\mathrm{wt}_2(j)+\mathrm{wt}_2(-jd)].$$

Definition 2 *Continuing the above notation, if $f(x)=\mu(x^d)$ let*

$$J_d = \{j\in\{1,2,\dots,2^n-2\} : \mathrm{wt}_2(j)+\mathrm{wt}_2(-jd)=M_d\}.$$

We refer to this set J_d as the J-set. Its importance is summarized in the following Lemma.

Lemma 1 *Continuing the above notation, if $f(x)=\mu(x^d)$ it holds that*

$$2^{M_d+1} | \widehat{f}(a) \Leftrightarrow \sum_{j\in J_d} a^{-jd} = 0.$$

Proof. Using the definition of J_d and Theorem 1,

$$\widehat{f}(a) \equiv 2^{M_d} \sum_{j\in J} \overline{\omega}^{jd}(a) \pmod{2^{M_d+1}}$$

by definition of the Teichmüller character $a\equiv\omega(a) \mod 2$. Thus the valuation of $\widehat{f}(a)$ is greater than M_d if and only if $\sum_{j\in J_d} a^{-jd}=0$.

□

Note that the J-set is closed under multiplication by 2, and so is a union of cyclotomic cosets. Let $\tilde{J}_d$ denote a set of cyclotomic representatives of J_d. Then, since n is odd,

$$\sum_{j\in J_d} a^{-jd} = \mathrm{tr}_L\Big(\sum_{j\in\tilde{J}_d} a^{-jd}\Big).$$

As the reader will see, determining the J-set leads to information about the Fourier spectrum. For the Kasami exponent $d=4^k-2^k+1$, where n is odd and $(k,n)=1$, we use the well-known fact (see [3] for example) that the Fourier spectrum is $\{0,\pm2^{(n+1)/2}\}$.

Corollary 1 *Continuing the above notation, if the Fourier spectrum of $f(x)=\mu(x^d)$ is $\{0,\pm2^{(n+1)/2}\}$ then $M_d=(n+1)/2$, and the support of the Fourier transform is $\{a\in L : \mathrm{tr}_L(\sum_{j\in\tilde{J}_d} a^{-jd})=1\}$.*

3. J-Sets for All Kasami Exponents

Let n be odd. Let $d = 4^k - 2^k + 1$ be a Kasami exponent, with $k > 1$ relatively prime to n. In this section we shall collect some results about the J-set.

We assume that j is an element of the J-set, with binary expansion

$$j = \sum_{i=0}^{n-1} j_i 2^i.$$

Let $s = jd \mod (2^n - 1)$. Then

$$s = \sum_{i=0}^{n-1} (j_{i-2k} - j_{i-k} + j_i) 2^i \mod (2^n - 1).$$

As usual, all subscripts are considered modulo n. First we recall the fundamental relation derived in [4]:

$$2c_i + s_i = j_{i-2k} - j_{i-k} + j_i + c_{i-1}, \quad i = 0, 1, \ldots, n-1, \tag{6}$$

where the *i-th carry* c_i is an integer in the set $\{-1, 0, 1\}$ and $\sum_i s_i 2^i$ is the binary expansion of s. Let c be the sequence $c_0, c_1, \ldots, c_{n-1}$. Overloading the notation, we write $\mathrm{wt}_2(c)$ for $\sum_i c_i$.

Lemma 2 $\mathrm{wt}_2(c) = -\frac{n-1}{2}$.

Proof. Summing equation (6) over all i gives

$$2\sum c_i + \sum s_i = \sum j_i - \sum j_i + \sum j_i + \sum c_i$$

or

$$\begin{aligned} \mathrm{wt}_2(c) &= \mathrm{wt}_2(j) - \mathrm{wt}_2(s) \\ &= \mathrm{wt}_2(j) - (n - \mathrm{wt}_2(-s)) \\ &= \mathrm{wt}_2(j) + \mathrm{wt}_2(-jd) - n \\ &= \frac{n+1}{2} - n \\ &= -\frac{n-1}{2}. \end{aligned}$$

□

Lemma 3 ([4]) *For all i we have $c_i + c_{i-k} \in \{-1, 0, 1\}$.*

Corollary 2 *c has exactly $(n-1)/2$ entries equal to -1. The remaining $(n+1)/2$ entries are equal to 0.*

Proof. By Lemma 3, c_i and c_{i-k} cannot both be -1. This means that c has at most $(n-1)/2$ entries equal to -1. But combining Lemma 2 with the fact that $c_i \in \{-1, 0, 1\}$, gives that c has at most $(n-1)/2$ entries equal to -1. This shows the result.

□

As $(k, n) = 1$ we may re-order the sequence c_i as

$$c_0, c_{-k}, c_{-2k}, c_{-3k}, \ldots, c_{-(n-1)k}. \tag{7}$$

In this ordering no two consecutive entries can be -1 by Lemma 3. We may assume $c_0 = c_1 = 0$ up to cyclotomic equivalence, and then in this re-ordering the sequence c_i is

$$0, 0, -1, 0, -1, 0, -1, 0, \ldots, 0, -1. \tag{8}$$

In other words, if we change the subscript variable to r where $i = -rk$, then $c_r = 0$ if $r \in \{0, 1, 3, 5, \ldots, n-2\}$, and $c_r = -1$ if $r \in \{2, 4, 6, \ldots, n-1\}$. In this new ordering, the fundamental relation (6) becomes

$$2c_r + s_r = j_{r+2} - j_{r+1} + j_r + c_{r+e} \tag{9}$$

where $ek \equiv 1 \pmod{n}$. We will assume wlog that e is odd (if e is even then replace k by $n-k$).

Note that if $r + e < n$ then the subscript $r + e$ has the opposite parity to r, whereas if $r + e \geq n$ then $r + e$ (reduced modulo n) has the same parity as r.

Let $\ell = n - e$, which we note is even. We now show that exactly $\ell + 3$ bits in j are uniquely determined. These bits are:

$$j_2, j_3, \ldots, j_{\ell+2}, j_{\ell+3}, j_{\ell+4}.$$

These are determined because for $0 < r \leq \ell$ and r even, the relation (9) becomes

$$-2 + s_r = j_{r+2} - j_{r+1} + j_r + 0 \tag{10}$$

and there is a unique solution $s_r = 1, j_{r+2} = 0, j_{r+1} = 1, j_r = 0$. The $r = \ell + 1$ equation is

$$s_{\ell+1} = j_{\ell+3} - j_{\ell+2} + j_{\ell+1}$$

which determines $j_{\ell+3} = 0$. Then the $r = \ell + 2$ equation is

$$-1 + s_{\ell+2} = j_{\ell+4} - j_{\ell+3} + j_{\ell+2}$$

which implies $j_{\ell+4} = 0$.

The remaining $n - \ell - 3 = e - 3$ bits of j are not determined uniquely, in general. The number of solutions is equal to the number of paths of length $(e-3)/2$ starting at a certain vertex in a certain directed graph, as we will explain in the next section. If $e = 3$ then j is completely determined and the J-set has one element (in this case $j_{\ell+3} = j_0$ and $j_{\ell+4} = j_1$). We will discuss particular values of e later.

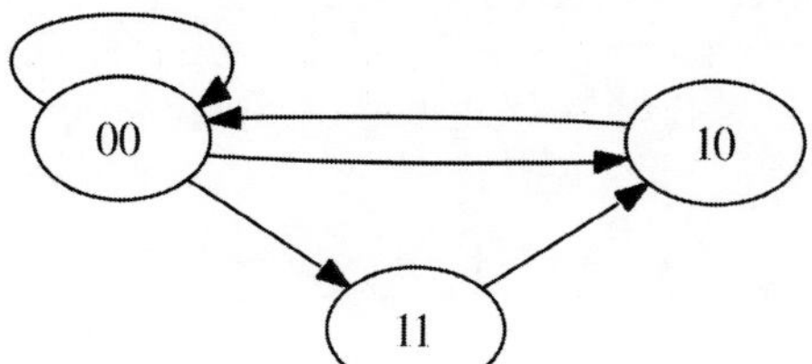

Figure 1. Graph G of Transitions $(j_r, j_{r+1}) \to (j_{r+2}, j_{r+3})$.

4. The Graph Describing the J-Sets

First we note that the equations for $r > \ell + 2$ are

$$-2 + s_r = j_{r+2} - j_{r+1} + j_r - 1, \; r \text{ even},$$

and

$$0 + s_r = j_{r+2} - j_{r+1} + j_r + 0, \; r \text{ odd},$$

which do not have a unique solution. For example, although $j_{\ell+3} = 0$, $j_{\ell+4} = 0$, the $r = \ell + 3$ equation is

$$s_{\ell+3} = j_{\ell+5} - j_{\ell+4} + j_{\ell+3}$$

which does not determine $j_{\ell+5}$ (assuming $e > 3$).

Equations for r between 2 and $\ell + 2$ have already been used. We will take the equations for $r > \ell + 2$ in pairs; the first corresponds to an odd r, the second to an even r. If r is odd, and $x = j_r$, $y = j_{r+1}$, $z = j_{r+2}$, $t = j_{r+3}$, the fact that the r-th equation has a solution is equivalent to

$$x - y + z \in \{0, 1\} \tag{11}$$

and the fact that the next equation for $r + 1$ has a solution is equivalent to

$$y - z + t + 1 \in \{0, 1\}. \tag{12}$$

We consider (11) and (12) as the restrictions in moving from one odd-even pair of bits (x, y) to the next odd-even pair (z, t). We note that these are the only restrictions on (z, t), and we further note that when (z, t) are chosen, they in turn will use (11) and (12) to determine the next odd-even pair. The next odd-even pair will not depend on x and y at all, only on z and t.

In this way, we can track how ordered pairs (j_r, j_{r+1}) (with r odd) of undetermined bits give rise to possibilities for the next pair of undetermined bits. We can express this situation very concisely in the form of the directed graph G given in Figure 1.

It is now evident that solutions for the undetermined bits correspond to paths in this graph G that start at 00. The length of the path should be one half the number of undetermined bits, i.e., $(e - 3)/2$. The last vertex in the path is the final pair of undetermined bits (j_0, j_1). We have proved:

Theorem 2 *There is a one-to-one correspondence between paths in G of length $(e-3)/2$ that start at 00, and representatives from cyclotomic cosets in the J-set.*

5. The Degree of the Support Function

The support of the Fourier transform is described by the function

$$a \mapsto \mathrm{tr}_L\Bigg(\sum_{j\in\tilde{J}_d} a^{-jd}\Bigg),$$

called the *support function.* An application of our results is that we can determine the degree of this function for any Kasami exponent.

Theorem 3 *Let n be odd. Let $d = 4^k - 2^k + 1$ be the Kasami exponent, with $ek \equiv 1 \pmod n$, e odd. Then the degree of the support function $a \mapsto \mathrm{tr}_L(\sum_{j\in\tilde{J}_d} a^{-jd})$ is $(e+1)/2$.*

Proof. The degree of the support function is determined by the maximum value of $\mathrm{wt}_2(-jd)$ where j is in the J-set. As

$$\mathrm{wt}_2(j) + \mathrm{wt}_2(-jd) = \frac{n+1}{2}$$

this maximum is achieved by minimising $\mathrm{wt}_2(j)$. By Theorem 2, elements of the J-set correspond to paths in the graph G. This minimum is achieved by choosing a path in G of smallest weight. It is clear that a path of weight 0 can always be chosen, simply by choosing the loop from 00 to itself for each edge. The resulting element of the J-set has weight $\ell/2$, so

$$\max_{j\in J_d} \mathrm{wt}_2(-jd) = \frac{n+1}{2} - \frac{\ell}{2} = \frac{n+1-(n-e)}{2} = \frac{e+1}{2}.$$

□

6. J-Sets for Particular Kasami Exponents

Experimental results in Langevin-Veron [8] showed that for $n < 39$ and not divisible by 3, there is exactly one Kasami exponent that has a J-set consisting of one cyclotomic coset. We are now able to prove this, and identify the exponent. It turns out that Dillon [2] and Dillon-Dobbertin [1] discussed this particular Kasami exponent before, and gave the trace description of the support of the Fourier transform. We give a different proof of this trace description as a simple corollary of our description of the J-set.

Theorem 4 *Let n be odd and not divisible by 3. Let $d = 4^k - 2^k + 1$ be the Kasami exponent with $3k \equiv 1 \pmod n$.*

(1) *There is exactly one element in the J-set up to cyclotomy, which is $1/(2^{-2k} - 1)$, i.e.,*

$$J_d = \left\{\frac{2^i}{2^{-2k}-1} : i = 0, 1, 2, \ldots, n-1\right\}.$$

(2) *This is the only Kasami exponent whose J-set consists of a single cyclotomic coset.*

(3) *The support of the Fourier transform is* $\{a \in L : \mathrm{tr}_L(a^{2^k+1}) = 1\}$.

Proof. (1) Here $e = 3$ and $\ell = n - e = n - 3$. By Theorem 2 the J-set has one element j up to cyclotomic equivalence. The element is completely determined as we explained in section 4. The bits of j in the re-ordering are

$$j_2 = 0, j_3 = 1, j_4 = 0, j_5 = 1, \ldots,$$

$$j_{\ell+1} = 1, j_{\ell+2} = 0, j_{\ell+3} = j_0 = 0, j_{\ell+4} = j_1 = 0.$$

Upon reverting to the original ordering, this gives j as

$$j = \sum_{s=1}^{(n-3)/2} 2^{-k(2s+1)} = 2^{-3k} \sum_{s=0}^{(n-5)/2} 2^{-2sk} = 2^{-3k} \frac{2^{-2k(n-3)/2} - 1}{2^{-2k} - 1} = \frac{1}{2^{-2k} - 1}.$$

(2) We assume $k \neq 1$, and thus $e \neq 1$ because $k = 1$ is the Gold case $d = 3$. As e is odd, any other value of e is at least 5 and there are at least three paths in G of the required type, and therefore the J-set contains at least three cyclotomic cosets.

(3) As we showed in section 2 the support of the Fourier transform is given by

$$\{a : \mathrm{tr}_L(\sum_{j \in \tilde{J}_d} a^{-jd}) = 1\}.$$

In this case $d = 2^{2k} - 2^k + 1 = (2^k + 1)(2^{2k} - 1)$ so $jd = -(2^k + 1)$.

□

Theorem 5 *Let n be odd and not divisible by 5. Let $d = 4^k - 2^k + 1$ be the Kasami exponent with $5k \equiv 1 \pmod{n}$.*

(1) *There are exactly three elements in the J-set up to cyclotomy, which are*

$$\frac{1}{2^{-2k} - 1}, 1 + \frac{2^{2k-1}}{2^{-2k} - 1}, 1 + 2^{-k} + \frac{2^{2k-1}}{2^{-2k} - 1}.$$

(2) *This is the only Kasami exponent whose J-set consists of three cyclotomic cosets.*

(3) *The support of the Fourier transform is the set of $a \in L$ such that*

$$\mathrm{tr}_L(a^{2^{3k}+2^k+1} + a^{2^{2k}+1} + a) = 1.$$

Proof. (1) Here $e = 5$ and $\ell = n - e = n - 5$. By Theorem 2 the J-set has three elements up to cyclotomic equivalence. All bits except two are completely determined. The bits of j in the re-ordering are

$$j_2 = 0, j_3 = 1, j_4 = 0, j_5 = 1, \ldots,$$

$$j_{\ell+1} = 1, j_{\ell+2} = 0, j_{\ell+3} = 0, j_{\ell+4} = 0.$$

Here $j_{\ell+5} = j_0$, $j_{\ell+6} = j_1$ are undetermined, the possibilities are determined by length one paths in the graph G starting at 00. There are three such paths, so the possibilities for (j_0, j_1) are 00, 10, and 11. Call the corresponding elements of the J-set $j^{(0)}, j^{(1)}, j^{(2)}$ respectively. Upon reverting to the original ordering, this gives $j^{(0)}$ as

$$j^{(0)} = \sum_{s=1}^{(n-5)/2} 2^{-k(2s+1)} = 2^{-3k} \sum_{s=0}^{(n-7)/2} 2^{-2sk} = 2^{-3k} \frac{2^{-2k(n-5)/2} - 1}{2^{-2k} - 1} = \frac{2^{2k-1}}{2^{-2k} - 1}.$$

Then $j^{(1)} = j^{(0)} + 1$ and $j^{(2)} = j^{(0)} + 1 + 2^{-k}$, by adding the appropriate powers of 2 according to the path.

(2) We assume $k \neq 1$, and thus $e \neq 1$, because $k = 1$ is the Gold case $d = 3$. We showed that the $e = 3$ case has one element in the J-set. As e is odd, any other value of e is at least 7 and there are more than three paths in G of the required type, and therefore the J-set contains more than three cyclotomic cosets.

(3) As we showed in section 2 the support of the Fourier transform is given by

$$\left\{ a : \mathrm{tr}_L \left(\sum_{j \in \tilde{J}_d} a^{-jd} \right) = 1 \right\}.$$

By part (1) this gives the support as the a such that

$$\mathrm{tr}_L \left(a^{-d[\frac{1}{2^{-2k}-1}]} + a^{-d[1+\frac{2^{2k-1}}{2^{-2k}-1}]} + a^{-d[1+2^{-k}+\frac{2^{2k-1}}{2^{-2k}-1}]} \right) = 1.$$

Note that, in this case, $d = (2^{3k} + 2^k + 1)(2^{2k} - 1)$ and thus

$$-d \left(\frac{1}{2^{-2k} - 1} \right) = (2^{3k} + 2^k + 1)2^{2k}$$

which is cyclotomic equivalent to $2^{3k} + 2^k + 1$. Furthermore,

$$-d \left(1 + \frac{2^{2k-1}}{2^{-2k} - 1} \right) = 2^{-k}(2^{2k} + 1)$$

which is cyclotomic equivalent to $2^{2k} + 1$ and finally

$$-d \left(1 + 2^{-k} + \frac{2^{2k-1}}{2^{-2k} - 1} \right) = 1.$$

Therefore, the support is given by

$$\mathrm{tr}_L \left(a^{2^{3k}+2^k+1} + a^{2^{2k}+1} + a \right) = 1$$

as claimed.

□

Theorem 6 *Let n be odd. For any $k > 1$ with $(k, n) = 1$, the J-set (for $d = 4^k - 2^k + 1$) contains the element*

$$\frac{1}{2^{-2k} - 1}.$$

Proof. This element is derived as in the proof of Theorem 5, by choosing the path in the graph G with all vertices 00.

□

7. The Case $d = (2^{5k} + 1)/(2^k + 1)$

In this section we shall extend the methods used to determine the J-set for the Kasami-Welch exponent $d = (2^{3k} + 1)/(2^k + 1) = 4^k - 2^k + 1$ to the exponent

$$d = \frac{2^{5k} + 1}{2^k + 1} = 16^k - 8^k + 4^k - 2^k + 1.$$

For this exponent it is known that the Fourier spectrum is 5-valued, and that the values are $\{0, \pm 2^{(n+1)/2}, \pm 2^{(n+3)/2}\}$. Thus we know that $M_d = (n+1)/2$ here, the same as in the Kasami-Welch case. However the analysis of the J-set is more complicated here. There are two graphs instead of one graph, and the number of solutions corresponds to paths in both graphs that link up correctly.

Using the same notations as in section 3, the fundamental relation of [4] becomes

$$2c_i + s_i = j_{i-4k} - j_{i-3k} + j_{i-2k} - j_{i-k} + j_i + c_{i-1}, \tag{13}$$

where $c_i \in \{-2, -1, 0, 1, 2\}$.

Lemma 4 $\mathrm{wt}_2(c) = -\frac{n-1}{2}$.

Proof. Same as in the Kasami-Welch case (Lemma 2).

□

Lemma 5 ([4]) *For all* i, $c_i + c_{i-k} \in \{-1, 0, 1\}$.

Corollary 3 *c has at least $(n-1)/2$ entries equal to -1.*

Proof. Applying the previous lemma gives the following information.

$$\begin{aligned} c_i = 2 &\implies c_{i+k} \in \{-2, -1\} \\ c_i = 1 &\implies c_{i+k} \in \{-2, -1, 0\} \\ c_i = 0 &\implies c_{i+k} \in \{-1, 0, 1\} \\ c_i = -1 &\implies c_{i+k} \in \{0, 1, 2\} \\ c_i = -2 &\implies c_{i+k} \in \{1, 2\}. \end{aligned}$$

¿From this it follows that if $c_i = -2$ then both c_{i-k} and c_{i+k} are $+1$ or $+2$. Therefore any c_i that is -2 cannot contribute anything towards $\mathrm{wt}_2(c)$. Since the 2-weight must reach $-(n-1)/2$, and any -2 entries contribute nothing, the only contribution can come from -1 entries.

□

Corollary 4 *c has exactly $(n-1)/2$ entries equal to -1. The remaining $(n+1)/2$ entries are equal to* 0.

Proof. By Lemma 5, c_i and c_{i+k} cannot both be -1. This means that c has at most $(n-1)/2$ entries equal to -1. Combining this with Corollary 3 gives the first statement.

If any c_i is positive, there are at least $(n+1)/2$ c_i's equal to -1 because of the 2-weight. This would imply there is an i with c_i and c_{i+k} both -1.

□

As we did in the Kasami-Welch case, we may re-order the sequence c_i as

$$c_0, c_{-k}, c_{-2k}, c_{-3k}, \ldots, c_{-(n-1)k} \tag{14}$$

which again wlog is

$$0, 0, -1, 0, -1, 0, -1, 0, \ldots, 0, -1. \tag{15}$$

In other words, if we change to the "r-ordering" where $i = -rk$, then $c_r = 0$ if $r \in \{0, 1, 3, 5, \ldots, n-2\}$, and $c_r = -1$ if $r \in \{2, 4, 6, \ldots, n-1\}$. In this new ordering, the fundamental relation (13) becomes

$$2c_r + s_r = j_{r+4} - j_{r+3} + j_{r+2} - j_{r+1} + j_r + c_{r+e} \tag{16}$$

where $ek \equiv 1 \pmod{n}$. We will assume wlog that e is odd (if e is even then replace k by $n-k$).

As before we let $\ell = n - e$, which is even. In the Kasami-Welch case, for some values of r the equation (16) had a unique solution. This does not happen here. There are four cases, depending on the parity of r and $r + e$:

$$r \in \{1, 2, \ldots \ell\}, \quad r \text{ odd}, \quad (c_r, c_{r+e}) = (0, -1) \tag{17}$$

$$r \in \{1, 2, \ldots \ell\}, \quad r \text{ even}, \quad (c_r, c_{r+e}) = (-1, 0) \tag{18}$$

$$r \in \{\ell+1, \ell+2, \ldots n-1\}, \quad r \text{ odd}, \quad (c_r, c_{r+e}) = (0, 0) \tag{19}$$

$$r \in \{\ell+1, \ell+2, \ldots n-1\}, \quad r \text{ even}, \quad (c_r, c_{r+e}) = (-1, -1). \tag{20}$$

The $r = n = 0$ equation also has $(c_r, c_{r+e}) = (0, 0)$.

We consider the bits of j in fours, starting at $r = 1$ although this is arbitrary. For any four bits $(j_r, j_{r+1}, j_{r+2}, j_{r+3})$, the transition to the next four bits $(j_{r+4}, j_{r+5}, j_{r+6}, j_{r+7})$ is determined by the four equations (16) for $r, r+1, r+2, r+3$. These four equations completely determine the transition. Therefore, we define two graphs. We consider the case $n \equiv 1 \pmod{4}$ and $\ell \equiv 0 \pmod{4}$, the other cases being similar but slightly different.

The first graph G_1 is for $r \in \{1, 2, \ldots \ell\}$. The four equations governing the transition will alternate between (17) and (18). The graph G_1 (Figure 2) shows all allowable transitions, and paths give us all possible solutions for $j_1, \ldots, j_{\ell+4}$.

The graph G_2 (Figure 3) does the same job for determining all solutions for $j_{\ell+5}, \ldots j_{n-1}$. The start vertex in G_2 must be the ending vertex of the path in G_1.

Finally, the ending vertex of the path in G_2 must link back to the starting vertex in G_1, satisfying the equations for $r = n-4, n-3, n-2, n-1, 0$. These are all the equations involving j_0.

Let $v_s(P)$ denote the first vertex of a path P, and let $v_e(P)$ denote the last vertex.

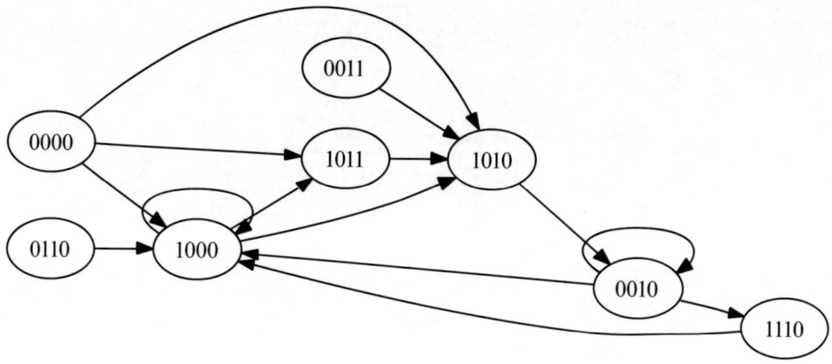

Figure 2. Graph G_1 of Transitions $(j_r, j_{r+1}, j_{r+2}, j_{r+3}) \rightarrow (j_{r+4}, j_{r+5}, j_{r+6}, j_{r+7})$ for $r \in \{1, 2, \ldots \ell\}$.

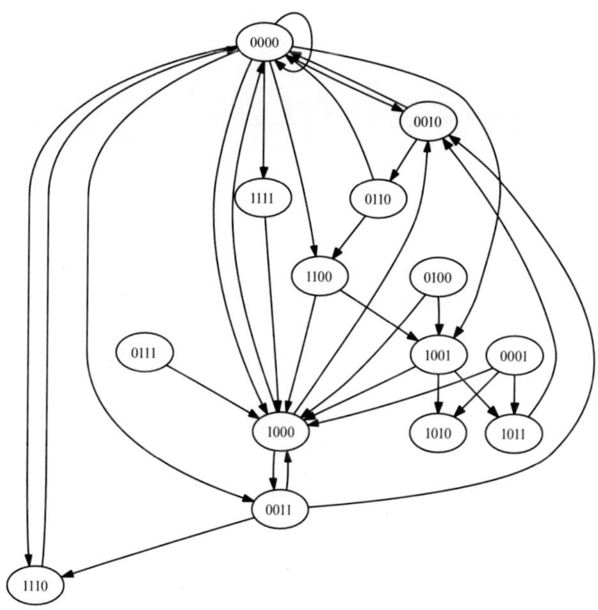

Figure 3. Graph G_2 of Transitions.

Theorem 7 *Let* $n \equiv 1 \pmod 4$ *and* $\ell = n - e \equiv 0 \pmod 4$. *There is a one-to-one correspondence between representatives from cyclotomic cosets in the* J*-set, and pairs* (P_1, P_2), *where* P_1 *is a path in* G_1 *of length* $\ell/4 - 1$ *and* P_2 *is a path in* G_2 *of length* $(n - 1 - \ell)/4$, *such that*

1. $v_e(P_1) = v_s(P_2)$
2. *there exists a path from* $v_e(P_2)$ *to* $v_s(P_1)$ *consistent with the* $r = n - 4, n - 3, n - 2, n - 1, 0$ *equations.*

Example Take $n = 13$, $e = k = 1$, and $d = 11$. Here $\ell = n - 1$ so we are only interested in paths of length 2 in G_1 and G_2 is not involved at all. There are three paths satisfying Theorem 7:

$$\begin{array}{ll} 0000 \longrightarrow 1000 \longrightarrow 1000 & (j_0 = 1) \\ 0000 \longrightarrow 1010 \longrightarrow 0010 & (j_0 = 1) \\ 0110 \longrightarrow 1000 \longrightarrow 1000 & (j_0 = 1). \end{array}$$

yielding three elements in the J-set, which up to cyclotomy are $\{273, 325, 557\}$.

8. Gross-Koblitz's Formula and the K-Set

Let us denote by ν the minimal dyadic valuation of the Fourier coefficient of the power mapping x^d. By section 2, we know that $\nu = M_d$. The goal of this section is to study the dyadic expansion of the Fourier coefficients, and thus to describe the Boolean functions f_0 and f_1 such that :

$$\widehat{f}(a) = 2^{\nu}\left(f_0(a) + f_1(a)2^1 + \cdots\right) \tag{21}$$

As we saw, ν and f_0 are connected to the J-set of d. The determination of f_1 will depend on the *K-set* of d :

$$K_d = \{j \mid \mathrm{wt}_2(j) + \mathrm{wt}_2(-jd) = M_d + 1\}.$$

The Gross-Koblitz formula [5,10] claims that, for any residue j modulo $2^n - 1$, the following equality holds:

$$\tau(\bar{\omega}^j) = (-2)^{\mathrm{wt}_2(j)} \prod_{i=0}^{n-1} \Gamma\left(1 - \left\langle \frac{2^i j}{2^n - 1} \right\rangle\right) \tag{22}$$

where $\langle x \rangle$ is the fractional part of x, and Γ the dyadic Gamma function. The dyadic Gamma function is defined over $\mathbf{N}$ the set of positive integers by

$$\forall k \in \mathbf{N}, \quad \Gamma(k) = (-1)^k \prod_{\substack{0<j<k \\ j \text{ odd}}} j. \tag{23}$$

If x and y are two positive integers such that $x \equiv y \pmod{2^k}$ then $\Gamma(x) \equiv \Gamma(y) \pmod{2^k}$. The dyadic Γ function is extended by continuity over the ring of dyadic numbers i.e.

$$\forall s \in \mathbf{Z}_2, \quad \Gamma(s) = \lim_{\mathbf{N} \ni k \to s} \Gamma(k).$$

In particular,

$$\Gamma(\left(1 - \left\langle \frac{j}{2^n - 1} \right\rangle\right) \equiv \Gamma(1 + j_0 + 2j_1) \pmod{4}, \tag{24}$$

where $j_0 + 2j_1 + \ldots$ in the dyadic expansion of j. Note that an empty product in (23) is equal to 1. The first values of the dyadic Gamma function are: $\Gamma(0) = 1, \Gamma(1) = -1, \Gamma(2) = +1,$

$\Gamma(3) = -1$, and $\Gamma(4) = 3 \equiv -1 \pmod 4$. In other words, for all bits $u, v \in \{0,1\}$, we have $\Gamma(1+u+2v) \equiv (-1)^{1+u+uv} \pmod 4$ and the following congruence holds

$$\Gamma\Big(\Big(1 - \Big\langle \frac{j}{2^n-1} \Big\rangle\Big)\Big) \equiv \Gamma(1+j_0+2j_1) \equiv (-1)^{1+j_0+j_0j_1} \pmod 4. \qquad (25)$$

It follows that

$$\begin{aligned}\tau(\bar{\omega}^j) &\equiv (-2)^{\mathrm{wt}_2(j)}(-1)^n(-1)^{\mathrm{wt}_2(j)}(-1)^{Q(j)} \\ &\equiv (-1)^n 2^{\mathrm{wt}_2(j)}(-1)^{Q(j)} \pmod{2^{\mathrm{wt}_2(j)+2}}\end{aligned}$$

where $Q(j) = j_0j_1 + j_1j_2 + \cdots + j_{n-1}j_0$ is the number of consecutive pairs of ones in the dyadic expansion of j.

Let us denote by g the Boolean function defined by

$$\sum_{j \in J_d} \bar{\omega}(a)^{jd} \equiv f_0(a) + 2g(a) \pmod 4. \qquad (26)$$

Then

$$g(a) \equiv \sum_{\substack{j,k \in J_d \\ j<k}} \bar{\omega}(a)^{(j+k)d} \pmod 2. \qquad (27)$$

Indeed, squaring (26) we obtain:

$$2\sum_{\substack{j,k \in J_d \\ j<k}} \bar{\omega}(a)^{(j+k)d} + \sum_{j \in J_d} \bar{\omega}(a)^{jd} \equiv f_0(a) \pmod 4 \qquad (28)$$

and so

$$2\sum_{\substack{j,k \in J_d \\ j<k}} \bar{\omega}(a)^{(j+k)d} + 2g(a) \equiv 0 \pmod 4. \qquad (29)$$

Now, if we introduce the set $J'_d = \{j \in J_d \mid Q(j) = 1\}$, we obtain the complete description of $f_1(a)$ as :

$$f_1(a) = \sum_{j \in K} \bar{\omega}^{jd}(a) + \sum_{j \in J'_d} \bar{\omega}^{jd}(a) + \sum_{\substack{j,k \in J_d \\ j<k}} \bar{\omega}^{(j+k)d}(a).$$

As a first consequence, the degree of f_0 is less than ν, and the degree of f_1 is less than $2\nu - 1$.

9. An Application to Gold Exponents

Now, we propose to use the theory developed in the preceding section to re-find a recent result of Lahtonen, McGuire and Ward [6] concerning the sign of the Fourier coefficient at 1 of the Gold exponent (the result for the Kasami-Welch exponent is also proved). This result for the Gold exponent is also stated in the appendix of [1].

Let n be odd, k be an integer coprime to n, and $d = 2^k + 1$. Let us denote by f the Gold mapping $f(x) = \mu(x^{2^k+1})$. The Fourier coefficient at one of f satisfies

$$\widehat{f}(1) = \begin{cases} 2^{(n+1)/2} & \text{if } n \equiv \pm 1 \pmod 8, \\ -2^{(n+1)/2}, & \text{if } n \equiv \pm 3 \pmod 8. \end{cases} \tag{30}$$

Note that

$$\widehat{f}(1) > 0 \iff \sharp K_d \equiv \sharp J'_d \pmod 2.$$

Indeed, the sign of $\widehat{f}(a)$ is completely determined by the value of $f_0(a) + 2f_1(a)$. In the case of $a = 1$, we get :

$$f_0(1) + 2f_1(1) \equiv (\sharp J_d)^2 + 2\sharp K_d + 2\sharp J'_d \pmod 4.$$

9.1. *J*-Set

For the computation of the J-set and K-set in the Gold case we restricted ourselves without loss of generality to the case where the inverse e of k modulo n is even.

The fundamental equation for the Gold exponent $d = 2^k + 1$ is

$$2c_i + s_i = j_i + j_{i-k} + c_{i-1}$$

where in this case the carries are usual bits, i.e. $c_i \in \{0, 1\}$. In particular we have the following relation

$$\mathrm{wt}_2(c) + \mathrm{wt}_2((2^k + 1)j) = 2\mathrm{wt}_2(j). \tag{31}$$

As it has been shown in [4] for the Gold case we get $M_d = (n + 1)/2$ and therefore the J-set is characterized by

$$j \in J_d \iff \mathrm{wt}_2(j) + \mathrm{wt}_2(-dj) = \frac{n+1}{2}.$$

In particular we have $\mathrm{wt}_2(j) < (n + 1)/2$. From equation (31) it follows that $\mathrm{wt}_2(dj) \leq 2\mathrm{wt}_2(j)$ with equality iff all carry bits are zero. Thus we get

$$(n + 1)/2 = \mathrm{wt}_2(j) + \mathrm{wt}_2(-dj) = \mathrm{wt}_2(j) + n - \mathrm{wt}_2(dj) \geq n - \mathrm{wt}_2(j).$$

This implies $\mathrm{wt}_2(j) \geq (n - 1)/2$, and therefore $\mathrm{wt}_2(j) = (n - 1)/2$. But this means nothing else than $\mathrm{wt}_2(-dj) = 1$. Thus in the Gold case the J-set consists of a single cyclotomic coset, or more precisely,

$$J_d = \left\{ -\frac{2^i}{2^k + 1} \; : \; i \in \{0, \ldots, n - 1\} \right\}.$$

In order to use the ideas developed in the previous section, we have to compute

$$\sharp J'_d = Q\left(-\frac{1}{2^k + 1}\right) \bmod 2.$$

For this note that, as we do not get any non-zero carry, in the k-ordering of the bits j_i we do not have any consecutive ones. Furthermore, due to cyclotomic equivalence, we can assume that in the k-ordering we have

$$j_0, j_k, j_{2k}, \ldots, j_{(n-1)k} = 0, 0, 1, 0, 1, 0, \ldots, 1, 0, 1.$$

In other words, we can assume that $j_i = 1$ if and only if there exist $t \in \{1, \ldots, (n-1)/2\}$ such that $i = 2kt$.

To compute $Q(j)$ we have to compute the number of consecutive ones in the normal ordering, i.e. the number of indices i such that $j_i = 1$ and $j_{i+1} = 1$. This corresponds to the number of solutions $t, t' \in \{1, \ldots, (n-1)/2\}$ such that

$$2kt \equiv 2kt' + 1 \pmod{n},$$

which implies

$$t - t' \equiv \frac{e}{2} \pmod{n}.$$

Due to the restriction $t, t' \in \{1, \ldots, (n-1)/2\}$ and the assumption that e is even, this number can easily be computed to be $(n-1)/2 - e/2$. The following lemma summarizes these results.

Lemma 6 *Let $d = 2^k + 1$ with $ke \equiv 1 \pmod{n}$ and e even. The J-set corresponds to the cyclotomic class of*

$$\widetilde{j} = -\frac{1}{d}$$

and we have

$$Q(\widetilde{j}) \equiv \frac{n-1}{2} - \frac{e}{2} \pmod{2}.$$

9.2. K-Set

Using the same arguments as for the J-set, it is easy to see that elements in the set

$$K = \left\{ j \ : \ \mathrm{wt}_2(j) + \mathrm{wt}_2(-(2^k+1)j) = \frac{n+3}{2} \right\}$$

can only have two possible weights and we split the set accordingly into two subsets: $K = K_1 \cup K_2$ where

$$K_1 = \left\{ j \in K \ : \ \mathrm{wt}_2(j) = \frac{n-3}{2} \right\}$$

and

$$K_2 = \left\{ j \in K \ : \ \mathrm{wt}_2(j) = \frac{n-1}{2} \right\}$$

which we treat separately.

Size of K_1 Elements of K_1 have weight $(n-3)/2$ and therefore we have $n-3 = \mathrm{wt}_2((2^k+1)j) = 2\mathrm{wt}_2(j)$, which implies that all carries c_i are zero (see equation (31)). Thus, in the k-ordering, we have to count the number of bit-strings of length n and weight $(n-3)/2$ with no consecutive ones. Using elementary combinatorics the size of K_1 can be shown to be

$$\sharp K_1 = \binom{\frac{n+1}{2}}{3} + \binom{\frac{n+3}{2}}{3}.$$

Size of K_2 Elements of K_2 have weight $(n-1)/2$, which implies that all but one carries c_i are zero and exactly one carry equals one. Again considering the k-ordering, we have exactly two consecutive bits. Due to cyclotomic equivalence we can assume that $j_0 = j_r = 1$. In this case having exactly one non-zero carry implies that $j_1 = j_{r+1} = 0$ and $j_{k(n-1)} = j_{2k} = 0$. The remaining positions have to be filled with $(n-5)/2$ ones and $(n-7)/2$ zeros with the restriction that there are no additional consecutive ones. This gives $(n-e-3)/2$ possibilities and therefore, including cyclotomic equivalent elements we get

$$\sharp K_2 = \frac{n-e-3}{2}n$$

Computing $(\sharp J'_d + \sharp K) \bmod 2$ we get

$$\begin{aligned}
\sharp J'_d + \sharp K &= Q(\widetilde{j}) + \sharp K = Q(\widetilde{j}) + \sharp K_1 + \sharp K_2 \\
&\equiv \frac{n-1-e}{2} + \sharp K_1 + \frac{n-e-3}{2}n \pmod 2 \\
&\equiv \sharp K_1 = \binom{\frac{n+1}{2}}{3} + \binom{\frac{n+3}{2}}{3} \pmod 2 \\
&\equiv \begin{cases} 0 & \text{if } n \equiv \pm 1 \pmod 8, \\ 1 & \text{if } n \equiv \pm 3 \pmod 8, \end{cases}
\end{aligned}$$

which is equivalent to (30).

References

[1] J. Dillon and H. Dobbertin, *New cyclic difference sets with Singer parameters*, Finite Fields Appl. 10(2004), 342–389.

[2] J. F. Dillon, *Multiplicative difference sets via additive characters*, Des. Codes Cryptogr. 17(1999), 225–235.

[3] H. Dobbertin, *Another proof of Kasami's theorem*, Des. Codes Cryptogr. 17(1999), 177–180.

[4] H. D. L. Hollmann and Q. Xiang, *A proof of the Welch and Niho conjectures on cross-correlations of binary m-sequences*, Finite Fields Appl. 7(2001), 253–286.

[5] N. Koblitz, *p-adic Analysis: a Short Course on Recent Work*, London Math. Soc. Lecture Note Series, vol. 46, Cambridge Univ. Press, 1980.

[6] J. Lahtonen, G. McGuire, and H. N. Ward, *Gold and Kasami-Welch functions, quadratic forms and bent functions*, Adv. Math. Commun. 1(2007), 243–250.

[7] P. Langevin and G. Leander, *Monomial bent functions and Stickelberger's theorem*, Finite Fields Appl. 14(2008), 727–742.

[8] P. Langevin and P. Véron, *On the non-linearity of power functions*, Des. Codes Cryptogr. 37(2005), 31–43.

[9] R. Lidl and H. Niederreiter, *Finite Fields*, Addison-Wesley, 1983.

[10] A. Robert, *The Gross-Koblitz formula revisited*, Rend. Sem. Mat. Univ. Padova 105(2001), 157–170.

In: Frontiers of Combinatorics and Number Theory, Volume 2 ISBN: 978-1-61942-918-5
Editor: Zhi-Wei Sun, pp. 79-84

Chapter 5

SOME NICE SUMS ARE ALMOST AS NICE IF YOU TURN THEM UPSIDE DOWN

Moa Apagodu[1] ***and Doron Zeilberger***[2]
[1]Department of Mathematics, Virginia Commonwealth University,
Richmond, VA, US
[2] Department of Mathematics, Rutgers University,
Hill Center-Busch Campus,
110 Frelinghuysen Rd., Piscataway,
New Jersey, US

Abstract

We represent the sums $\sum_{k=0}^{n-1} \binom{n}{k}^{-2}$, $\sum_{k=0}^{m} \binom{m}{k}^{-1} \binom{a}{n-k}^{-1}$, $\sum_{k=0}^{n-1} \frac{q^{-k(k-1)}}{\left[n \atop k \right]_q}$, and the sum of the reciprocals of the summands in Dixon's identity, each as a product of an *indefinite hypergeometric sum* times a (closed form) *hypergeometric sequence*.

In [3], Rockett proved the formula

$$\sum_{k=0}^{n} \binom{n}{k}^{-1} = \frac{n+1}{2^n} \sum_{j=0}^{n} \frac{2^j}{j+1} \ , \qquad (Rockett)$$

for all nonnegative integers n. Since then, several authors (e.g. [1][4][5]) used different techniques to re-derive (*Rockett*) and considered analogous, more complicated, sums.

As we all know,

$$\sum_{k=0}^{n} \binom{n}{k} = 2^n \ , \qquad (Binomial)$$

is the *simplest* example of a **definite** binomial coefficient sum that evaluates in *closed-form*. Such sums have the form

$$\sum_{k=0}^{n} F(n,k) = A(n) \ ,$$

where $A(n)$ is *hypergeometric* (i.e. $A(n+1)/A(n)$ is a rational function of n), and $F(n,k)$ is *bi-hypergeometric* (i.e $F(n+1,k)/F(n,k)$ and $F(n,k+1)/F(n,k)$ are both rational functions of n and k). Other famous examples, are

$$\sum_{k=0}^{n} \binom{n}{k}\binom{a}{k} = \binom{n+a}{a} , \qquad (Chu - Vandermonde)$$

and

$$\sum_{k=-n}^{n} (-1)^k \binom{a+n}{n+k}\binom{b+n}{b+k}\binom{a+b}{a+k} = \frac{(a+b+n)!}{a!b!n!} . \qquad (Dixon)$$

The *best* thing that can happen to a binomial coefficients sum is to be evaluable in *closed form*, i.e. as a hypergeometric sequence. The *next best* thing is to be "almost closed-form", i.e. of the form

$$ClosedForm(n) \cdot \left(\sum_{j=1}^{n} h(j) \right) ,$$

where $h(j)$ is hypergeometric in the *single* variable j.

Inspired by $(Rockett)$, that has this "almost perfect" form, and that came out from considering the sum of the reciprocals of the summands of the *simplest* binomial coefficient sum known, we searched for other cases where one starts from a well-known binomial coefficients sum that evaluates in closed form, and looks at what happens if one considers the sum of the reciprocals of the summands. To our great surprise and delight, it worked for the Chu-Vandermonde summand and for the Dixon summand (see below). To our disappointment, it didn't work for the so-called Pfaff-Saalschutz identity (see, e.g. [2]).

We use the WZ method([2][7][8]), and the reader is assumed to be familiar with it. In particular we used Zeilberger's Maple packages `EKHAD` (procedure zeillim) and `qEKHAD` accompanying [2], available from

`http://www.math.rutgers.edu/~zeilberg/tokhniot/[EKHAD,qEKHAD]` .

In most applications of the WZ method, the summation limits are *natural*, i.e. the summand is well-defined and vanishes at $k = -1$ and $k = n+1$ (or in the case of $(Dixon)$, at $k = -n-1$ and $k = n+1$) so one can get a *homogeneous* first-order linear recurrence for the definite sum, that leads to a closed-form solution. In the present cases, this is no longer the case, and we only get *inhomogeneous* first-order linear recurrences, that lead to the above-mentioned "almost-perfect" kind of solutions.

Consider the sum

$$f(n) := \sum_{k=\alpha}^{n-\beta} F(n,k) ,$$

where α and β are constants and $[\alpha, n-\beta]$ is properly contained in $[0,n]$, the support of $F(n,k)$. Assume also that there exists a function $G(n,k)$, (the so-called certificate) such that

$$a(n)F(n+1,k) + b(n)F(n,k) = G(n,k+1) - G(n,k) . \qquad (WZeqn)$$

Take $\alpha = 0$ and $\beta = 1$. Then, if we add both sides of (*WZeqn*) from $k = 0$ to $k = n$ and rewrite the result we get a nonhomogeneous recurrence relation satisfied by $f(n)$:

$$a(n)f(n+1) + b(n)f(n) = G(n,n) - G(n,0) + a(n)F(n+1,n) . \quad (nonhomorec)$$

Finally, we solve (*nonhomorec*) and get the following representation of $f(n)$ as an indefinite sum:

$$f(n) = g(n) \sum_{i=0}^{n-1} \frac{C(i)}{g(i)a(i)} ,$$

where

$$C(i) = G(i,i) - G(i,0) + a(i)F(i+1,i) .$$

(i.e. the right side of $(nonhomorec)$ with n replaced by i), and $g(n)$ is a solution to the associated homogeneous equation $a(n)f(n+1) + b(n)f(n) = 0$.

We note that in case $\alpha \neq 0$ and $\beta \neq 1$, the left-hand side of (*nonhomrec*) remains the same. The only change is on the right-hand side.

Theorem 1 [Upside Down Binomials Squared]:

$$\sum_{k=0}^{n-1} \binom{n}{k}^{-2} = \frac{(n+1)!(n+1)^2}{(n+\frac{3}{2})!4^n} \sum_{j=0}^{n-1} \frac{(3j^3+12j^2+18j+10)4^j(j+3/2)!}{(j+1)^2(j+1)!(j+2)^3} .$$

(Here $n! := \Gamma(n+1)$).
Proof: We construct the function

$$G(n,k) = \frac{(2k-3n-6)(k-n-1)^2}{\binom{n}{k}^2}$$

such that $(4n+10)(n+1)^2F(n+1,k) - (n+2)^2F(n,k) = G(n,k+1) - G(n,k)$, where $F(n,k)$ is the summand on the left-hand side sum.
Next add both sides from $k = 0$ to $k = n$ and rearrange to get the nonhomogeneous recurrence relation satisfied by the sum on the left-hand side, S_n :

$$(4n+10)(n+1)^2 S_{n+1} - (n+2)^2 S_n = 3n^3 + 12n^2 + 18n + 10 .$$

Finally the theorem follows by solving this recurrence with the initial condition $S_1 = 1$.
Next we give q-analog of (*Rockett*). Let $(a;q)_n = (1-a)(1-aq)\ldots(1-aq^{n-1})$.

Theorem 2 [Upside Down q-binomial]:

$$\sum_{k=0}^{n-1} \frac{q^{-k(k-1)/2}}{\begin{bmatrix} n \\ k \end{bmatrix}_q} = \frac{(q^2;q)_n}{(q^2;q^2)_n} \sum_{i=0}^{n-1} \frac{C(i)(q^2;q^2)_i}{(q^2;q)_i(q^{i+2}-1)} ,$$

where

$$C(i) =$$
$$= \frac{q^{i+1} + q^{3(i+1)} + q^{-i(i-1)/2} + q^{-(i+1)(i-4)/2} - q^{(2+3i-i^2)/2} - q^{-(i+1)(i-2)/2} - 2q^{2(i+1)}}{q^{i+1}-1} .$$

Proof: We construct the function

$$G(n,k) = \frac{q^{n+1}(1-q^{n-k+1})q^{-k(k-1)/2}}{\begin{bmatrix} n \\ k \end{bmatrix}_q}$$

such that $(q^{n+1}-1)(q^{n+1}+1)F(n+1,k)-(q^{n+2}-1)F(n,k)=G(n,k+1)-G(n,k)$, where $F(n,k)$ is the summand on the left-hand side.

Add both sides from $k=0$ to $k=n$ and rearrange to get the nonhomogeneous recurrence relation satisfied by the sum on the left-hand side, S_n :

$$(q^{n+1}-1)(q^{n+1}+1)S_{n+1}-(q^{n+2}-1)S_n=C(n)\ .$$

Finally the theorem follows by solving this recurrence with the initial condition $S_1=1$.

Next we consider the reciprocal of the summand in the Chu-Vandermonde and the Dixon classical identities.

Theorem 3 [Upside Down Chu-Vandermonde]

$$\sum_{k=0}^{m-1}\binom{m}{k}^{-1}\binom{a}{n-k}^{-1}=g(m)\sum_{i=0}^{m-1}\frac{C(i)}{g(i)(i+2)(a+i-n+2)}$$

where

$$g(m)=\frac{(a+m-n+1)!(a+4)!(m+1)}{2(a+m+3)!(a-n+2)!}$$

and

$$C(i)=(n+i+in+1)\binom{a}{n}^{-1}+(2i+a-n+3)\binom{a}{n-i}^{-1}.$$

Proof: The proof follows similarly from the recurrence relation

$$(a+m+4)(m+1)S_{m+1}-(m+2)(a+m-n+2)S_m=C(m)\ .$$

where S_m is the sum on left-hand side.

Theorem 4 [Upside Down Dixon]

$$\sum_{k=0}^{n-1}(-1)^k\binom{n+b}{n+k}^{-1}\binom{n+c}{c+k}^{-1}\binom{b+c}{b+k}^{-1}=$$

$$=g(n)\sum_{i=0}^{n-1}\frac{C(i)}{g(i)(2i+2)(c+i+2)(b+i+2)}$$

where

$$g(n)=\frac{(b+n+1)(c+n+1)(b+c+2)!n!}{(b+1)(c+1)(b+c+n+2)!}$$

and

$C(i) = (-1)^i(3bi + 3ci + bc + 3b + 5i^2 + 12i + 3c + 7)\binom{i+b}{2i}^{-1}\binom{b+c}{b+i}^{-1} - (b+1)(c+1)(i+1)\binom{i+b}{i}^{-1}\binom{i+c}{c}^{-1}\binom{b+c}{b}^{-1}$.

Proof: The proof follows similarly from the recurrence relation

$$2(c+n+1)(b+n+1)(b+c+n+3)S_{n+1} - 2(n+1)(c+n+2)(b+n+2)S_n = C(n)\ .$$

where S_n is the sum on left-hand side.

Remarks a) From the recurrence in the proof of theorem 1,

$$2S_n = \frac{(n+1)^3}{2n^3+3n^2}S_{n-1} + \frac{3n^3+3n^2+3n+1}{2n^3+3n^2}$$

that implies

$$\lim_{n\to\infty}\sum_{k=0}^{n}\binom{n}{k}^{-2} = 2\ .$$

b) Identity (*Rockett*) is a special case of the following identity with $x = y = 1$ in

$$\sum_{k=0}^{n}\binom{n}{k}^{-1}x^k y^{n-k} = x^n + \left(\frac{xy}{x+y}\right)^n (n+1)\sum_{j=0}^{n-1}\frac{((j+1)y^{j+2} + yx^{j+1})(x+y)^j}{(xy)^{j+1}(j+1)(j+2)}$$

which follows from the recurrence [S_n the left-hand side sum]:

$$(x+y)(n+1)S_{n+1} - (n+2)xyS_n = (n+1)y^{n+2} + yx^{n+1}\ .$$

c) T. Mansour [1] derived the following representation for the reciprocals of binomial coefficients to any power *m* as:

$$\sum_{k=0}^{n-1}\binom{n}{k}^{-m} = (n+1)^m\sum_{k=0}^{n}\left[\sum_{j=0}^{k}\frac{(-1)^j}{n-k+1+j}\binom{k}{j}\right]^m\ .$$

One of the referees kindly pointed out that (b) and (c) can also be deduced from a general identity derived in [6]. We thank them for this remark and for numerous corrections and improvements.

References

[1] T. Mansour, Combinatorial identities and inverse binomial coefficients, *Adv. in Appl. Math.* 28 (2002), 196–202.

[2] M. Petkovsek, H.S. Wilf and D. Zeilberger, $A = B$, A.K. Peters Ltd., 1996.

[3] A. M. Rockett, Sum of the inverse of binomial coefficients, *Fibonacci Quart.* 19(1981), 433–437.

[4] R. Sprugnoli, Sums of reciprocals of the central binomial coefficients, *Integers* 6(2006), #A27.

[5] B. Sury, Sum of the reciprocals of the binomial coefficients, *European J. Combin.* 14(1993), 351–353.

[6] B. Sury, T. Wang and F. Zhao, Identities involving reciprocals of binomial coefficients, *J. Integer Seq.* 7(2007), Article 04.2.8.

[7] H. Wilf and D. Zeilberger, Rational functions certify combinatorial identities, *J. Amer. Math. Soc.* 3(1990), 147–158.

[8] D. Zeilberger, The method of creative telescoping, *J. Symbolic Comput.* 11(1991), 195–204.

In: Frontiers of Combinatorics and Number Theory, Volume 2 ISBN: 978-1-61942-918-5
Editor: Zhi-Wei Sun, pp. 85-96

Chapter 6

A SUMSET PROBLEM

***Javier Cilleruelo*[1,*], *Manuel Silva*[2,†] *and Carlos Vinuesa*[3,‡]**
[1]Instituto de Ciencias Matemáticas (CSIC-UAM-UC3M-UCM)
and Departamento de Matemáticas.
Universidad Autónoma de Madrid, Madrid, Spain
[2]Departamento de Matemática, Universidade Nova de Lisboa,
Caparica, Portugal
[3]Departamento de Matemáticas.
Universidad Autónoma de Madrid, Madrid, Spain

Abstract

We study the sumset $A + k \cdot A$ for the first non trivial case, $k = 3$, where $k \cdot A = \{k \cdot a,\ a \in A\}$. We prove that $|A + 3 \cdot A| \geq 4|A| - 4$ and that the equality holds only for $A = \{0, 1, 3\}$, $A = \{0, 1, 4\}$, $A = 3 \cdot \{0, \ldots, n\} \cup (3 \cdot \{0, \ldots, n\} + 1)$ and all the affine transforms of these sets.

2000 Mathematics Subject Classification: 11B75

1. Introduction

We address here the question of how large is the sumset $A + k \cdot A$ where $k \cdot A = \{k \cdot a,\ a \in A\}$ and A is a finite set of integers. It is well known that $|A + A| \geq 2|A| - 1$ and that equality only holds when A is an arithmetic progression. The case $k = 2$ (see [2]) can be studied easily by splitting A in the two residue classes modulo 2 and then obtaining $|A + 2 \cdot A| \geq 3|A| - 2$. It is not difficult to check that for any arithmetic progression with at least k elements, we have $|A + k \cdot A| = (k + 1)|A| - k$, so it might be expected that arithmetic progressions are extremal cases for this problem, as when $k = 1$. Indeed this is the case for $k = 2$ as we will prove in section §2.

*E-mail address: franciscojavier.cilleruelo@uam.es
†E-mail address: mnas@fct.unl.pt
‡E-mail address: c.vinuesa@uam.es

Theorem 1 *For any set A we have $|A + 2 \cdot A| \geq 3|A| - 2$. Furthermore, if $|A + 2 \cdot A| = 3|A| - 2$, then A is an arithmetic progression or a singleton.*

For $k = 3$ Bukh [1] has proved in a recent paper that $|A + 3 \cdot A| \geq 4|A| - O(1)$ for any set A. Our main theorem gives, using a different argument, a sharp lower bound and a complete description of the extremal sets. We observe that these minimal sets for $k = 3$ are not arithmetic progressions anymore, as in the previous cases $k = 1, 2$. We observe also that the problem in consideration here is affine invariant.

Theorem 2 *For any set A we have $|A + 3 \cdot A| \geq 4|A| - 4$. Furthermore if $|A + 3 \cdot A| = 4|A| - 4$ then $A = 3 \cdot \{0, \dots, n\} \cup (3 \cdot \{0, \dots, n\} + 1)$ or $A = \{0, 1, 3\}$ or $A = \{0, 1, 4\}$ or A is an affine transform of one of these sets.*

The general sums of dilated sets, $\lambda_1 \cdot A + \cdots + \lambda_k \cdot A$, have been studied by Bukh in [1]. The main theorem there says that for coprime integers $\lambda_1, \dots, \lambda_k$,

$$|\lambda_1 \cdot A + \cdots + \lambda_k \cdot A| \geq (|\lambda_1| + \cdots + |\lambda_k|)|A| - o(|A|).$$

In particular it gives $|A + k \cdot A| \geq (k+1)|A| - o(|A|)$. As we will prove in section §5, there exist arbitrarily large sets A such that $|A + k \cdot A| = (k+1)|A| - \left\lceil \frac{k^2+2k}{4} \right\rceil$. We conjecture that this lower bound is sharp for large $|A|$.

2. Case $k = 2$ and Preliminary Lemmas

The next lemma is folklore, and we state it without proof.

Lemma 1 *For arbitrary non empty finite sets A, B, we have*

i) $|A + B| \geq |A| + |B| - 1$.

ii) Furthermore, if equality holds, then A and B are arithmetic progressions with the same difference unless one of them is a singleton.

We generalize this lemma for any k. For that purpose, it is natural to divide A into residue classes modulo k. We define $\hat{A}$ to be the projection of A into $\mathbb{Z}/k\mathbb{Z}$.

Lemma 2 *For arbitrary non empty sets B and $A = \bigcup_{i \in \hat{A}} (k \cdot A_i + i)$ we have*

i) $|A + k \cdot B| = \sum_{i \in \hat{A}} |A_i + B|$

ii) $|A + k \cdot B| \geq |A| + |\hat{A}|(|B| - 1)$.

iii) Furthermore, if equality holds in ii), then either $|B| = 1$ or $|A_i| = 1$ for all $i \in \hat{A}$ or B and all the sets A_i with more than one element are arithmetic progressions with the same difference.

Proof 3 *For i)* $|A + k \cdot B| = |\cup_{i \in \hat{A}} (k \cdot A_i + i + k \cdot B)| = \sum_{i \in \hat{A}} |k \cdot (A_i + B) + i| = \sum_{i \in \hat{A}} |A_i + B|$. *To prove ii) we use i) and Lemma 1-i). To prove iii) we observe that Lemma 1-ii) implies that* A_i *and* B *are arithmetic progressions with the same difference except for the degenerate cases.*

Next we prove Theorem 1 as a direct application of Lemma 2.

Proof 4 (Proof of Theorem 1) *If* $|A| = 1$ *then* $|A + 2 \cdot A| = 3|A| - 2$, *and these sets are described in Theorem 1, so the inverse part is also proved.*

So we assume $|A| \geq 2$. *If* $|\hat{A}| = 1$ *then we can write* $A = 2 \cdot A_i + i$ *for some* $i \in \{0, 1\}$ *and* $|A + 2 \cdot A| = |2 \cdot A_i + i + 4 \cdot A_i + 2i| = |A_i + 2 \cdot A_i|$. *Now, if* $|\hat{A}_i| = 1$, *we can repeat this process and it is clear that we will eventually obtain a set* A' *with* $|\hat{A}'| = 2$ *(since it cannot be greater for* $k = 2$*) that is an affine transformation of the set* A.

Thus we can also assume that $|\hat{A}| = 2$. *By Lemma 2-ii) we conclude that* $|A + 2 \cdot A| \geq |A| + 2(|A| - 1) = 3|A| - 2$. *For the inverse part, if the equality holds, Lemma 2-iii) implies that either* $|A| = 1$ *or* $|A_0| = |A_1| = 1$ *or* A *is an arithmetic progression. We finish by observing that* $|A| = 1$ *is impossible since we assumed* $|A| \geq 2$ *and* $|A_0| = |A_1| = 1$ *implies that* $|A| = 2$, *so it is an arithmetic progression.*

For the case $k = 3$ we will need some preliminary lemmas. In the rest of this section and in the next one, k will be equal to 3 and so $\hat{A}$ will always be the projection of A into $\mathbb{Z}/3\mathbb{Z}$.

Lemma 3 *If* $A = 3 \cdot A_0 \cup (3 \cdot A_1 + 1)$ *with* A_0 *and* A_1 *non empty sets, then*

i) $|A + 3 \cdot A| \geq |A_0 + 3 \cdot A_0| + |A_1 + 3 \cdot A_1| + 2$.

ii) $|A + 3 \cdot A| \geq |A_0 + 3 \cdot A_1| + |A_1 + 3 \cdot A_0| + 2$.

Proof 5 *To prove i) we write*

$$\begin{aligned}
|A + 3 \cdot A| &= |A_0 + A| + |A_1 + A| \\
&= |(A_0 + 3 \cdot A_0) \cup (A_0 + 3 \cdot A_1 + 1)| + |(A_1 + 3 \cdot A_0) \cup (A_1 + 3 \cdot A_1 + 1)| \\
&= |A_0 + 3 \cdot A_0| + |A_1 + 3 \cdot A_1 + 1| \\
&+ |(A_0 + 3 \cdot A_1 + 1) \setminus (A_0 + 3 \cdot A_0)| + |(A_1 + 3 \cdot A_0) \setminus (A_1 + 3 \cdot A_1 + 1)|
\end{aligned}$$

Then we only need to check that the last line above is at least 2. If $|A_0| = 1$ *and* $|A_1| = 1$, *we write* $A_0 = \{a_0\}$ *and* $A_1 = \{a_1\}$. *Then* $a_0 + 3a_1 + 1 \neq a_0 + 3a_0$ *and* $a_1 + 3a_0 \neq a_1 + 3a_1 + 1$ *because they are different modulo 3, so we have two extra elements. If not, let* m_i *and* M_i *be the minimum and the maximum of* A_i, $i = 0, 1$, *and we know that for at least one* i, $m_i \neq M_i$.

If $M_0 \leq M_1$ *then* $M_0 + 3M_1 + 1 \in (A_0 + 3 \cdot A_1 + 1) \setminus (A_0 + 3 \cdot A_0)$ *because* $M_0 + 3M_1 + 1$ *is greater than* $M_0 + 3M_0$, *which is the maximum of* $A_0 + 3 \cdot A_0$. *On the other hand, if* $M_0 > M_1$, *then* $M_1 + 3M_0 \in (A_1 + 3 \cdot A_0) \setminus (A_1 + 3 \cdot A_1 + 1)$.

If $m_0 \leq m_1$ *then* $m_1 + 3m_0 \in (A_1 + 3 \cdot A_0) \setminus (A_1 + 3 \cdot A_1 + 1)$ *and if* $m_0 > m_1$ *then* $m_0 + 3m_1 + 1 \in (A_0 + 3 \cdot A_1 + 1) \setminus (A_0 + 3 \cdot A_0)$.

We obtain one extra element in each case. To see that the elements in the same set are distinct, observe that if $M_0 + 3M_1 + 1 = m_0 + 3m_1 + 1$*, then we must have* $M_0 = m_0$ *and* $M_1 = m_1$*, a contradiction. The same thing happens if* $M_1 + 3M_0 = m_1 + 3m_0$*. The proof of ii) is similar.*

Lemma 4 *If* $A = 3 \cdot A_0 \cup (3 \cdot A_1 + 1)$ *with* A_0 *and* A_1 *non empty sets, then*

i) If $|\hat{A}_0| = 2$*, we have* $|A_0 + A| \geq 2|A| - 2$*.*

ii)
- *If* $|\hat{A}_0| \leq 2$ *and* $|A_0 + 3 \cdot A_0| \geq 4|A_0| - 4$*, we have* $|A_0 + A| \geq 4|A_0| + |A_1| - 4$*.*
- *If* $|\hat{A}_1| \leq 2$ *and* $|A_1 + 3 \cdot A_1| \geq 4|A_1| - 4$*, we have* $|A_1 + A| \geq 4|A_1| + |A_0| - 4$*.*

Proof 6 *For i), let* $\hat{A}_0 = \{u, u+1\}$*. We can write* $A_0 = A_0^u \cup A_0^{u+1}$*, where* $A_0^u = \{x \in A_0,\ x \equiv u \pmod 3\}$*. Then*

$$\begin{aligned}
|A_0 + A| &= |(A_0 + 3 \cdot A_0) \cup (A_0 + 3 \cdot A_1 + 1)| \\
&= |(A_0^u + 3 \cdot A_0) \cup (A_0^{u+1} + 3 \cdot A_0) \cup (A_0^u + 3 \cdot A_1 + 1) \cup (A_0^{u+1} + 3 \cdot A_1 + 1)| \\
&\geq |A_0^u + 3 \cdot A_0| + |A_0^u + 3 \cdot A_1 + 1| + |A_0^{u+1} + 3 \cdot A_1 + 1| \\
&\geq |A_0^u| + |A_0| - 1 + |A_1| + |A_0^{u+1}| + |A_1| - 1 = 2|A| - 2,
\end{aligned}$$

where we have used Lemma 1-i) twice.

For part ii), we again write $A_0 = A_0^u \cup A_0^{u+1}$ *if* $|\hat{A}_0| = 2$*, or* $A_0 = A_0^{u+1}$ *if* $|\hat{A}_0| = 1$*. Then*

$$\begin{aligned}
|A_0 + A| &= |(A_0 + 3 \cdot A_0) \cup (A_0 + 3 \cdot A_1 + 1)| \geq |(A_0 + 3 \cdot A_0) \cup (A_0^{u+1} + 3 \cdot A_1 + 1)| \\
&= |A_0 + 3 \cdot A_0| + |A_0^{u+1} + 3 \cdot A_1 + 1| \geq 4|A_0| - 4 + |A_1|.
\end{aligned}$$

The same argument works for A_1 *instead of* A_0*.*

Lemma 5

i) If $|A| = 2$ *then* $|A + 3 \cdot A| = 4|A| - 4 = 4$*.*

ii) If $|A| = 3$ *then* $|A + 3 \cdot A| \geq 4|A| - 4$*. Furthermore, if* $|A| = 3$ *and* $|A + 3 \cdot A| = 4|A| - 4$ *then* A *is an affine transform of* $\{0, 1, 3\}$ *or* $\{0, 1, 4\}$*.*

Proof 7 *i) By an affine transformation we may assume* $A = \{0, 1\}$*. Then* $A + 3 \cdot A = \{0, 1, 3, 4\}$*.*

ii) By an affine transformation we may assume that $A = \{0, 1, a\}$*, where* $a > 1$ *is a rational number. We have that*

$$A + 3 \cdot A = \{0, 1, a, 3, 4, 3 + a, 3a, 3a + 1, 4a\}.$$

If $A + 3 \cdot A$ *has 8 or less elements then there is some repeated element in the sumset. The possible repetitions come from* $a = 3,\ a = 4,\ 4 = 3a,\ 3 + a = 3a$ *which provide the sets* $\{0, 1, 3\}$*,* $\{0, 1, 4\}$*,* $\{0, 3, 4\}$*,* $\{0, 2, 3\}$*. All these sets have* $|A + 3 \cdot A| = 8 = 4|A| - 4$*.*

3. Proof of Theorem 2: The Inequality

In this section, we will prove the lower bound in Theorem 2 for $|A + 3 \cdot A|$. That is, we will prove that for every set A, we have $|A+3 \cdot A| \geq 4|A| - 4$. In the next section, we will prove the inverse part which is more involved.

We distinguish three cases according to the different values of $|\hat{A}|$.

If $|\hat{A}| = 3$ then (by Lemma 2-ii)) we have that $|A + 3 \cdot A| \geq 4|A| - 3$, a better lower bound than that we want to prove in Theorem 2.

If $|\hat{A}| = 1$ and $|A| = 1$, then $4|A| - 4 = 0$ and the theorem is trivial. If $|\hat{A}| = 1$ and $|A| > 1$, then we write $A = 3 \cdot A_i + i$ and we have $|A + 3 \cdot A| = |A_i + 3 \cdot A_i|$. If $|\hat{A}_i| = 1$, we repeat the process with A replaced by A_i until we find a set with $|\hat{A}_i|$ equal to 2 or 3. Cardinality 3 has been treated above.

So we can now assume that $|\hat{A}| = 2$ and write $A = (3 \cdot A_i + i) \cup (3 \cdot A_{i+1} + i + 1)$. We can assume that $|\hat{A}_i| \leq |\hat{A}_{i+1}|$. If not the set $B = -A$ could be written as $B = (3 \cdot B_j + j) \cup (3 \cdot B_{j+1} + j + 1)$ where $B_j = -A_{i+1} - 1,\ B_{j+1} = -A_i - 1,\ j = 2 - i$, and in this case we would have $|\hat{B}_j| \leq |\hat{B}_{j+1}|$. Finally, by translation we can assume

- $A = 3 \cdot A_0 \cup (3 \cdot A_1 + 1)$
- $\min A_0 = 0$
- $|\hat{A}_0| \leq |\hat{A}_1|$.

Assuming all this, we prove $|A+3 \cdot A| \geq 4|A| - 4$ by induction on $|A|$. The inequality clearly holds for $|A| = 1$. Suppose we have proved it for any set with fewer elements than A, in particular for A_0 and A_1. We distinguish three cases:

Case $|\hat{A}_0| = |\hat{A}_1| = 3$. We use Lemma 3-i) and Lemma 2-ii) to obtain

$$|A+3 \cdot A| \geq |A_0 + 3 \cdot A_0| + |A_1 + 3 \cdot A_1| + 2 \geq 4|A_0| - 3 + 4|A_1| - 3 + 2 = 4|A| - 4.$$

Case $|\hat{A}_1| = 3,\ |\hat{A}_0| < 3$. We apply Lemma 4-ii) (using the induction hypothesis) and Lemma 2-ii) to obtain

$$\begin{aligned}|A + 3 \cdot A| &= |A_0 + A| + |A_1 + A| \geq |A_0 + A| + |A_1 + 3 \cdot A_1| \geq \\ 4|A_0| &+ |A_1| - 4 + 4|A_1| - 3 = 4|A| - 4 + |A_1| - 3 \geq 4|A| - 4.\end{aligned}$$

In the last inequality we have used that $|A_1| \geq |\hat{A}_1| = 3$.

Case $|\hat{A}_1| < 3$. We apply Lemma 4-ii) to A_0 and A_1 (again, using the induction hypothesis) to obtain

$$|A+3 \cdot A| = |A_0+A|+|A_1+A| \geq 4|A_0|+|A_1|-4+4|A_1|+|A_0|-4 = 4|A|-4+|A|-4.$$

If $|A| \geq 4$ this gives the bound. If not, we use Lemma 5. This completes the proof.

4. Proof of Theorem 2: The Cases of Equality

Now we turn to the proof of the inverse part of Theorem 2. Once we know that $|A+3\cdot A| \geq 4|A|-4$ for every set - and also we have some examples where the equality holds (see Lemma 5) - we characterize all the sets that satisfy the equality $|A+3\cdot A| = 4|A|-4$.

As in the previous section we can assume $A = 3\cdot A_0 \cup (3\cdot A_1+1)$, $|\hat{A}_0| \leq |\hat{A}_1|$ and $\min A_0 = 0$.

Case $|\hat{A}_0| = |\hat{A}_1| = 3$. We use Lemma 3-i) and 3-ii) and Lemma 2-ii) to obtain

$$\begin{aligned} 4|A|-4 = |A+3\cdot A| &\geq |A_0+3\cdot A_0| + |A_1+3\cdot A_1| + 2 \\ &\geq |A_0| + 3|A_0| - 3 + |A_1| + 3|A_1| - 3 + 2 = 4|A|-4. \end{aligned}$$

and

$$\begin{aligned} 4|A|-4 = |A+3\cdot A| &\geq |A_0+3\cdot A_1| + |A_1+3\cdot A_0| + 2 \\ &\geq |A_0| + 3|A_1| - 3 + |A_1| + 3|A_0| - 3 + 2 = 4|A|-4. \end{aligned}$$

Then, the inequalities are, indeed, equalities. So $|A_0+3\cdot A_0| = |A_0|+3|A_0|-3$, $|A_1+3\cdot A_1| = |A_1|+3|A_1|-3$, $|A_0+3\cdot A_1| = |A_0|+3|A_1|-3$ and $|A_1+3\cdot A_0| = |A_1|+3|A_0|-3$. Now we apply Lemma 2-iii) to conclude that (since $|A_0| \geq |\hat{A}_0| = 3$ and $|A_1| \geq |\hat{A}_1| = 3$)

a) either $A_0 = \{x_0, x_1, x_2\}$ and $A_1 = \{y_0, y_1, y_2\}$ with $x_i, y_i \equiv i \pmod 3$

b) or A_0 and A_1 are arithmetic progressions with the same difference, d.

a) In this subcase, $|A| = 6$ and $4|A|-4 = 20$, and we know by Lemma 2-i) that $20 = |A+3\cdot A| = |A_0+A|+|A_1+A|$. Then, $|A_0+A| \leq 10$ or $|A_1+A| \leq 10$. We suppose that $|A_0+A| \leq 10$ (the other case is identical) and, because $A_0 = \{x_0, x_1, x_2\}$ with $x_i \equiv i \pmod 3$, we have

$$\begin{aligned} 10 &\geq |(A_0+3\cdot A_0)\cup(A_0+3\cdot A_1+1)| \\ &= |(x_0+3\cdot A_0)\cup(x_2+3\cdot A_1+1)| \\ &+ |(x_1+3\cdot A_0)\cup(x_0+3\cdot A_1+1)| \\ &+ |(x_2+3\cdot A_0)\cup(x_1+3\cdot A_1+1)| \\ &= \left|A_0\cup\left(A_1+\frac{x_2-x_0+1}{3}\right)\right| \\ &+ \left|A_0\cup\left(A_1+\frac{x_0-x_1+1}{3}\right)\right| \\ &+ \left|A_0\cup\left(A_1+\frac{x_1-x_2+1}{3}\right)\right| \end{aligned}$$

and we can observe that each addend gives us at least 4 elements unless the two members of the union are equal (in this case we have only 3 elements). But, because the sum of the three addends is less than or equal to 10, we must have at least two equalities, for example:

$$A_0 = A_1 + \frac{x_2-x_0+1}{3} \text{ and } A_0 = A_1 + \frac{x_0-x_1+1}{3}.$$

Then, we have $x_2 - x_0 = x_0 - x_1$, so A_0 is an arithmetic progression and also A_1 is an arithmetic progression with the same difference, since it is a translation of A_0. The other possibilities are identical.

b) In this case both A_0 and A_1 are arithmetic progressions with difference d, and because $\min A_0 = 0$ we can write $A_0 = d\cdot[0, n_0-1]$, $A_1 = d\cdot[0, n_1-1]+e$. Since $n_0, n_1 \geq 3$, we have that $[0, n_i - 1] + 3 \cdot [0, n_j - 1] = [0, 3n_j + n_i - 4]$ for any $i, j \in \{0, 1\}$. Thus

$$\begin{aligned}&|A + 3 \cdot A| = |A_0 + A| + |A_1 + A|\\&= |d \cdot ([0, n_0 - 1] + 3 \cdot [0, n_0 - 1]) \cup\\&(d \cdot ([0, n_0 - 1] + 3 \cdot [0, n_1 - 1]) + 3e + 1)|\\&+|(d \cdot ([0, n_1 - 1] + 3 \cdot [0, n_0 - 1]) + e) \cup\\&(d \cdot ([0, n_1 - 1] + 3 \cdot [0, n_1 - 1]) + 4e + 1)|\\&= |d \cdot [0, 4n_0 - 4] \cup (d \cdot [0, 3n_1 + n_0 - 4] + 3e + 1)|\\&+|d \cdot [0, 4n_1 - 4] \cup (d \cdot [0, 3n_0 + n_1 - 4] - 3e - 1)|.\end{aligned}$$

If $n_1 > n_0$ then

$$|A + 3 \cdot A| \geq 3n_1 + n_0 - 3 + 4n_1 - 3 = 4(n_0 + n_1) + 3(n_1 - n_0) - 6 \geq 4|A| - 3,$$

which is a contradiction. So $n_1 \leq n_0$. For the same reason (interchanging n_0 and n_1) we have that $n_0 \leq n_1$ and then $n_0 = n_1$. Now we can write

$$\begin{aligned}|A + 3 \cdot A| &= |d \cdot [0, 4n_0 - 4] \cup (d \cdot [0, 4n_0 - 4] + 3e + 1)|\\&+ |d \cdot [0, 4n_0 - 4] \cup (d \cdot [0, 4n_0 - 4] - 3e - 1)|.\end{aligned}$$

If $3e + 1 \not\equiv 0 \pmod d$ then the unions are disjoint and we have $4|A| - 4 = |A + 3 \cdot A| = 2(4n_0 - 3) + 2(4n_0 - 3) = 8|A| - 12$. That implies that $|A| = 2$ and this is impossible since $|A| = |A_0| + |A_1| \geq |\hat{A}_0| + |\hat{A}_1| = 6$. If $3e + 1 \equiv 0 \pmod d$ we write $3e + 1 = de'$ and then

$$\begin{aligned}|A + 3 \cdot A| &= |[0, 4n_0 - 4] \cup ([0, 4n_0 - 4] + e')|\\&+ |[0, 4n_0 - 4] \cup ([0, 4n_0 - 4] - e')|.\end{aligned}$$

If $|e'| \geq 2$ then the cardinality of each union is greater than or equal to $4n_0 - 1$, and $|A+3\cdot A| \geq 4n_0 - 1 + 4n_0 - 1 = 4|A| - 2$. So, since $e' \neq 0$ (remember that de' is an integer different from 0) then $e' = \pm 1$, so $3e+1 = \pm d$ and $A = 3 \cdot A_0 \cup 3 \cdot A_1 + 1 = d \cdot (3 \cdot [0, n_0 - 1] \cup 3 \cdot [0, n_0 - 1] \pm 1)$. These sets are listed as extremal sets in Theorem 2.

Case $|\hat{A}_0| = 2,\ |\hat{A}_1| = 3$. We write

$$|A_1 + A| = |(A_1 + 3\cdot A_0) \cup (A_1 + 3\cdot A_1 + 1)| = |A_1 + 3\cdot A_1| + |(A_1 + 3\cdot A_0) \setminus (A_1 + 3\cdot A_1 + 1)|.$$

Lemma 2-i), Lemma 4-ii) and the equality above imply that

$$\begin{aligned} |A+3\cdot A| &= |A_0+A|+|A_1+A| \\ &\geq 4|A_0|+|A_1|-4+4|A_1|-3+(|A_1+3\cdot A_1|-4|A_1|+3) \\ &+ |(A_1+3\cdot A_0)\setminus(A_1+3\cdot A_1+1)|. \end{aligned}$$

Then

$$4|A|-4=|A+3\cdot A|\geq 4|A|-4+(|A_1|-3)+(|A_1+3\cdot A_1|-$$

$$-4|A_1|+3)+|(A_1+3\cdot A_0)\setminus(A_1+3\cdot A_1+1)|.$$

Using that $|A_1|\geq|\hat{A}_1|=3$ and Lemma 2-ii) we see that the three last addends are non negative. But the inequality implies that all of them are indeed 0. Then,

i) $|A_1|=3$.

ii) By Lemma 2-iii),

a) either $A_1=\{y_0,y_1,y_2\}$ with $y_i\equiv i \pmod 3$

b) or A_1 is an arithmetic progression, say $A_1=d\cdot[0,2]+e$.

iii) $3\cdot A_0\subset A_1-A_1+3\cdot A_1+1$ (because $A_1+3\cdot A_0\subset A_1+3\cdot A_1+1$).

Now we claim that also $|A_0|=3$. To see that we will obtain a lower and an upper bound.

To prove $|A_0|\leq 3$ we use Lemma 2-i), Lemma 4-ii), Lemma 2-ii) and the fact that $|A_1|=3$ to have

$$\begin{aligned} |A+3\cdot A| &= |A_0+A|+|A_1+A|\geq|A_0+A|+|A_1+3\cdot A_0|\geq \\ & \quad 4|A_0|+|A_1|-4+|A_1|+3(|A_0|-1)=4|A|-4+3|A_0|-9. \end{aligned}$$

Since we have assumed that $|A+3\cdot A|=4|A|-4$ then $|A_0|\leq 3$.

To prove $|A_0|\geq 3$ we use Lemma 2-i), Lemma 4-i) and Lemma 5-ii) (for a set A of three elements that covers the three classes modulo 3 we must have $|A+3\cdot A|=9$) to obtain

$$4|A|-4=|A+3\cdot A|=|A_0+A|+|A_1+A|\geq 2|A|-2+|A_1+3\cdot A_1|=2|A|-2+9,$$

so $|A|\geq 11/2$. And since $|A_1|=3$ we have that $|A_0|\geq 5/2$, so $|A_0|\geq 3$.

So we have proved that $|A|=6$.

Next we will see that if we are in case ii)-a), that is if $A_1=\{y_0,y_1,y_2\}$ with $y_i\equiv i \pmod 3$, then A_1 is an arithmetic progression. As in a) of the case $|\hat{A}_0|=|\hat{A}_1|=3$ we have, $20=4|A|-4=|A+3\cdot A|=|A_0+A|+|A_1+A|$. Again, one of them is less or equal than 10. If $|A_1+A|\leq 10$ then we proceed exactly as we did in that case and we have that A_1 is an arithmetic progression. If $|A_0+A|\leq 10$ then $A_0=\{x_0,y_0,x_1\}$ or $A_0=\{x_0,y_0,x_2\}$ where $x_i\equiv y_i\equiv i \pmod 3$ except for translations. In the first case $10\geq|A_0+A|=|(A_0+3\cdot A_0)\cup(A_0+3\cdot A_1+1)|\geq|(x_0+3\cdot A_0)\cup(y_0+3\cdot A_0)|+$

$|(x_0 + 3 \cdot A_1 + 1) \cup (y_0 + 3 \cdot A_1 + 1)| + |x_1 + 3 \cdot A_1 + 1| \geq 4 + 4 + 3 = 11$, which is a contradiction. The second case is similar.

Thus, the only possibility is ii)-b), that is, A_1 is an arithmetic progression, say $A_1 = d \cdot [0, 2] + e$, and then $A_1 + 3 \cdot A_1 + 1 - A_1 = d \cdot [-2, 8] + 3e + 1$, so by iii) we have that

$$3 \cdot A_0 \subset d \cdot [-2, 8] + 3e + 1. \tag{1}$$

Inclusion (1) implies that $(d, 3) = 1$.

Suppose $d \equiv 1 \pmod 3$. Then $3 \cdot A_0 \subset d \cdot \{-1, 2, 5, 8\} + 3e + 1$. In this case $A = d \cdot (S \cup \{0, 3, 6\}) + 3e + 1$, where $S = \{-1, 2, 8\}$ or $S = \{-1, 5, 8\}$. Observe that these sets are the only subsets of three elements of $\{-1, 2, 5, 8\}$ satisfying that $|\frac{1}{3}\widehat{(S+1)}| = 2$. Since the problem is invariant by translations and dilations we only have to check the sets $A = \{-1, 0, 2, 3, 6, 8\}$ and $A = \{-1, 0, 3, 5, 6, 8\}$.

If $d \equiv 2 \pmod 3$ the sets we have to check are $A = \{-2, 0, 1, 3, 6, 7\}$ and $A = \{-2, 0, 3, 4, 6, 7\}$. The four sets we have described satisfy $|A + 3 \cdot A| = 24 \neq 4|A| - 4$.

Case $|\hat{A}_0| = 1$, $|\hat{A}_1| = 3$. Since $|\hat{A}_0| = 1$ we have $|A_0 + A| = |(A_0 + 3 \cdot A_0) \cup (A_0 + 3 \cdot A_1 + 1)| = |A_0 + 3 \cdot A_0| + |A_0 + 3 \cdot A_1| \geq 4|A_0| - 4 + |A_0| + |A_1| - 1 = 5|A_0| + |A_1| - 5$. Also we have that $|A_1 + A| \geq |A_1 + 3 \cdot A_1| \geq 4|A_1| - 3$. Then

$$4|A| - 4 = |A + 3 \cdot A| = |A_0 + A| + |A_1 + A| \geq 5|A_0| + |A_1| - 5 + 4|A_1| - 3 = 5|A| - 8,$$

thus $|A| \leq 4$. But since $|\hat{A}_0| = 1$ and $|\hat{A}_1| = 3$ we have that $|A_0| = 1$ and $|A_1| = 3$. In this case we get $|A_0 + A| = |A_0 + 3 \cdot A_0| + |A_0 + 3 \cdot A_1| = 1 + |A_1| = 4$. Then

$$12 = 4|A| - 4 = |A + 3 \cdot A| = |A_0 + A| + |A_1 + A| \geq 4 + 4|A_1| - 3 = 13$$

and we get a contradiction.

Case $|\hat{A}_0| = 2$, $|\hat{A}_1| = 2$. We can write, as in the proof of Lemma 4, $A_0 = A_0^u \cup A_0^{u+1}$ and $A_1 = A_1^v \cup A_1^{v+1}$, where $A_i^j = \{x \in A_i,\ x \equiv j \pmod 3\}$.

$$|A_0 + A| \geq |A_0 + 3 \cdot A_0| + |A_0^{u+1} + 3 \cdot A_1 + 1| \geq 4|A_0| - 4 + |A_1|.$$

Similary $|A_1 + A| \geq 4|A_1| - 4 + |A_0|$. Then $4|A| - 4 = |A + 3 \cdot A| = |A_0 + A| + |A_1 + A| \geq 4|A_0| - 4 + |A_1| + 4|A_1| - 4 + |A_0| = 5|A| - 8$ and thus $|A| \leq 4$. Since $|\hat{A}_0| = |\hat{A}_1| = 2$ we have that $|A_0| = |A_1| = 2$. Write $A_0 = \{a_0, b_0\}$, $A_1 = \{a_1, b_1\}$ with $b_i \equiv a_i + 1 \pmod 3$, $i = 0, 1$, and

$$\begin{aligned} |A_0 + A| &= |a_0 + 3 \cdot A_0| + |(b_0 + 3 \cdot A_0) \cup (a_0 + 1 + 3 \cdot A_1)| + |b_0 + 1 + 3 \cdot A_1| \\ &\geq 4 + |3 \cdot A_0 \cup (a_0 - b_0 + 1 + 3 \cdot A_1)|, \\ |A_1 + A| &= |a_1 + 3 \cdot A_0| + |(b_1 + 3 \cdot A_0) \cup (a_1 + 1 + 3 \cdot A_1)| + |b_1 + 1 + 3 \cdot A_1| \\ &\geq 4 + |3 \cdot A_0 \cup (a_1 - b_1 + 1 + 3 \cdot A_1)|. \end{aligned}$$

Hence

$$\begin{aligned} 12 &= 4|A| - 4 = |A + 3 \cdot A| = |A_0 + A| + |A_1 + A| \\ &\geq 8 + |3 \cdot A_0 \cup (a_0 - b_0 + 1 + 3 \cdot A_1)| + |3 \cdot A_0 \cup (a_1 - b_1 + 1 + 3 \cdot A_1)|. \end{aligned}$$

We claim that $3 \cdot A_0 = a_1 - b_1 + 1 + 3 \cdot A_1$. Otherwise we would obtain more than 2 elements in the last sum and we would get a contradiction. Then $3 \cdot A_0 = \{3a_1 + a_1 - b_1 + 1, 3b_1 + a_1 - b_1 + 1\} = \{4a_1 - b_1 + 1, a_1 + 2b_1 + 1\}$, so we obtain a set A described in Theorem 2,

$$\begin{aligned} A &= 3 \cdot A_0 \cup (3 \cdot A_1 + 1) = \{4a_1 - b_1, a_1 + 2b_1, 3a_1, 3b_1\} + 1 \\ &= 3b_1 + 1 + (a_1 - b_1) \cdot \{0, 1, 3, 4\}. \end{aligned}$$

Case $|\hat{A}_0| = 1,\ |\hat{A}_1| = 2$. In this case we have

$$|A_0 + A| = |A_0 + 3 \cdot A_0| + |A_0 + 3 \cdot A_1| \geq 4|A_0| - 4 + |A_0| + |A_1| - 1 = 5|A_0| + |A_1| - 5$$

and we apply Lemma 4-ii) to obtain $|A_1 + A| \geq 4|A_1| + |A_0| - 4$. Then

$$4(|A_0| + |A_1|) - 4 = |A + 3 \cdot A| = |A_0 + A| + |A_1 + A| \geq 5|A_0| + |A_1| - 5 + 4|A_1| + |A_0| - 4,$$

so $5 \geq 2|A_0| + |A_1|$. Since $|\hat{A}_1| = 2$ then $|A_1| \geq 2$ and $|A_0| \leq 3/2$; so $|A_0| = 1$. But in this case we have that $|A_0 + A| = |A|$ and $|A_1 + A| \geq 4|A_1| - 3$. Then $4|A_1| = 4|A| - 4 \geq |A| + 4|A_1| - 3$, so $|A| \leq 3$. Indeed, since $|\hat{A}_1| = 2$ and $|\hat{A}_0| = 1$ we have that $|A| = 3$. These cases are analyzed in Lemma 5.

Case $|\hat{A}_0| = 1,\ |\hat{A}_1| = 1$. As above we have $|A_0 + A| \geq 5|A_0| + |A_1| - 5$ and also we have $|A_1 + A| \geq 5|A_1| + |A_0| - 5$. Then $4|A| - 4 = |A + 3 \cdot A| = |A_0 + A| + |A_1 + A| \geq 6|A| - 10$, so $|A| \leq 3$ and the result again follows from Lemma 5.

5. Small sumsets $A + k \cdot A$

Now we show some constructions that give a small sumset, $A + k \cdot A$, for general $k \in \mathbb{N}$.

Proposition 8 *For any $k \in \mathbb{Z}_{>0}$*

i) there exist arbitrarily large sets A such that

$$|A + k \cdot A| = (k+1)|A| - \left\lceil \frac{k^2 + 2k}{4} \right\rceil$$

ii) there exists a set A such that

$$|A + k \cdot A| = (k+1)|A| - \frac{k^3 + 6k^2 + 9k + \delta_k}{27}$$

where

$$\delta_k = \begin{cases} 3k + 8 & \text{if } k \equiv 1 \pmod 3 \\ 4 & \text{if } k \equiv 2 \pmod 3 \\ 0 & \text{if } k \equiv 0 \pmod 3 \end{cases}.$$

We conjecture that, for a fixed k, the constructions given in i) are the best possible, in the sense that for a large set A we always have $|A + k \cdot A| \geq (k+1)|A| - \left\lceil \frac{k^2+2k}{4} \right\rceil$. But the construction given in ii) says that there are small sets that achieve a smaller lower bound.

Proof 9 *Following the examples we obtained in the inverse problem for k=3, we consider sets that are unions of arithmetic progressions with difference* k*. We write*

$$A = \bigcup_{i \in I} (k \cdot [0, m-1] + i)$$

where I *is an interval,* $I = [0, |I| - 1] \subseteq [0, k-1]$*. Then* $|A| = |I|m$*. As in Lemma 2, we have (with* $A_i = [0, m-1]$ *for all i)*

$$|A + k \cdot A| = \sum_{i \in I} |A_i + A|.$$

and

$$A_i + A = [0, m-1] + \bigcup_{i \in I} (k \cdot [0, m-1] + i) = [0, m-1] + k \cdot [0, m-1] + I.$$

Now, we try to find the sets of this shape that give us the smallest sumset, $A + k \cdot A$*.*

i) If $m \geq k$

$$A_i + A = [0, (k+1)(m-1) + |I| - 1]$$

so

$$|A + k \cdot A| = |I|((k+1)(m-1) + |I|) = (k+1)|A| - |I|(k+1-|I|).$$

We want to maximize $|I|(k+1-|I|)$ *in order to get an* A *with small sumset. If we think of* $|I|$ *as a real number we can look at the derivative to see that this happens when* $|I| = \frac{k+1}{2}$*. If* k *is odd everything works and if* k *is even we take* $|I| = \frac{k}{2}$ *or* $|I| = \frac{k+2}{2}$ *and in any case we have the formula of the proposition.*

ii) If $0 < m < k$*, then* $A_i + A$ *is the union of* m *intervals of lenght* $m + |I| - 1$ *starting on* $0, k, 2k, \ldots$ *and* $(m-1)k$*. If we do not want this intervals to overlap, then we must impose* $m + |I| - 1 \leq k$*, i. e.* $|I| \leq k + 1 - m$*. Then*

$$|A_i + A| = (m + |I| - 1)m$$

and

$$|A + k \cdot A| = |I|m(m + |I| - 1) = (k+1)|A| - m|I|(k + 2 - m - |I|).$$

We want to maximize $m|I|(k+2-m-|I|)$*. If we think of* m *and* $|I|$ *as real numbers, we can look at the gradient to conclude that the maximum occurs for* $m = |I| = \frac{k+2}{3}$*. If* $k \equiv 1$ (3)*, everything works and we have* $|A + k \cdot A| = (k+1)|A| - \left(\frac{k+2}{3}\right)^3$ *as in ii) of the theorem. If* $k \equiv 2$ (3)*, we can take* $m = |I| = \frac{k+1}{3}$ *or one of them equal to* $\frac{k+1}{3}$ *and the other to* $\frac{k+4}{3}$ *and we have* $|A + k \cdot A| = (k+1)|A| - \left(\frac{k+1}{3}\right)^2 \left(\frac{k+4}{3}\right)$*. Finally, if* $k \equiv 0$ (3)*, we take* $m = |I| = \frac{k+3}{3}$ *or one equal to* $\frac{k+3}{3}$ *and the other to* $\frac{k}{3}$ *and we have* $|A + k \cdot A| = (k+1)|A| - \left(\frac{k+3}{3}\right)^2 \left(\frac{k}{3}\right)$*. This proves ii).*

References

[1] Boris Bukh, *Sums of dilates*, Combinatorics, Probability and Computing, vol. 17, 627-639 (2008).

[2] Melvin B. Nathanson, *Inverse problems for linear forms over finite sets of integers*, arXiv:0708.2304 (2007).

In: Frontiers of Combinatorics and Number Theory, Volume 2 ISBN: 978-1-61942-918-5
Editor: Zhi-Wei Sun, pp. 97-112

Chapter 7

On the Correlations, Selberg Integral and Symmetry of Sieve Functions in Short Intervals

Giovanni Coppola*
DIIMA, Salerno University,
Fisciano(SA), Italy.

Abstract

We study the arithmetic (real) function $f = g * \mathbf{1}$, with g "essentially bounded" and supported over the integers of $[1, Q]$. In particular, we obtain non-trivial bounds, through f "correlations", for the "Selberg integral" and the "symmetry integral" of f in almost all short intervals $[x-h, x+h], N \le x \le 2N$, beyond the "classical" level, up to level of distribution, say, $\lambda = \log Q / \log N < 2/3$ ($h \gg N^{1/3+\varepsilon}$, $\varepsilon > 0$ fixed). This time we don't apply Large Sieve inequality, as in our paper [CS]. Precisely, our method is completely elementary.

2010 Mathematics Subject Classification: Primary 11N37; Secondary 11N25.

1. Introduction

We study "*sieve functions*", i.e. real arithmetic functions $f = g * \mathbf{1}$ (see the hypotheses on g in the sequel), in almost all the short intervals $[x-h, x+h]$ (i.e., almost all stands $\forall x \in [N, 2N]$, except $o(N)$ of them and short means, say, $h \to \infty$ and $h = o(N)$, as $N \to \infty$). Here $\mathbf{1}(n) = 1$ is the constant-1 arithmetic function and $*$ is the Dirichlet product (esp., [T]). In order to study the sum of f in a.a. (abbreviates almost all, now on) $[x-h, x+h]$, we define (in analogy with the classical Selberg integral [CS]) the "*Selberg integral*" of f as:

$$J_f(N,h) \overset{def}{=} \int_N^{2N} \Big| \sum_{0<|n-x|\le h} f(n) - M_f(2h) \Big|^2 \, dx,$$

*E-mail address: gcoppola@diima.unisa.it; Website: http://www.giovannicoppola.name/

where (from heuristics in accordance with the classic case) we expect the "mean-value" to be $M_f(2h) \stackrel{def}{=} 2h\sum_d g(d)/d$ (that converges, from $d \le 2N+h$).

Actually, this definition comes from what the "natural" choice of $M_f(2h)$ is (recall $[\ ] =$ *integer part*):

$$2h\Big(\frac{1}{x}\sum_{n\le x} f(n)\Big) = \frac{2h}{x}\sum_d g(d)\left[\frac{x}{d}\right] = 2h\sum_d \frac{g(d)}{d} + \mathcal{O}\Big(\frac{h}{x}\sum_{d\le Q}|g(d)|\Big),$$

since $f = g * \mathbf{1}$, $g(q) = 0$, $\forall q > Q$. Assuming Q much "smaller" than N (as we may), we get $M_f(2h)$.

Selberg integral counts the values of f in a.a. $[x-h, x+h]$.

We study their symmetry through the "*symmetry integral*" of f (we write $\text{sgn}(0) \stackrel{def}{=} 0$, $\text{sgn}(r) \stackrel{def}{=} \frac{|r|}{r}$, $\forall r \ne 0$):

$$I_f(N,h) \stackrel{def}{=} \int_N^{2N} \Big| \sum_{|n-x|\le h} \text{sgn}(n-x) f(n)\Big|^2 dx.$$

We'll generalize the results given in [CS] for these integrals, applying the Large Sieve inequality, in the case $g = \mathbf{1}$ of the divisor function $d = \mathbf{1} * \mathbf{1}$. We point out that the procedure given there works, as well, for more general g to bound I_f; but fails in the case of J_f, whenever g is not constant (i.e., the Dirichlet "flipping" of the divisors can't be applied). Here, we give another approach valid for both integrals, even for non-constant g. It is based on the "correlations" of f. The *correlation* of f *is* defined as ($\forall a \in \mathbb{Z}, a \ne 0$)

$$\mathcal{C}_f(a) \stackrel{def}{=} \sum_{n\sim N} f(n)f(n-a) = \sum_{\ell|a}\sum_{(d,q)=1}\sum g(\ell d)g(\ell q)\frac{1}{q}\left(\left[\frac{2N}{\ell d}\right] - \left[\frac{N}{\ell d}\right]\right) + R_f(a)$$

(hereon $x \sim X$ is $X < x \le 2X$), where the orthogonality of additive characters [V] (see Lemma 2.3) gives (as usual, $e(\theta) \stackrel{def}{=} e^{2\pi i\theta}$ and $e_q(m) \stackrel{def}{=} e(m/q)$, here)

$$R_f(a) \stackrel{def}{=} \sum_{\ell|a}\sum_{(d,q)=1}\sum g(\ell d)g(\ell q)\frac{1}{q}\sum_{j\ne 0} e_q(-ja/\ell)\sum_{m\sim \frac{N}{\ell d}} e_q(jdm)$$

(hereon, $j \ne 0$ means that j describes exactly once all classes $(\text{mod } q)$, except $j \equiv 0(q)$); and I_f is a sum (see Lemma 2.1) of these correlations, weighted with W (name from the shape), W *even*,

$$W(a) \stackrel{def}{=} \begin{cases} 2h - 3a & \text{if } 0 \le a \le h \\ a - 2h & \text{if } h \le a \le 2h \\ 0 & \text{if } a > 2h \end{cases} \quad \Longrightarrow \quad \sum_{a\in\mathbb{Z}} W(a) = 0.$$

In complete analogy, Lemma 2.2 gives the Selberg integral $J_f(N,h)$ as a weighted sum of correlations, with (Selberg) weight $S(a) \stackrel{def}{=} \max(2h - |a|, 0)$. Notice that S is always non-negative (while W oscillates in sign).

(Here, as usual, $F = o(G) \stackrel{def}{\Longleftrightarrow} \lim F/G = 0$ and $F = \mathcal{O}(G) \stackrel{def}{\Longleftrightarrow} \exists c > 0 : |F| \le c\,G$ are Landau's notation. When c depends on ε, write $F = \mathcal{O}_\varepsilon(G)$ or, like Vinogradov, $F \ll_\varepsilon G$).

We call an arithmetical function *essentially bounded* when, $\forall \varepsilon > 0$, its $n-$th value is $\mathcal{O}_\varepsilon(n^\varepsilon)$ and we'll write $\lll 1$; i.e.,

$$F(N) \lll G(N) \stackrel{def}{\Longleftrightarrow} \forall \varepsilon > 0 \;\; F(N) \ll_\varepsilon N^\varepsilon G(N) \quad (\text{as } N \to \infty)$$

e.g., the divisor function $d(n)$ is essentially bounded (like many other number-theoretic f) and we remark that $f = g * \mathbf{1}$ is essentially bounded if and only if g is (from Möbius inversion, see [D]). From Lemma 2.2, applying Lemma 2.3 to f correlations, together with $\sum_a S(a\ell) = 4h^2/\ell + \mathcal{O}(h)$, uniformly $\forall \ell \in \mathbb{N}$ (like in (2.1), see Lemma 2.4 proof), we get (see section 3 for explicit calculations)

$$J_f(N,h) = \sum_{a \ne 0} S(a) R_f(a) + \mathcal{O}_\varepsilon\left(N^\varepsilon\left(Nh + h^3 + Qh^2\right)\right)$$

In fact, (compare the considerations about $M_f(2h)$ with section 3 calculations), since the hypothesis $f \lll 1$ gives $g \lll 1$, we have $M_f(2h) \lll h$ and

$$\sum_{n \sim N} f(n) = \sum_d g(d)\left(\left[\frac{2N}{d}\right] - \left[\frac{N}{d}\right]\right) = N \sum_d \frac{g(d)}{d} + \mathcal{O}_\varepsilon\left(N^\varepsilon Q\right).$$

Abbreviate $n \equiv a(q)$ for $n \equiv a (\bmod q)$; $\|r\| \stackrel{def}{=} \min_{n \in \mathbb{Z}} |r - n|$ is the DISTANCE TO $\mathbb{Z}$.

We give our main result.

Theorem 1.1. *Let $N, h, Q \in \mathbb{N}$, be such that $h \to \infty$, $Q \ll N$ and $h = o(N)$, as $N \to \infty$. Let $f : \mathbb{N} \to \mathbb{R}$ be essentially bounded, with $f = g * \mathbf{1}$ and $g(q) = 0 \; \forall q > Q$. Then*

$$J_f(N,h) \lll Nh + h^3 + Q^2 h + Qh^2; \qquad I_f(N,h) \lll Nh + h^3 + Q^2 h + Qh^2.$$

Also, only for the symmetry integral $I_f(N,h)$,

$$I_f(N,h) = 2 \sum_a S(a)\left(\mathfrak{C}_f(a) - \mathfrak{C}_f(a+h)\right) + \mathcal{O}_\varepsilon\left(N^\varepsilon (Nh + h^3)\right). \qquad (1.0)$$

Remark 1.1. We explicitly point out that our Theorem 1.1 implies non-trivial estimates $J_f(N,h) \ll \frac{Nh^2}{N^\varepsilon}$ and $I_f(N,h) \ll \frac{Nh^2}{N^\varepsilon}$ for both integrals, with LEVEL OF DISTRIBUTION, say, $\frac{\log Q}{\log N} \stackrel{def}{=} \lambda < \frac{1+\theta}{2}$, where, say, $\theta \stackrel{def}{=} (\log h)/(\log N)$ is the WIDTH; hence, level up to $2/3$, when the width is above $1/3$. (The same result can also be achieved with the method of [CS], but only for I_f.)

In fact, an immediate consequence of our Theorem 1.1 is the following

Corollary 1.1. *Let $0 < \theta < 1$, $0 \le \lambda < \frac{1+\theta}{2}$ and $N, h, Q \in \mathbb{N}$, be such that $N^\theta \ll h \ll N^\theta$, $N^\lambda \ll Q \ll N^\lambda$, as $N \to \infty$. Let $f : \mathbb{N} \to \mathbb{R}$ be essentially bounded, with $f = g * \mathbf{1}$ and $g(q) = 0 \; \forall q > Q$. Then $\exists \varepsilon_0 = \varepsilon_0(\theta, \lambda) > 0$ (depending only on θ, λ) such that*

$$J_f(N,h) \ll_{\varepsilon_0} Nh^2 N^{-\varepsilon_0}, \qquad I_f(N,h) \ll_{\varepsilon_0} Nh^2 N^{-\varepsilon_0}.$$

The paper is organized as follows:

◇ we will give our Lemmas in the next section;

◇ we will calculate Selberg integral in section 3;

◇ we will prove our Theorem 1.1 in section 4;

◇ then, we give our final remarks in section 5.

Acknowledgement. The author thanks Professor Alberto Perelli for friendly comments and helpful discussions.

2. Lemmas

Lemma 2.1. *Let $N, h \in \mathbb{N}$, with $h \to \infty$ and $h = o(N)$ as $N \to \infty$. If $f : \mathbb{N} \to \mathbb{R}$ has* $\|f\|_\infty = \max_{n \le 2N+h} |f(n)|$,

$$\int_N^{2N} \Big| \sum_{|n-x| \le h} \mathrm{sgn}(n-x) f(n) \Big|^2 dx = \sum_a W(a) \mathcal{C}_f(a) + \mathcal{O}\left(h^3 \|f\|_\infty^2\right).$$

Proof. This is a kind of dispersion method, but the main term "vanishes".
Use f REAL to write the integral above $I_f(N,h)$ as $D_f(N,h)$ plus

$$2 \sum\sum_{N-h<n_1<n_2\le 2N+h} f(n_1)f(n_2) \int_{x\sim N, |x-n_1|\le h, |x-n_2|\le h} \mathrm{sgn}(x-n_1)\mathrm{sgn}(x-n_2)dx,$$

where

$$D_f(N,h) := \sum_{N-h<n\le 2N+h} f^2(n) \int_{N<x\le 2N, 0<|x-n|\le h} dx =$$

$$= \sum_{N+h<n\le 2N-h} f^2(n) \int_{|x-n|\le h} dx + \mathcal{O}\left(h\|f\|_\infty^2 \left(\sum_{|n-N|\le h} 1 + \sum_{|n-2N|\le h} 1 \right)\right)$$

$$= W(0)\mathcal{C}_f(0) + \mathcal{O}\left(h^2 \|f\|_\infty^2\right)$$

is the DIAGONAL. The remainder is (negligible) in the second one. Since, $\forall a > 0$,

$$\mathcal{C}_f(-a) = \sum_{N+a<m\le 2N+a} f(m-a)f(m) = \mathcal{C}_f(a) + \mathcal{O}\left(a\|f\|_\infty^2\right), \text{ whence}$$

$$W \text{ EVEN}, W(a) \ll h \Rightarrow \sum_{a\le 2h} W(a)\mathcal{C}_f(-a) = \sum_{a\le 2h} W(a)\mathcal{C}_f(a) + \mathcal{O}\left(h^3\|f\|_\infty^2\right)$$

(left limits into sums, if implicit, mean positive integers), we confine to:

$$I_f(N,h) - D_f(N,h) = W(0)\mathcal{C}_f(0) + 2\sum_{0<a\le 2h} W(a)\mathcal{C}_f(a) + E_f(N,h),$$

with, SAY,

$$E_f(N,h) \ll h^3\|f\|_\infty^2.$$

The left-hand side, changing variables, namely $n = n_1$, $a = n_2 - n_1$, $s = x - n_1$, is (introducing the remainders which shall take part of the final $E_f(N,h)$, here)

$$I_f(N,h) - D_f(N,h) = 2 \sum_{N-h<n<2N+h} f(n) \sum_{0<a\le 2h, a\le 2N+h-n} f(n+a)\times$$

$$\times \int_{N-n<s\le 2N-n, |s|\le h, |s-a|\le h} \mathrm{sgn}(s)\mathrm{sgn}(s-a)ds =$$

$$= 2 \sum_{N-h<n\le 2N-h} f(n) \sum_{a\le 2h} f(n+a) \int_{s>N-n, |s|\le h, |s-a|\le h} \mathrm{sgn}(s)\mathrm{sgn}(s-a)ds + E_1 =$$

$$= 2 \sum_{N+h<n\le 2N-h} f(n) \sum_{a\le 2h} f(n+a) \int_{|s|\le h, |s-a|\le h} \mathrm{sgn}(s)\mathrm{sgn}(s-a)ds + E_1 + E_2 =$$

$$= 2 \sum_{N<n\le 2N} f(n) \sum_{a\le 2h} f(n+a)W(a) + E_1 + E_2 + E_3,$$

where, SAY, the following agrees with W definition (see the introduction) :

$$W(a) := \int_{|s|\le h, |s-a|\le h} \mathrm{sgn}(s)\mathrm{sgn}(s-a)ds \ll h,$$

whence $E_3 \ll \left(\sum_{N<n\le N+h} + \sum_{2N-h<n\le 2N} \right) |f(n)| \sum_{0<a\le 2h} |f(n+a)|h \ll h^3\|f\|_\infty^2$ is a "TAIL", like:

$$E_1 \ll \sum_{|n-2N|\le h} |f(n)| \sum_{0<a\le 2h} |f(n+a)|h,$$

$$E_2 \ll \sum_{|n-N|\le h} |f(n)| \sum_{0<a\le 2h} |f(n+a)|h \text{ are } \ll h^3\|f\|_\infty^2.$$

□

Lemma 2.2. *Let $N, h \in \mathbb{N}$, with $h \to \infty$ and $h = o(N)$ as $N \to \infty$. If $f : \mathbb{N} \to \mathbb{R}$ has* $\|f\|_\infty = \max_{n\le 2N+h} |f(n)|$,

$$\int_N^{2N} \Big| \sum_{0<|n-x|\le h} f(n) - M_f(2h) \Big|^2 dx = \sum_a S(a)\mathfrak{C}_f(a) - 4M_f(2h)h \sum_{n\sim N} f(n) + M_f^2(2h)N$$

$$+ O\left(h^3 \|f\|_\infty^2 + h^2\|f\|_\infty |M_f(2h)|\right).$$

Proof. This is a direct application of the dispersion method [L]. Use f REAL :

$$J_f(N,h)=D_f(N,h)+2\sum\sum_{N-h<n_1<n_2\le 2N+h} f(n_1)f(n_2)\int_{x\sim N,|x-n_1|\le h,|x-n_2|\le h} dx$$

$$-2M_f(2h)\sum_{N-h<n\le 2N+h} f(n)\int_{x\sim N,|x-n|\le h} dx+M_f^2(2h)\int_N^{2N} dx=$$

$$=D_f(N,h)+2\sum\sum_{N-h<n_1<n_2\le 2N+h} f(n_1)f(n_2)\int_{x\sim N,|x-n_1|\le h,|x-n_2|\le h} dx$$

$$-4hM_f(2h)\sum_{n\sim N} f(n)+M_f^2(2h)N,$$

save an error which is $\mathcal{O}(|M_f(2h)|h^2\|f\|_\infty)$; here ($J_f$ is the integral above and)

$$D_f(N,h)=\sum_{N-h<n\le 2N+h} f^2(n)\int_{x\sim N,0<|x-n|\le h} dx=$$

$$=\sum_{N<n\le 2N} f^2(n)\int_{0<|x-n|\le h} dx+\mathcal{O}\left(h\|f\|_\infty^2\left(\sum_{|n-N|\le h}1+\sum_{|n-2N|\le h}1\right)\right)=$$

$$=S(0)\mathfrak{C}_f(0)+\mathcal{O}\left(h^2\|f\|_\infty^2\right)$$

is the same diagonal (with same negligible remainder) of Lemma 2.1. In fact, we closely follow its proof; due to:

$$S\ \text{EVEN},\ S(a)\ll h\ \Rightarrow\ \sum_{a\le 2h} S(a)\mathfrak{C}_f(-a)=\sum_{a\le 2h} S(a)\mathfrak{C}_f(a)+\mathcal{O}\left(h^3\|f\|_\infty^2\right),$$

this time we confine to

$$E'_f(N,h):=\sum\sum_{N-h<n_1<n_2\le 2N+h} f(n_1)f(n_2)\int_{x\sim N,|x-n_1|\le h,|x-n_2|\le h} dx-\sum_{a\le 2h} S(a)\mathfrak{C}_f(a)$$

is given by TAILS. The left-hand side, changing $n=n_1$, $a=n_2-n_1$, $s=x-n_1$, is (see Lemma 2.1 proof)

$$\sum_{N-h<n<2N+h} f(n)\sum_{0<a\le 2h,a\le 2N+h-n} f(n+a)\int_{N-n<s\le 2N-n,|s|\le h,|s-a|\le h} ds=$$

$$=\sum_{N<n\le 2N} f(n)\sum_{a\le 2h} f(n+a)S(a)+E'_1+E'_2+E'_3,$$

WITH

$$S(a):=\int_{\substack{|s|\le h\\|s-a|\le h}} ds\ll h,$$

whence, say,

$$E_3' \ll \left(\sum_{N<n\le N+h} + \sum_{2N-h<n\le 2N} \right) |f(n)| \sum_{0<a\le 2h} |f(n+a)| h \ll h^3 \|f\|_\infty^2$$

is a "TAIL", like:

$$E_1' \ll \sum_{|n-2N|\le h} |f(n)| \sum_{a\le 2h} |f(n+a)| h \ll h^3 \|f\|_\infty^2,$$

$$E_2' \ll \sum_{|n-N|\le h} |f(n)| \sum_{a\le 2h} |f(n+a)| h \ll h^3 \|f\|_\infty^2.$$

□

Lemma 2.3. *Let $N, h, Q \in \mathbb{N}$, where $h \to \infty$, $h = o(N)$ and $Q \ll N$, as $N \to \infty$. Let $f = g * \mathbf{1}$, where $g : \mathbb{N} \to \mathbb{R}$, with $q > Q \;\Rightarrow\; g(q) = 0$. Then*

$$a \ne 0 \Rightarrow \mathcal{C}_f(a) = \sum_{\ell|a} \sum_{(d,q)=1} \sum g(\ell d) g(\ell q) \frac{1}{q} \left(\left[\frac{2N}{\ell d} \right] - \left[\frac{N}{\ell d} \right] \right) + R_f(a),$$

where, say, as in the introduction

$$R_f(a) = \sum_{\ell|a} \sum_{(d,q)=1} \sum g(\ell d) g(\ell q) \frac{1}{q} \sum_{j\ne 0} e_q(-ja/\ell) \sum_{m\sim \frac{N}{\ell d}} e_q(jdm), \qquad \forall a \ne 0.$$

Also, every weight function $K : \mathbb{N} \to \mathbb{C}$, K EVEN, with $K(0) = 2h$, gives

$$\sum_a K(a) R_f(a) = 2h \mathcal{C}_f(0) +$$

$$+ \sum_{\ell\le 2h} \sum_{(d,q)=1} \sum g(\ell d) \frac{g(\ell q)}{q} \sum_{j\ne 0} \sum_{m\sim \frac{N}{\ell d}} \cos \frac{2\pi j d m}{q} \sum_{a\ne 0} K(a\ell) e_q(ja).$$

Proof. We'll always assume a non-zero. First of all, we start from the correlation:

$$\mathcal{C}_f(a) = \sum_{n\sim N} f(n) f(n-a) = \sum_d \sum_q g(d) g(q) \sum_{\substack{N<n\le 2N \\ n\equiv 0(d) \\ n\equiv a(q)}} 1 = \sum_{(d,q)|a} \sum g(d) g(q) \sum_{\substack{\frac{N}{d}<m\le \frac{2N}{d} \\ md\equiv a(q)}} 1$$

since last congruence is solveable if and only if the GCD (greatest common divisor) (d, q) divides a; changing variables, this is

$$\sum_{\ell|a} \sum_{(d,q)=1} \sum g(\ell d) g(\ell q) \sum_{\substack{\frac{N}{\ell d}<m\le \frac{2N}{\ell d} \\ md\equiv \frac{a}{\ell}(q)}} 1 = R_f(a) + \sum_{\ell|a} \sum_{(d,q)=1} \sum g(\ell d) g(\ell q) \frac{1}{q} \left(\left[\frac{2N}{\ell d} \right] - \left[\frac{N}{\ell d} \right] \right);$$

use now the orthogonality of additive characters [V]: here $R_f(a)$ is as above; summing on a with K,

$$\sum_{a\neq 0} K(a)R_f(a) = \sum_{\ell\le 2h}\sum_{(d,q)=1}\sum g(\ell d)g(\ell q)\frac{1}{q}\sum_{j\neq 0}\sum_{m\sim\frac{N}{\ell d}} e_q(jdm)\sum_{b\neq 0} K(b\ell)e_q(jb) =$$

(using K even, here)

$$= \sum_{\ell\le 2h}\sum_{(d,q)=1}\sum g(\ell d)g(\ell q)\frac{1}{q}\sum_{j\neq 0}\sum_{m\sim\frac{N}{\ell d}} e_q(jdm)\sum_{a\neq 0} K(a\ell)\cos\frac{2\pi ja}{q} =$$

$$= \sum_{\ell\le 2h}\sum_{(d,q)=1}\sum g(\ell d)g(\ell q)\frac{1}{q}\sum_{j\neq 0}\sum_{m\sim\frac{N}{\ell d}} \cos\frac{2\pi jdm}{q}\sum_{a\neq 0} K(a\ell)\cos\frac{2\pi ja}{q} =$$

$$= \sum_{\ell\le 2h}\sum_{(d,q)=1}\sum g(\ell d)g(\ell q)\frac{1}{q}\sum_{j\neq 0}\sum_{m\sim\frac{N}{\ell d}} \cos\frac{2\pi jdm}{q}\sum_{a\neq 0} K(a\ell)e_q(ja).$$

(We used once more K even, here.) Then, add the term $K(0)\mathfrak{C}_f(0) = 2h\mathfrak{C}_f(0)$.

□

Remark 2.1. We explicitly point out : for REAL f

$$f \lll 1 \Rightarrow 2h\mathfrak{C}_f(0) = 2h\sum_{n\sim N} f^2(n) \lll Nh,$$

a trivial estimate which will be useful in future occurrences.

Lemma 2.4. *Defining, $\forall h\in\mathbb{N}$, W as above, we have, $\forall \ell, q\in\mathbb{N}$, $\forall\beta\notin\mathbb{Z}$, $\forall\alpha\in\mathbb{R}$,*

$$\sum_{\substack{a\\ a\equiv 0(q)}} W(a) = 2q\left\|\frac{h}{q}\right\|;\ \sum_a W(a)e(a\beta) = \frac{4\sin^4(\pi h\beta)}{\sin^2(\pi\beta)};\ \sum_b W(\ell b)e(b\alpha)\ge 0. \quad (2.1)$$

More in general (in the same hypotheses), letting $E_X(\beta) \stackrel{def}{=} \sum_{0\le|a|\le X} e(a\beta)$,

$$\frac{1}{\ell}\sum_a W(a\ell)e(a\beta) = \frac{4\sin^2\pi\beta[\frac{h}{\ell}] - \sin^2\pi\beta[\frac{2h}{\ell}]}{\sin^2\pi\beta} + \quad (2.2)$$

$$+4\left\{\frac{h}{\ell}\right\}E_{\frac{h}{\ell}}(\beta) - \left\{\frac{2h}{\ell}\right\}E_{\frac{2h}{\ell}}(\beta) =$$

$$= 2\left(1-\cos\left(2\pi\beta\left[\frac{h}{\ell}\right]\right)\right)\sum_{0\le|a|\le\frac{h}{\ell}}\left(\left[\frac{h}{\ell}\right]-|a|\right)e(a\beta)$$

$$-\left(2\left\{\frac{h}{\ell}\right\}-\left\{\frac{2h}{\ell}\right\}\right)E_{2[\frac{h}{\ell}]}(\beta) + 4\left\{\frac{h}{\ell}\right\}E_{\frac{h}{\ell}}(\beta) - \left\{\frac{2h}{\ell}\right\}E_{\frac{2h}{\ell}}(\beta).$$

Proof. We will prove (2.1), even if it's a special case of (2.2);

$$\sum_{\substack{a \\ a\equiv 0 (\bmod q)}} W(a) = 2h + 4h\left(\left[\frac{h}{q}\right] - \left[\frac{2h}{q}\right] + \left[\frac{h}{q}\right]\right) + 2q\left(\sum_{\frac{h}{q}<b\le\frac{2h}{q}} b - 3\sum_{b\le\frac{h}{q}} b\right)$$

$$= q\left(\left\{\frac{2h}{q}\right\}^2 - 4\left\{\frac{h}{q}\right\}^2 - \left\{\frac{2h}{q}\right\} + 4\left\{\frac{h}{q}\right\}\right).$$

We get the first using, $\forall\alpha\in\mathbb{R}$,

$$\{2\alpha\} = \{2\{\alpha\}\} = \begin{cases} 2\{\alpha\} & \text{if } \ 0\le\{\alpha\}<1/2 \\ 2\{\alpha\}-1 & \text{if } \ 1/2\le\{\alpha\}<1 \end{cases}.$$

We come, now, to the second:

$$\sum_{0\le|a|\le 2h} W(a)e(a\beta) = 2h + 2\sum_{a\le 2h} W(a)\cos 2\pi a\beta = 2h + 2\Sigma,$$

SAY; then, partial summation gives

$$\Sigma = 4\sum_{a\le h} C_a(\beta) - 4C_h(\beta) - \sum_{a\le 2h} C_a(\beta) + C_{2h}(\beta),$$

SAY, where $\forall X\in\mathbb{N}, \forall\theta\notin\mathbb{Z}$

$$C_X(\theta) \stackrel{def}{=} \sum_{n\le X}\cos(2\pi n\theta) = \frac{\sin(2\pi\theta X)}{2\tan(\pi\theta)} - \frac{1-\cos(2\pi\theta X)}{2}$$

(a well-known formula) to get that Σ is

$$2\cot(\pi\beta)\sum_{a\le h}\sin(2\pi a\beta) - 2h - 2C_h(\beta) - \frac{1}{2}\cot(\pi\beta)\sum_{a\le 2h}\sin(2\pi a\beta) + \frac{1}{2}C_{2h}(\beta) + h =$$

$$= \cot(\pi\beta)\left(2S_h(\beta) - \frac{1}{2}S_{2h}(\beta)\right) - h - 2C_h(\beta) + \frac{1}{2}C_{2h}(\beta),$$

SAY, $\forall X\in\mathbb{N}, \forall\theta\notin\mathbb{Z}, \ S_X(\theta) \stackrel{def}{=} \sum_{n\le X}\sin(2\pi n\theta) = \frac{\sin^2(\pi\theta X)}{\tan(\pi\theta)} + \frac{\sin(2\pi\theta X)}{2}.$

Then, since, both for $X = h$ and $X = 2h$, $\frac{1-\cos(2\pi\beta X)}{2} = \sin^2(\pi\beta X)$,

$$\Sigma = \cot^2(\pi\beta)\left(2\sin^2(\pi\beta h) - \frac{1}{2}\sin^2(2\pi\beta h)\right) + 2\sin^2(\pi\beta h) - \frac{1}{2}\sin^2(2\pi\beta h) - h =$$

$$= 2\cot^2(\pi\beta)\left(1-\cos^2(\pi h\beta)\right)\sin^2(\pi h\beta) + 2\sin^2(\pi h\beta)\left(1-\cos^2(\pi h\beta)\right) - h =$$

$$= 2\left(\cot^2(\pi\beta)+1\right)\sin^4(\pi h\beta) - h = \frac{2\sin^4(\pi h\beta)}{\sin^2(\pi\beta)} - h.$$

This gives the second. Finally, the third follows from: $\forall \ell \in \mathbb{N}$

$$\sum_b W(\ell b)e(b\alpha) \geq 0 \ \forall \alpha \in \mathbb{R} \Longleftrightarrow \sum_{a \equiv 0 (\mathrm{mod}\ \ell)} W(a)e(a\beta) \geq 0 \ \forall \beta \in \mathbb{R}$$

which, using again orthogonality [V] and $\sum_a W(a)e(a\beta) \geq 0 \ \forall \beta \in \mathbb{R}$, is

$$\sum_{a \equiv 0 (\mathrm{mod}\ \ell)} W(a)e(a\beta) = \frac{1}{\ell}\sum_{j \leq \ell}\sum_a W(a)e(a\beta)e_\ell(ja) =$$

$$= \frac{1}{\ell}\sum_{j \leq \ell}\sum_a W(a)e\left(a\left(\beta + \frac{j}{\ell}\right)\right) \geq 0.$$

(We explicitly remark that this last property isn't "visible" from (2.2): not an immediate consequence.)

We manage (2.2): $\sum_a W(a\ell)e(a\beta) = 2h + 2\sum_{a \leq \frac{2h}{\ell}} W(a\ell)\cos 2\pi a\beta = 2h + 2\Sigma_\ell$, say;

then,

$$\frac{1}{\ell}\Sigma_\ell = 4\sum_{a \leq [\frac{h}{\ell}]} C_a(\beta) - 4C_{[\frac{h}{\ell}]}(\beta) - \sum_{a \leq [\frac{2h}{\ell}]} C_a(\beta) + C_{[\frac{2h}{\ell}]}(\beta) +$$

$$+\left(4\left\{\frac{h}{\ell}\right\}C_{[\frac{h}{\ell}]}(\beta) - \left\{\frac{2h}{\ell}\right\}C_{[\frac{2h}{\ell}]}(\beta)\right),$$

from partial summation (the term in brackets isn't present whenever $\ell = 1$); then, (see above formulas)

$$\frac{1}{\ell}\Sigma_\ell = \cot(\pi\beta)\left(2S_{[\frac{h}{\ell}]}(\beta) - \frac{1}{2}S_{[\frac{2h}{\ell}]}(\beta)\right) - \left(2\left[\frac{h}{\ell}\right] - \frac{1}{2}\left[\frac{2h}{\ell}\right]\right)$$

$$-2C_{[\frac{h}{\ell}]}(\beta) + \frac{1}{2}C_{[\frac{2h}{\ell}]}(\beta) + \left(4\left\{\frac{h}{\ell}\right\}C_{[\frac{h}{\ell}]}(\beta) - \left\{\frac{2h}{\ell}\right\}C_{[\frac{2h}{\ell}]}(\beta)\right),$$

i.e.

$$\Sigma_\ell = \frac{2\sin^2 \pi\beta\left[\frac{h}{\ell}\right] - \frac{1}{2}\sin^2 \pi\beta\left[\frac{2h}{\ell}\right]}{\sin^2(\pi\beta)}\ell - h +$$

$$+\left(2\left\{\frac{h}{\ell}\right\}\left(1 + 2C_{[\frac{h}{\ell}]}(\beta)\right) - \frac{1}{2}\left\{\frac{2h}{\ell}\right\}\left(1 + 2C_{[\frac{2h}{\ell}]}(\beta)\right)\right)\ell;$$

hence,

$$\frac{1}{\ell}\sum_a W(a\ell)e(a\beta) = \frac{4\sin^2 \pi\beta\left[\frac{h}{\ell}\right] - \sin^2 \pi\beta\left[\frac{2h}{\ell}\right]}{\sin^2(\pi\beta)} +$$

$$+\left(4\left\{\frac{h}{\ell}\right\}\sum_{0 \leq |a| \leq \frac{h}{\ell}} e(a\beta) - \left\{\frac{2h}{\ell}\right\}\sum_{0 \leq |a| \leq \frac{2h}{\ell}} e(a\beta)\right);$$

we distinguish two cases: first, $0 \le \left\{\frac{h}{\ell}\right\} < \frac{1}{2}$ and, then, $\frac{1}{2} \le \left\{\frac{h}{\ell}\right\} < 1$. In the first, we have $\left[\frac{2h}{\ell}\right] = 2\left[\frac{h}{\ell}\right]$:

$$\frac{1}{\ell}\sum_a W(a\ell)e(a\beta) = \frac{4\sin^4 \pi\beta\left[\frac{h}{\ell}\right]}{\sin^2(\pi\beta)} +$$

$$+\left(4\left\{\frac{h}{\ell}\right\}\sum_{0\le|a|\le\frac{h}{\ell}} e(a\beta) - \left\{\frac{2h}{\ell}\right\}\sum_{0\le|a|\le\frac{2h}{\ell}} e(a\beta)\right),$$

while in the second case we have $\left[\frac{2h}{\ell}\right] = 2\left[\frac{h}{\ell}\right] + 1$, so join (only for $2\left\{\frac{h}{\ell}\right\} - \left\{\frac{2h}{\ell}\right\} = 1$) the term

$$-\cos 4\pi\beta\left[\frac{h}{\ell}\right] - \sin 4\pi\beta\left[\frac{h}{\ell}\right]\cot(\pi\beta) = \left(\text{use the formula for } C_{2\left[\frac{h}{\ell}\right]}(\beta), \text{ here}\right)$$

$$= -2\left(\sum_{a\le 2\left[\frac{h}{\ell}\right]}\cos(2\pi a\beta) + \frac{1}{2}\right) = -\sum_{0\le|a|\le 2\left[\frac{h}{\ell}\right]} e(a\beta) =$$

$$= -\left(2\left\{\frac{h}{\ell}\right\} - \left\{\frac{2h}{\ell}\right\}\right)\sum_{0\le|a|\le 2\left[\frac{h}{\ell}\right]} e(a\beta).$$

□

3. Selberg Integral Details

Adapting Lemma 2.2 to the case f ESSENTIALLY BOUNDED,

$$J_f(N,h) = \sum_a S(a)\mathfrak{C}_f(a) - 4hM_f(2h)\sum_{n\sim N} f(n) + M_f^2(2h)N +$$

$$+\mathcal{O}_\varepsilon\left(N^\varepsilon\left(h^3 + h^2\left|M_f(2h)\right|\right)\right).$$

We need the following elementary calculation (use only $g := f * \mu \lll 1$, and that f is supported up to $\ll Q$):

$$\sum_{n\sim N} f(n) = \sum_d g(d)\sum_{\substack{n\sim N\\ n\equiv 0(d)}} 1 = \sum_d g(d)\left(\left[\frac{2N}{d}\right] - \left[\frac{N}{d}\right]\right) =$$

$$= \sum_d g(d)\left(\frac{N}{d} + \mathcal{O}(1)\right) = N\sum_d \frac{g(d)}{d} + \mathcal{O}_\varepsilon\left(N^\varepsilon Q\right)$$

and (recall $f \lll 1 \Leftrightarrow g \lll 1$, here), since $g(q) = 0, \forall q > Q$, (here $Q \ll N$)

$$M_f(2h) = 2h\sum_q \frac{g(q)}{q} \ll_\varepsilon N^\varepsilon h\sum_{q\ll N}\frac{1}{q} \lll h,$$

so to obtain from the previous (assuming only $f \lll 1$ here), see the Remark 2.1 (after Lemma 2.3),

$$J_f(N,h) = \sum_{a\neq 0} S(a)\mathfrak{C}_f(a) - M_f^2(2h)N + \mathcal{O}_\varepsilon\left(N^\varepsilon\left(Nh + h^3 + Qh^2\right)\right).$$

From Lemma 2.3, actually, $M_f^2(2h)N$ is implicit in the sum over $a \neq 0$:

$$\mathfrak{C}_f(a) = B_f(a) + R_f(a), \ \ \forall a \in \mathbb{Z}^*,$$

SAY,

$$B_f(a) := \sum_{\ell|a}\sum_{(d,q)=1}\sum g(\ell d)\frac{g(\ell q)}{q}\sum_{m\sim\frac{N}{\ell d}} 1 = \sum_{\ell|a}\sum_{(d,q)=1}\sum g(\ell d)\frac{g(\ell q)}{q}\left(\frac{N}{\ell d} + \mathcal{O}(1)\right),$$

(while $R_f(a)$ defined above) whence, changing variables,

$$\sum_{a\neq 0} S(a)B_f(a) = N\sum_{\ell\le 2h}\frac{1}{\ell}\left(\sum_{a\neq 0} S(a\ell)\right)\sum_{(d,q)=1}\sum \frac{g(\ell d)}{d}\frac{g(\ell q)}{q} + \mathcal{O}_\varepsilon\left(N^\varepsilon Qh^2\right) =$$

$$= N\sum_{\ell\le 2h}\frac{4h^2}{\ell^2}\sum_{(d,q)=1}\sum \frac{g(\ell d)}{d}\frac{g(\ell q)}{q} + \mathcal{O}_\varepsilon\left(N^\varepsilon\left(Nh + Qh^2\right)\right),$$

having used (first the fact that $g \lll 1$ is supported up to $\mathcal{O}(Q)$ and then) the following bound (uniform $\forall \ell \le 2h$):

$$\sum_{a\neq 0} S(a\ell) = 2\sum_{a\le\frac{2h}{\ell}}(2h - a\ell) = 4h\left[\frac{2h}{\ell}\right] - \ell\left[\frac{2h}{\ell}\right]^2 - \ell\left[\frac{2h}{\ell}\right] = \frac{4h^2}{\ell} + \mathcal{O}(h);$$

in fact (using also $\sum_{\ell>2h} 1/\ell^2 \ll 1/h$), we rather get $M_f^2(2h)N$, here :

$$\sum_{a\neq 0} S(a)B_f(a) = 4Nh^2\sum_{\ell=1}^{\infty}\frac{1}{\ell^2}\sum_{(d,q)=1}\sum \frac{g(\ell d)}{d}\frac{g(\ell q)}{q} + \mathcal{O}_\varepsilon\left(N^\varepsilon\left(Nh + Qh^2\right)\right);$$

and, finally, another change of variables gives

$$4h^2\sum_{\ell=1}^{\infty}\frac{1}{\ell^2}\sum_{(d,q)=1}\sum \frac{g(\ell d)}{d}\frac{g(\ell q)}{q} = 4h^2\sum_{\ell=1}^{\infty}\sum_{(d,q)=\ell}\sum \frac{g(d)}{d}\frac{g(q)}{q} = \left(2h\sum_d \frac{g(d)}{d}\right)^2,$$

which is $M_f^2(2h)$:

$$\sum_{a\neq 0} S(a)B_f(a) = M_f^2(2h)N + \mathcal{O}_\varepsilon\left(N^\varepsilon\left(Nh + Qh^2\right)\right).$$

Resuming, we can say that, provided $f : \mathbb{N} \to \mathbb{R}$ is ESSENTIALLY BOUNDED, where $g := f * \mu$ is supported up to $\mathcal{O}(Q)$, when our definitions are applied to the Dispersion Method, give (as stated above, in the introduction)

$$J_f(N,h) = \sum_{a\neq 0} S(a)R_f(a) + \mathcal{O}_\varepsilon\left(N^\varepsilon\left(Nh + h^3 + Qh^2\right)\right),$$

with $R_f(a)$ as previously defined ("REMAINDER OF CORRELATIONS" $\mathfrak{C}_f(a)$, see the introduction).

4. Proof of Theorem 1.1

Proof of Theorem 1.1. We will ignore the $R_g(N,h)$ that are $\lll Nh + h^3 + Qh^2$ (Good remainders!). Linking the Lemmas,

$$I_f(N,h) = \sum_{\ell \le 2h} \sum_{(d,q)=1} \sum g(\ell d) g(\ell q) \frac{1}{q} \sum_{j \ne 0} \sum_{m \sim \frac{N}{\ell d}} \cos \frac{2\pi jdm}{q} \sum_{a \ne 0} W(a\ell) e_q(ja)$$

(save $\lll R_g(N,h)$, hereon); and using Lemma 2.2 instead of Lemma 2.1, see previous section,

$$J_f(N,h) = \sum_{\ell \le 2h} \sum_{(d,q)=1} \sum g(\ell d) g(\ell q) \frac{1}{q} \sum_{j \ne 0} \sum_{m \sim \frac{N}{\ell d}} \cos \frac{2\pi jdm}{q} \sum_{a \ne 0} S(a\ell) e_q(ja)$$

Then, we'll show, for each K like in Lemma 2.3, supported in $[-2h, 2h]$, where uniformly bounded as $K \ll h$,

$$(*) \qquad T_g(N,h) \stackrel{def}{=} \sum_{\ell \le 2h} \sum_{(d,q)=1} \sum g(\ell d) g(\ell q) \frac{1}{q} \sum_{j \ne 0} \sum_{m \sim \frac{N}{\ell d}} \cos \frac{2\pi jdm}{q} \sum_{a} K(a\ell) e_q(ja)$$

$$\lll Q^2 h + R_g(N,h)$$

In fact, reintroduce $a = 0$–terms ($K(0) = 2h$), with contributes ($\sum^* =$ coprime to d)

$$2h \sum_{\ell \le 2h} \sum_{(d,q)=1} \sum g(\ell d) \frac{g(\ell q)}{q} \sum_{j \ne 0} \sum_{m \sim \frac{N}{\ell d}} e_q(jdm) =$$

$$= 2h \sum_{\ell \le 2h} \sum_{d} g(\ell d) \sum_{m \sim \frac{N}{\ell d}} \left(\sum_{q|dm}^{*} g(\ell q) - \sum_{q}^{*} \frac{g(\ell q)}{q} \right) \lll Nh.$$

(Once more from orthogonality, see above) We'll prove now (1.0). We may also join $j = 0$ whenever $K = W$:

$$\sum_{\ell \le 2h} \sum_{(d,q)=1} \sum g(\ell d) g(\ell q) \frac{1}{q} \sum_{m \sim \frac{N}{\ell d}} \sum_{a} W(a\ell) \lll \sum_{\ell \le 2h} \sum_{(d,q)=1} \sum \frac{1}{q} \left(\frac{N}{\ell d} + 1 \right) h \lll Nh$$

and using (2.2), see Lemma 2.4, we get (only for $K = W$)

$$T_g(N,h) = 2 \sum_{\ell \le 2h} \ell \sum_{(d,q)=1} \sum g(\ell d) g(\ell q) \frac{1}{q} \sum_{j} \sum_{m \sim \frac{N}{\ell d}} \cos \frac{2\pi jdm}{q} \times$$

$$\times \left(1 - \cos \frac{2\pi j}{q} [\frac{h}{\ell}] \right) \sum_{0 \le |a| \le \frac{h}{\ell}} \left(\left[\frac{h}{\ell} \right] - |a| \right) e_q(ja) +$$

$$+\sum_{\ell\le 2h}\ell\sum\sum_{(d,q)=1} g(\ell d)g(\ell q)\frac{1}{q}\sum_{j}\sum_{m\sim\frac{N}{\ell d}}\cos\frac{2\pi jdm}{q}\times B\left(\frac{h}{\ell}\right)\sum_{0\le|a|\le\frac{2h}{\ell}}U_a\left(\frac{h}{\ell}\right)e_q(-ja),$$

(plus negligible remainders), where B and U_a (uniformly on a) are bounded functions. From orthogonality,

$$\frac{1}{q}\sum_{j}\sum_{m\sim\frac{N}{\ell d}}\cos\frac{2\pi jdm}{q}B\left(\frac{h}{\ell}\right)\sum_{0\le|a|\le\frac{2h}{\ell}}U_a\left(\frac{h}{\ell}\right)e_q(-ja)=$$

$$=B\left(\frac{h}{\ell}\right)\sum_{0\le|a|\le\frac{2h}{\ell}}U_a\left(\frac{h}{\ell}\right)\sum_{\substack{m\sim\frac{N}{\ell d}\\ md\equiv a(q)}}1,$$

whence we get (1.0), applying orthogonality (and the Lemmas) also on the main term, since for remainders we obtain:

$$\sum_{\ell\le 2h}\ell\sum\sum_{(d,q)=1} g(\ell d)g(\ell q)\frac{1}{q}\sum_{j}\sum_{m\sim\frac{N}{\ell d}}\cos\frac{2\pi jdm}{q}B\left(\frac{h}{\ell}\right)\sum_{0\le|a|\le\frac{2h}{\ell}}U_a\left(\frac{h}{\ell}\right)e_q(-ja)$$

$$\lll Nh.$$

We pass to the two bounds for our integrals. Now, from (2.1) and the well-known formula (Fejér kernel)

$$\sum_{a}S(a)e_q(ja)=\sum_{0\le|a|\le 2h}(2h-|a|)e_q(ja)=\frac{\sin^2\frac{2\pi jh}{q}}{\sin^2\frac{\pi j}{q}},$$

which gives (like in Lemma 2.4 proof) $\sum_a S(a\ell)e_q(ja)\ge 0$, $\forall j\ne 0$, we have, say, (for $K=W,S$)

$$\widehat{K}\left(\frac{j}{q}\right)\overset{def}{=}\sum_{a}K(a\ell)e_q(ja)\ge 0\quad\forall j\not\equiv 0(q)$$

($\widehat{W}(0)\ge 0$ and $\widehat{S}(0)=\frac{4h^2}{\ell}+\mathcal{O}(h)$, trivially); whence (apart from $\lll R_g(N,h)$), writing "$*$" for $(d,q)=1$,

$$T_g(N,h)=\sum_{\ell\le 2h}\sum\sum_{(d,q)=1} g(\ell d)g(\ell q)\frac{1}{q}\sum_{j\ne 0}\sum_{m\sim\frac{N}{\ell d}}\cos\frac{2\pi jdm}{q}\widehat{K}\left(\frac{j}{q}\right)$$

$$\lll \max_{D\le Q}\sum_{\ell\le 2h}\sum_{q\sim\frac{D}{\ell}}\frac{1}{q}\sum_{d\le 2q}{}^{*}\sum_{j\ne 0}\frac{1}{\left\|\frac{jd}{q}\right\|}\widehat{K}\left(\frac{j}{q}\right)$$

due to [D, ch. 26]

$$\sum_{m\sim\frac{N}{\ell d}}e_q(jdm)\ll\frac{1}{\left\|\frac{jd}{q}\right\|},$$

having used a dissection argument (over q). Changing variables (with $\overline{d}d \equiv 1(q)$, here) into

$$\sum_{j\neq 0} \frac{1}{\left\|\frac{jd}{q}\right\|}\widehat{K}\left(\frac{j}{q}\right) = \sum_{0<|j|\leq \frac{q}{2}} \frac{1}{\left\|\frac{j}{q}\right\|}\widehat{K}\left(\frac{j\overline{d}}{q}\right) \overset{K\ \mathrm{EVEN}}{=\!=} 2q\sum_{j\leq q/2}\frac{1}{j}\widehat{K}\left(\frac{j\overline{d}}{q}\right)$$

gives

$$T_g(N,h) \lll \max_{D\leq 2Q}\sum_{\ell\leq 2h}\sum_{q\sim\frac{D}{\ell}}\sum_{j\leq q/2}\frac{1}{j}\sum_{n\leq 2q}{}^{*}\widehat{K}\left(\frac{jn}{q}\right)$$

which, since

$$\sum_{n\leq 2q}{}^{*}\widehat{K}\left(\frac{jn}{q}\right) \ll \sum_{n\leq 2q}\widehat{K}\left(\frac{jn}{q}\right) = \sum_{a}K(a\ell)\sum_{n\leq 2q}e_q(jan) = 2q\sum_{\substack{a\\ ja\equiv 0(q)}}K(a\ell)$$

and "flipping" the divisors (say, F, i.e., change t into q/t) in the following

$$2q\sum_{j\leq q/2}\frac{1}{j}\sum_{\substack{a\\ ja\equiv 0(q)}}K(a\ell) = 2\sum_{\substack{t|q\\ t<q}}q\sum_{\substack{j\leq q/2\\ (j,q)=t}}\frac{1}{j}\sum_{a\equiv 0(q/t)}K(a\ell) \overset{\mathrm{F}}{=}$$

$$\overset{\mathrm{F}}{=} 2\sum_{\substack{t|q\\ t>1}}t\sum_{\substack{j\leq t/2\\ (j,t)=1}}\frac{1}{j}\sum_{a\equiv 0(t)}K(a\ell) = 4h\sum_{\substack{t|q\\ t>1}}t\sum_{\substack{j\leq t/2\\ (j,t)=1}}\frac{1}{j} + 2\sum_{\substack{t|q\\ 1<t\leq 2h}}t\sum_{\substack{j\leq t/2\\ (j,t)=1}}\frac{1}{j}\sum_{\substack{a\neq 0\\ a\equiv 0(t)}}K(a\ell)$$

$$\lll qh + h^2,$$

finally entails

$$T_g(N,h) \lll Nh + \max_{D\leq 2Q}\sum_{\ell\leq 2h}\sum_{q\sim\frac{D}{\ell}}(qh+h^2) \lll Nh + Q^2h + Qh^2 \Rightarrow (*).$$

□

5. Conclusions

We remark that our arguments are completely elementary (not even a Cauchy inequality !), relying on the calculations of our Lemmas; in particular, we are able to exploit the positivity (namely, non-negativity) of the exponential sums of S (say, "SELBERG WEIGHT") and W (say, the $0-$mean value weight).

This, in turn, resembles the positivity arguments, used in order to prove the Large Sieve inequality; actually, we don't apply neither this last (classic) bound for L^2 estimates of exponential sums. Instead, our proof of the present Theorem 1.1 applies an L^1 estimate of the transforms of our weights (i.e., $\hat{S}$ and $\hat{W}$), coupled with an L^∞ estimate for the exponential sum of f (on FAREY FRACTIONS, i.e. j/q, with j, q coprime).

We hope, for the future, to cope the case of level larger than $(1+\theta)/2$; however, this seems to set the limit of our present methods.

References

[CS] G. Coppola and S. Salerno, On the symmetry of the divisor function in almost all short intervals, *Acta Arith.* 113 (2004), no.2, 189–201.

[D] H. Davenport, Multiplicative Number Theory, 3rd Edition, Graduate Texts in Math., Vol. 74, *Springer,* New York, 2000.

[L] Ju. V. Linnik, The Dispersion Method in Binary Additive Problems, translated by S. Schuur, *Amer. Math. Soc.*, Providence, RI, 1963.

[T] G. Tenenbaum, *Introduction to Analytic and Probabilistic Number Theory*, Cambridge Studies in Adv. Math., Vol. 46, Cambridge Univ. Press, 1995.

[V] I. M. Vinogradov, *The Method of Trigonometrical Sums in the Theory of Numbers*, Interscience Publishers LTD, London, 1954.

In: Frontiers of Combinatorics and Number Theory, Volume 2 ISBN: 978-1-61942-918-5
Editor: Zhi-Wei Sun, pp. 113-134

Chapter 8

ON THE BRAUER–MANIN OBSTRUCTION FOR CUBIC SURFACES*

Andreas-Stephan Elsenhans[1, #] ***and Jörg Jahnel***[2, ‡]
[1]Math. Inst., Universität Bayreuth, Bayreuth, Germany
[2]FB6 Mathematik, Siegen, Germany

Abstract

We describe a method to compute the Brauer-Manin obstruction for smooth cubic surfaces over $\mathbb{Q}$ such that $\mathrm{Br}(S)/\mathrm{Br}(\mathbb{Q})$ is of order two or four. This covers the vast majority of the cases when this group is non-zero. Our approach is to associate a Brauer class with every Galois invariant double-six. We show that all order two Brauer classes may be obtained in this way. We also recover Sir Peter Swinnerton-Dyer's result that $\mathrm{Br}(S)/\mathrm{Br}(\mathbb{Q})$ may take only five values.

2010 Mathematics Subject Classification: Primary 11D25; Secondary 11D85, 11G35

1. Introduction

1.1. — For cubic surfaces, weak approximation and even the Hasse principle are not always fulfilled. The first example of a cubic surface violating the Hasse principle was constructed by Sir Peter Swinnerton-Dyer [SD1]. A series of examples generalizing that of Swinnerton-Dyer is due to L. J. Mordell [Mo]. An example of a different sort was given by J. W. S. Cassels and M. J. T. Guy [CG].

A way to explain these examples in a unified manner was provided by Yu. I. Manin in his book [Ma]. This is what today is called the Brauer-Manin obstruction. Manin's idea is that a non-trivial Brauer class may be responsible for the failure of weak approximation. We will recall the Brauer-Manin obstruction in some detail in section 2.

---

*The computer part of this work was executed on the Sun Fire V20z Servers of the Gauß Laboratory for Scientific Computing at the Göttingen Mathematical Institute. Both authors are grateful to Prof. Y. Tschinkel for the permission to use these machines as well as to the system administrators for their support.

#E-mail: Stephan.Elsenhans@uni-bayreuth.de; Website: http://www.staff.uni-bayreuth.de/~btm216
‡E-mail: jahnel@mathematik.uni-siegen.de; Website: http://www.uni-math.gwdg.de/jahnel

An important point is that only the factor group $\mathrm{Br}(S)/\mathrm{Br}(\mathbb{Q})$ of the Grothendieck-Brauer group of the cubic surface S is relevant. That is isomorphic to the Galois cohomology group $H^1(\mathrm{Gal}(\overline{\mathbb{Q}}/\mathbb{Q}),\mathrm{Pic}(S_{\overline{\mathbb{Q}}}))$. A theorem of Sir Peter Swinnerton-Dyer [SD2] states that, for this group, there are only five possibilities. It may be isomorphic to 0, $\mathbb{Z}/2\mathbb{Z}$, $\mathbb{Z}/2\mathbb{Z}\times\mathbb{Z}/2\mathbb{Z}$, $\mathbb{Z}/3\mathbb{Z}$, or $\mathbb{Z}/3\mathbb{Z}\times\mathbb{Z}/3\mathbb{Z}$. We observed that, today, Swinnerton-Dyer's theorem from 1993 may easily be established by a script in GAP.

The effect of the Brauer-Manin obstruction has been studied by several authors. For example, for diagonal cubic surfaces, the computations were carried out by J.-L. Colliot-Thélène and his coworkers in [CKS]. In this case, $\mathrm{Br}(S)/\mathrm{Br}(\mathbb{Q})=\mathbb{Z}/3\mathbb{Z}$. The same applies to the examples of Mordell or Cassels-Guy which were explained by the Brauer-Manin obstruction in [Ma].

1.2. — It seems that, for the cases that $H^1(\mathrm{Gal}(\overline{\mathbb{Q}}/\mathbb{Q}),\mathrm{Pic}(S_{\overline{\mathbb{Q}}}))\cong\mathbb{Z}/2\mathbb{Z}$ or $\mathbb{Z}/2\mathbb{Z}\times\mathbb{Z}/2\mathbb{Z}$, no computations have been done up to now. The goal of the present paper is to fill this gap.

Our starting point is a somewhat surprising observation. It turns out that $H^1(\mathrm{Gal}(\overline{\mathbb{Q}}/\mathbb{Q}),\mathrm{Pic}(S_{\overline{\mathbb{Q}}}))$ is of order two or four only in cases when, on S, there is a Galois invariant double-six. This reduces the possibilities for the action of $\mathrm{Gal}(\overline{\mathbb{Q}}/\mathbb{Q})$ on the 27 lines. In general, the automorphism group of the configuration of the 27 lines is the Weyl group $W(E_6)$ [Ma, Theorem 23.9]. Among the 350 conjugacy classes of subgroups in $W(E_6)$, exactly 158 stabilize a double-six.

In a previous paper [EJ2], we described a method to construct smooth cubic surfaces with a Galois invariant double-six. Our method is based on the hexahedral form of L. Cremona and Th. Reye and an explicit Galois descent. It is able to produce examples for each of the 158 conjugacy classes.

Among them, however, there are 56 which even stabilize a sixer. Those may be constructed by blowing up six points in $\mathbf{P}^2$ and, thus, certainly fulfill weak approximation. There are 26 further conjugacy classes which lead to $H^1(\mathrm{Gal}(\overline{\mathbb{Q}}/\mathbb{Q}),\mathrm{Pic}(S_{\overline{\mathbb{Q}}}))=0$.

1.3. — In this article, we compute the Brauer-Manin obstruction for each of the 76 cases such that $H^1(\mathrm{Gal}(\overline{\mathbb{Q}}/\mathbb{Q}),\mathrm{Pic}(S_{\overline{\mathbb{Q}}}))\cong\mathbb{Z}/2\mathbb{Z}$ or $\mathbb{Z}/2\mathbb{Z}\times\mathbb{Z}/2\mathbb{Z}$. We start with two "model cases" for the Brauer groups $\mathbb{Z}/2\mathbb{Z}$ and $\mathbb{Z}/2\mathbb{Z}\times\mathbb{Z}/2\mathbb{Z}$. These are the maximal subgroup $U_1\subset W(E_6)$ stabilizing a double-six and the maximal subgroup $U_3\subset W(E_6)$ stabilizing a triple of azygetic double sixes [Ko].

In both cases, we compute the Brauer group explicitly. This means, we produce representatives which we describe as Azumaya algebras. We then show that every subgroup $H\subset W(E_6)$ which leads to a Brauer group of order four is actually contained in U_3. Recall that every subgroup $H\subset W(E_6)$ leading to a Brauer group of order two is contained in U_1. Finally, we prove the main result that the restriction map is bijective in each of the cases.

1.4. — The article is concluded by examples showing the effect of the Brauer-Manin obstruction. It turns out that, unlike the situation described in [CKS] where a Brauer class of order three typically excludes two thirds of the adelic points, various fractions are possible.

2. The Brauer-Manin Obstruction – Generalities

2.1. — For cubic surfaces, all known counterexamples to the Hasse principle or weak approximation are explained by the following observation.

2.2. Definition. — Let X be a projective variety over $\mathbb{Q}$ and $\mathrm{Br}(X)$ its Grothendieck-Brauer group. Then, we will call

$$\mathrm{ev}_\nu\colon \mathrm{Br}(X) \times X(\mathbb{Q}_\nu) \longrightarrow \mathbb{Q}/\mathbb{Z}, \quad (\alpha, \xi) \mapsto \mathrm{inv}_\nu(\alpha|_\xi)$$

the *local evaluation map*. Here, $\mathrm{inv}_\nu\colon \mathrm{Br}(\mathbb{Q}_\nu) \to \mathbb{Q}/\mathbb{Z}$ (and $\mathrm{inv}_\infty\colon \mathrm{Br}(\mathbb{R}) \to \frac{1}{2}\mathbb{Z}/\mathbb{Z}$) denote the canonical isomorphisms.

2.3. Observation (Manin). — *Let* $\pi\colon X \to \mathrm{Spec}(\mathbb{Q})$ *be a projective variety over* $\mathbb{Q}$. *Choose an element* $\alpha \in \mathrm{Br}(X)$. *Then, every* $\mathbb{Q}$*-rational point* $x \in X(\mathbb{Q})$ *gives rise to an adelic point* $(x_\nu)_\nu \in X(\mathbf{A}_\mathbb{Q})$ *satisfying the condition*

$$\sum_{\nu \in \mathrm{Val}(\mathbb{Q})} \mathrm{ev}_\nu(\alpha, x_\nu) = 0\,.$$

2.4. Remarks — i) It is obvious that altering $\alpha \in \mathrm{Br}(X)$ by some Brauer class $\pi^*\rho$ for $\rho \in \mathrm{Br}(\mathbb{Q})$ does not change the obstruction defined by α. Consequently, it is only the factor group $\mathrm{Br}(X)/\pi^*\mathrm{Br}(\mathbb{Q})$ which is relevant for the Brauer-Manin obstruction.

ii) The local evaluation map $\mathrm{ev}_\nu\colon \mathrm{Br}(X) \times X(\mathbb{Q}_\nu) \to \mathbb{Q}/\mathbb{Z}$ is continuous in the second variable.

iii) Further, for every projective variety X over $\mathbb{Q}$ and every $\alpha \in \mathrm{Br}(X)$, there exists a finite set $S \subset \mathrm{Val}(\mathbb{Q})$ such that $\mathrm{ev}(\alpha, \xi) = 0$ for every $\nu \notin S$ and $\xi \in X(\mathbb{Q}_\nu)$.

These facts imply that the Brauer-Manin obstruction, if present, is an obstruction to the principle of weak approximation.

2.5. Lemma. — *Let* $\pi\colon S \to \mathrm{Spec}\,\mathbb{Q}$ *be a non-singular cubic surface. Then, there is a canonical isomorphism*

$$\delta\colon H^1(\mathrm{Gal}(\overline{\mathbb{Q}}/\mathbb{Q}), \mathrm{Pic}(S_{\overline{\mathbb{Q}}})) \longrightarrow \mathrm{Br}(S)/\pi^*\mathrm{Br}(\mathbb{Q})$$

making the diagram

$$\begin{array}{ccc}
H^1(\mathrm{Gal}(\overline{\mathbb{Q}}/\mathbb{Q}), \mathrm{Pic}(S_{\overline{\mathbb{Q}}})) & \xrightarrow{\ \delta\ } & \mathrm{Br}(S)/\pi^*\mathrm{Br}(\mathbb{Q}) \\
\downarrow d & & \\
H^2(\mathrm{Gal}(\overline{\mathbb{Q}}/\mathbb{Q}), \overline{\mathbb{Q}}(S)^*/\overline{\mathbb{Q}}^*) & & \downarrow \mathrm{res} \\
\| & & \\
H^2(\mathrm{Gal}(\overline{\mathbb{Q}}/\mathbb{Q}), \overline{\mathbb{Q}}(S)^*)/\pi^*\mathrm{Br}(\mathbb{Q}) & \xrightarrow{\ \mathrm{inf}\ } & \mathrm{Br}(\mathbb{Q}(S))/\pi^*\mathrm{Br}(\mathbb{Q})
\end{array}$$

commute. Here, d is induced by the short exact sequence

$$0 \to \overline{\mathbb{Q}}(S)^*/\overline{\mathbb{Q}}^* \to \mathrm{Div}(S_{\overline{\mathbb{Q}}}) \to \mathrm{Pic}(S_{\overline{\mathbb{Q}}}) \to 0$$

and the other morphisms are the canonical ones.

Proof. The equality at the lower left corner comes from the fact [Ta, section 11.4] that $H^3(\mathrm{Gal}(\overline{\mathbb{Q}}/\mathbb{Q}), \overline{\mathbb{Q}}^*) = 0$. The main assertion is [Ma, Lemma 43.1.1]. □

2.6. Remark — The group $H^1(\mathrm{Gal}(\overline{\mathbb{Q}}/\mathbb{Q}), \mathrm{Pic}(S_{\overline{\mathbb{Q}}}))$ is always finite. Hence, by Remark 2.4.iii), we know that only finitely many primes are relevant for the Brauer-Manin obstruction.

3. One Double-Six

3.1. — Recall that, on a non-singular cubic surface, there are exactly 27 lines. Their configuration was studied intensively by the geometers of the 19th century. They showed that the 27 lines contain 72 *sixers* of lines which are mutually skew. For each sixer, there is a complementary one formed by the lines meeting exactly five lines of the sixer. A pair of complementary sixers is called a *double-six*.

The automorphism group $W(E_6)$ acts transitively on double-sixes. In the blown-up model, one double-six may easily be written down explicitly. For this, we refer to [Ha, Remark V.4.9.1] from which we also adopt the notation for the lines.

3.2. Lemma. — *Let S be a non-singular cubic surface over* $\mathbb{Q}$*. Suppose that, under the operation of* $\mathrm{Gal}(\overline{\mathbb{Q}}/\mathbb{Q})$*, the 27 lines on S decompose into orbits one of which is of size 15. Then, the complementary twelve lines form a double-six.*

Proof. The Galois group $\mathrm{Gal}(\overline{\mathbb{Q}}/\mathbb{Q})$ operates via a certain subgroup $G \subseteq W(E_6)$. Our assumption implies that $5|\#G$. I.e., G contains the 5-Sylow subgroup of $W(E_6)$.

The operation of this is easily described in the blown-up model. The cyclic group $\langle(12345)\rangle \subset S_6$ acts on the indices. The two lines E_0 and G_0 are stationary while the others form five orbits of size five each. These are $E := \{E_1, \ldots, E_5\}$, $G := \{G_1, \ldots, G_5\}$, $F_0 := \{F_{01}, \ldots, F_{05}\}$, $F_1 := \{F_{12}, F_{23}, F_{34}, F_{45}, F_{15}\}$,and, finally,$F_2 := \{F_{13}, F_{24}, F_{35}, F_{14}, F_{25}\}$. The intersection matrix of the five latter orbits turns out to be

$$\begin{pmatrix} -5 & 20 & 5 & 10 & 10 \\ 20 & -5 & 5 & 10 & 10 \\ 5 & 5 & -5 & 15 & 15 \\ 10 & 10 & 15 & 5 & 5 \\ 10 & 10 & 15 & 5 & 5 \end{pmatrix}.$$

We assert that a size 15 orbit must be formed by F_0, F_1, and F_2.

Indeed, three orbits of size five may be put together to form an orbit only if, for the corresponding divisors, $D(D+D'+D'') = D'(D+D'+D'') = D''(D+D'+D'')$. This excludes all combinations, except for $E \cup G \cup F_1$, ($E \cup G \cup F_2$, and $F_0 \cup F_1 \cup F_2$). The set $E \cup G \cup F_1$ contains, however, only two fivers of skew lines, namely E and G. As the lines in F_1 are missing, this is a contradiction. □

3.3. Remark — U_1, the largest subgroup of $W(E_6)$ stabilizing a double-six is isomorphic to $S_6 \times \mathbb{Z}/2\mathbb{Z}$ of order 1440.

3.4. Notation. — Let S be a non-singular cubic surface. Assume that twelve of the 27 lines on S form a double-six which is $\mathrm{Gal}(\overline{\mathbb{Q}}/\mathbb{Q})$-invariant. Choose such a double-six. Then, there are two kinds of tritangent planes. We have 15 tritangent planes which meet S only within the 15 complementary lines. The other 30 tritangent planes contain one of the 15 lines and two from the double-six.

We write F_{30} for a product over the linear forms defining the 30 tritangent planes and F_{15} for a product over the linear forms defining the 15 others. Note that $F_{30}/F_{15}^2 \in \mathbb{Q}(S)$.

3.5. Theorem. — *Let* $\pi\colon S \to \operatorname{Spec}\mathbb{Q}$ *be a non-singular cubic surface such that the 27 lines have orbit structure* $[12,15]$ *under the operation of* $\mathrm{Gal}(\overline{\mathbb{Q}}/\mathbb{Q})$.

i) *Then,* $\mathrm{Br}(S)/\pi^*\mathrm{Br}(\mathbb{Q}) = \mathbb{Z}/2\mathbb{Z}$.

ii) *For* $0 \neq c \in \mathrm{Br}(S)/\pi^*\mathrm{Br}(\mathbb{Q})$, *a representative* $\underline{c}$ *of* $\mathrm{res}(c) \in \mathrm{Br}(\mathbb{Q}(S))/\pi^*\mathrm{Br}(\mathbb{Q})$ *is given as follows.*

Consider the quadratic number field $\mathbb{Q}(\sqrt{D})$ *splitting the double-six into two sixers. Then, apply to the class*

$$(F_{30}/F_{15}^2) \in \widehat{H}^0(\mathrm{Gal}(\mathbb{Q}(\sqrt{D})/\mathbb{Q}), \mathbb{Q}(\sqrt{D})(S)^*) = \mathbb{Q}(S)^*/N\mathbb{Q}(\sqrt{D})(S)^*$$

the periodicity isomorphism to H^2 *and the inflation map.*

Proof. *First step.* Inflation.

We have the isomorphism $\delta\colon H^1(\mathrm{Gal}(\overline{\mathbb{Q}}/\mathbb{Q}), \mathrm{Pic}(S_{\overline{\mathbb{Q}}})) \to \mathrm{Br}(S)/\pi^*\mathrm{Br}(\mathbb{Q})$ and will work with the group on the left.

An element of the group $\mathrm{Gal}(\overline{\mathbb{Q}}/\mathbb{Q})$ may either flip the two sixers forming the twelve lines or not. Therefore, there is an index two subgroup stabilizing the sixers. This group corresponds to the quadratic number field $\mathbb{Q}(\sqrt{D})$. By Fact 3.6 below, we know H^1 $(\mathrm{Gal}(\overline{\mathbb{Q}}/\mathbb{Q}(\sqrt{D})), \mathrm{Pic}(S_{\overline{\mathbb{Q}}})) = 0$. The inflation-restriction-sequence yields that

$$\mathrm{inf}\colon H^1(\mathrm{Gal}(\mathbb{Q}(\sqrt{D})/\mathbb{Q}), \mathrm{Pic}(S_{\overline{\mathbb{Q}}})^{\mathrm{Gal}(\overline{\mathbb{Q}}/\mathbb{Q}(\sqrt{D}))}) \longrightarrow H^1(\mathrm{Gal}(\overline{\mathbb{Q}}/\mathbb{Q}), \mathrm{Pic}(S_{\overline{\mathbb{Q}}}))$$

is an isomorphism.

Second step. Divisors.

The orbit structure of the 27 lines under the operation of $\mathrm{Gal}(\overline{\mathbb{Q}}/\mathbb{Q}(\sqrt{D}))$ is $[6,6,15]$. Indeed, when going over to an index two subgroup an orbit of odd size must not split. Denote by E, G, and F the divisors formed by summing over the first, second, and third orbit, respectively. E, G, and F clearly define elements in $\mathrm{Pic}(S_{\overline{\mathbb{Q}}})^{\mathrm{Gal}(\overline{\mathbb{Q}}/\mathbb{Q}(\sqrt{D}))}$. Write P for the subgroup generated by these three divisors.

P is of finite index in $\mathrm{Pic}(S_{\overline{\mathbb{Q}}})^{\mathrm{Gal}(\overline{\mathbb{Q}}/\mathbb{Q}(\sqrt{D}))}$. Indeed, every element of $\mathrm{Pic}(S_{\overline{\mathbb{Q}}})$ is an integral linear combination of the divisors given by the 27 lines. Therefore, every element in $\mathrm{Pic}(S_{\overline{\mathbb{Q}}})^{\mathrm{Gal}(\overline{\mathbb{Q}}/\mathbb{Q}(\sqrt{D}))}$ is a $\mathbb{Q}$-linear combination of E, G, and F and the denominators are at most six or 15.

We claim that the index of P is a divisor of 15. In fact, we have the relation $5E + 5G - 4F \sim 0$. Further, the discriminant of the lattice spanned by E and F is

$$\begin{vmatrix} -6 & 30 \\ 30 & 75 \end{vmatrix} = -1350 = (-6)\cdot 15^2.$$

Consequently, P is of odd index in $\mathrm{Pic}(S_{\overline{\mathbb{Q}}})^{\mathrm{Gal}(\overline{\mathbb{Q}}/\mathbb{Q}(\sqrt{D}))}$. This implies that the natural homomorphism

$$H^1(\mathrm{Gal}(\mathbb{Q}(\sqrt{D})/\mathbb{Q}),P) \longrightarrow H^1(\mathrm{Gal}(\mathbb{Q}(\sqrt{D})/\mathbb{Q}),\mathrm{Pic}(S_{\overline{\mathbb{Q}}})^{\mathrm{Gal}(\overline{\mathbb{Q}}/\mathbb{Q}(\sqrt{D}))})$$

is a bijection.

Third step. The fundamental class.
As $\mathrm{Gal}(\mathbb{Q}(\sqrt{D})/\mathbb{Q})$ is a cyclic group of order two, we have

$$H^2(\mathrm{Gal}(\mathbb{Q}(\sqrt{D})/\mathbb{Q}),\mathbb{Z}) \cong \mathbb{Z}/2\mathbb{Z}\,.$$

Write u for the non-zero element. Then, the periodicity isomorphism is given by

$$\cup u\colon \widehat{H}^{-1}(\mathrm{Gal}(\mathbb{Q}(\sqrt{D})/\mathbb{Q}),P) \longrightarrow H^1(\mathrm{Gal}(\mathbb{Q}(\sqrt{D})/\mathbb{Q}),P)\,.$$

Observe that, for a cyclic group of order two, this isomorphism is canonical as there is no ambiguity in the choice of u.

Fourth step. Computing $\widehat{H}^{-1}$.
We have $P = S/S_0$ for $S := \mathbb{Z}E \oplus \mathbb{Z}G \oplus \mathbb{Z}F$ and S_0 the group of the principal divisors contained in S. The relation $\widehat{H}^{-1}(\mathrm{Gal}(\mathbb{Q}(\sqrt{D})/\mathbb{Q}),S) = 0$ follows immediately from the definition. Hence, the short exact sequence

$$0 \to S_0 \to S \to P \to 0$$

yields

$$\begin{aligned}
&\widehat{H}^{-1}(\mathrm{Gal}(\mathbb{Q}(\sqrt{D})/\mathbb{Q}),P) = \\
&\qquad = \ker(\widehat{H}^0(\mathrm{Gal}(\mathbb{Q}(\sqrt{D})/\mathbb{Q}),S_0) \to \widehat{H}^0(\mathrm{Gal}(\mathbb{Q}(\sqrt{D})/\mathbb{Q}),S)) \\
&\qquad = \ker(S_0^{\mathrm{Gal}(\mathbb{Q}(\sqrt{D})/\mathbb{Q})}/NS_0 \to S^{\mathrm{Gal}(\mathbb{Q}(\sqrt{D})/\mathbb{Q})}/NS) \\
&\qquad = (S_0^{\mathrm{Gal}(\mathbb{Q}(\sqrt{D})/\mathbb{Q})} \cap NS)/NS_0\,.
\end{aligned}$$

Here, the norm map acts by the rule $N\colon aE + bG + cF \mapsto (a+b)E + (a+b)G + 2cF$. Hence, $NS = \langle E+G, 2F\rangle$. Principal divisors are characterized by the property that all intersection numbers are zero. A direct calculation shows

$$S_0^{\mathrm{Gal}(\mathbb{Q}(\sqrt{D})/\mathbb{Q})} \cap NS = \langle 5E+5G-4F\rangle\,.$$

The generator is the norm of any divisor of the form $aE + (5-a)G - 2F$. None of these is principal. Indeed, the intersection number with E is equal to $-6a + 30(5-a) - 60 = -36a + 90$ and this terms does not vanish for $a \in \mathbb{Z}$. Assertion i) is proven.

Fifth step. The representative.
We actually constructed a non-zero element $c' \in H^1(\mathrm{Gal}(\overline{\mathbb{Q}}/\mathbb{Q}),\mathrm{Pic}(S_{\overline{\mathbb{Q}}}))$. The calculations given above show that

$$d(c') = (F_{30}/F_{15}^2) \cup u\,.$$

Indeed, it is easy to see that $\mathrm{div}(F_{30}/F_{15}^2) = 5E+5G-4F$. The assertion now follows from the commutative diagram given in Lemma 2.5. □

3.6. Fact — *Let S be a non-singular cubic surface over a field K obtained by blowing up $\mathbf{P}^2_K$ in six $\overline{K}$-rational points which form a Galois invariant set.*
Then, $H^1(\mathrm{Gal}(\overline{K}/K), \mathrm{Pic}(S_{\overline{K}})) = 0$.

Proof. This is, of course, a particular case of the very general [Ma, Theorem 29.1]. Let us give an elementary proof here.

According to Shapiro's lemma, we may replace $\mathrm{Gal}(\overline{K}/K)$ by a finite quotient G. We have $\mathrm{Pic}(S_{\overline{K}}) = \mathbb{Z}H \oplus \mathbb{Z}E_1 \oplus \ldots \oplus \mathbb{Z}E_6$ for H the hyperplane section of $\mathbf{P}^2$ and $E_1, \ldots, E_6$ the exceptional divisors. Therefore, as a G-module,

$$\mathrm{Pic}(S_{\overline{K}}) = \mathbb{Z} \oplus \mathbb{Z}[G/H_1] \oplus \ldots \oplus \mathbb{Z}[G/H_l]$$

for l the number of Galois orbits and certain subgroups $H_1, \ldots, H_l$. Clearly, we have $H^1(G, \mathbb{Z}) = \mathrm{Hom}(G, \mathbb{Z}) = 0$.

On the other hand, for any subgroup H, the G-module $\mathbb{Z}[G/H]$ is equipped with a non-degenerate pairing. Hence, $\mathbb{Z}[G/H] \cong \mathrm{Hom}(\mathbb{Z}[G/H], \mathbb{Z})$ and

$$\begin{aligned} H^1(G, \mathbb{Z}[G/H]) &\cong H^1(G, \mathrm{Hom}(\mathbb{Z}[G/H], \mathbb{Z})) \\ &\cong \widehat{H}^0(G, \mathrm{Hom}(\mathbb{Z}[G/H], \mathbb{Q}/\mathbb{Z})) \\ &\cong \mathrm{Hom}\big(\widehat{H}^{-1}(G, \mathbb{Z}[G/H]), \mathbb{Q}/\mathbb{Z}\big) \end{aligned}$$

by the duality theorem for cohomology of finite groups [CE, Chap. XII, Corollary 6.5]. Finally, $\widehat{H}^{-1}(G, \mathbb{Z}[G/H])$ vanishes as is seen immediately from the definition. □

4. Triples of Azygetic Double-sixes

4.1. Definition. — Let $\pi \colon S \to \operatorname{Spec} \mathbb{Q}$ be a smooth cubic surface and $\mathscr{D}$ be a Galois invariant double-six. This induces a group homomorphism $\mathrm{Gal}(\overline{\mathbb{Q}}/\mathbb{Q}) \to U_1$, given by the operation on the 27 lines.

i) Then, the image of the non-zero element under the natural homomorphism

$$\mathbb{Z}/2\mathbb{Z} \cong H^1(U_1, \mathrm{Pic}(S_{\overline{\mathbb{Q}}})) \longrightarrow H^1(\mathrm{Gal}(\overline{\mathbb{Q}}/\mathbb{Q}), \mathrm{Pic}(S_{\overline{\mathbb{Q}}})) \xrightarrow{\ \delta\ } \mathrm{Br}(S)/\pi^*\mathrm{Br}(\mathbb{Q})$$

is called *the Brauer class associated with* the double-six $\mathscr{D}$. We denote it by $\mathrm{cl}(\mathscr{D})$.

ii) This defines a map

$$\mathrm{cl} \colon \Phi_S^{\mathrm{Gal}(\overline{\mathbb{Q}}/\mathbb{Q})} \longrightarrow \mathrm{Br}(S)/\pi^*\mathrm{Br}(\mathbb{Q})$$

from the set of all Galois invariant double-sixes which we will call the *class map*.

4.2. — Two double-sixes may have either four or six lines in common. In the former case, the two are called syzygetic, in the latter azygetic. A pair of azygetic double-sixes is built as follows.

$$\begin{pmatrix} E_0 & E_1 & E_2 & E_3 & E_4 & E_5 \\ G_0 & G_1 & G_2 & G_3 & G_4 & G_5 \end{pmatrix}, \quad \begin{pmatrix} E_0 & E_1 & E_2 & F_{45} & F_{35} & F_{34} \\ F_{12} & F_{02} & F_{01} & G_3 & G_4 & G_5 \end{pmatrix}.$$

The twelve lines which appear only once form a third double-six

$$\begin{pmatrix} F_{12} & F_{02} & F_{01} & E_3 & E_4 & E_5 \\ G_0 & G_1 & G_2 & F_{45} & F_{35} & F_{34} \end{pmatrix}$$

azygetic to the other two.

4.3. — The largest group U_3 stabilizing a triple of azygetic double-sixes is isomorphic to $(S_3 \times S_3) \ltimes \mathbb{Z}/2\mathbb{Z}$ of order 72. The induced orbit structure is $[6,6,6,9]$. The orbits themselves are, in the notation above, $\{E_0,E_1,E_2,G_3,G_4,G_5\}$, $\{G_0,G_1,G_2,E_3,E_4,E_5\}$, $\{F_{01},F_{02},F_{12},F_{34},F_{35},F_{45}\}$, and $\{F_{03},F_{04},F_{05},F_{13},F_{14},F_{15},F_{23},F_{24},F_{25}\}$.

The quadratic extension $\mathbb{Q}(\sqrt{D})$ splitting one of the three double-sixes into two sixers automatically splits the others, too. The operation of $\mathrm{Gal}(\overline{\mathbb{Q}}/\mathbb{Q}(\sqrt{D}))$ yields the orbits $\{E_0,E_1,E_2\}$, $\{E_3,E_4,E_5\}$, $\{G_0,G_1,G_2\}$, $\{G_3,G_4,G_5\}$, $\{F_{01},F_{02},F_{12}\}$, $\{F_{34},F_{35},F_{45}\}$, and $\{F_{03},F_{04},F_{05},F_{13},F_{14},F_{15},F_{23},F_{24},F_{25}\}$.

4.4. Theorem. — *Let* $\pi\colon S \to \operatorname{Spec}\mathbb{Q}$ *be a non-singular cubic surface. Assume that* $\mathrm{Gal}(\overline{\mathbb{Q}}/\mathbb{Q})$ *stabilizes a triple* $\{\mathscr{D}_1,\mathscr{D}_2,\mathscr{D}_3\}$ *of azygetic double-sixes and that the 27 lines have orbit structure* $[6,6,6,9]$.

i) *Then,* $\mathrm{Br}(S)/\pi^*\mathrm{Br}(\mathbb{Q}) = \mathbb{Z}/2\mathbb{Z} \times \mathbb{Z}/2\mathbb{Z}$.

ii) *The three non-zero elements are* $\mathrm{cl}(\mathscr{D}_1)$, $\mathrm{cl}(\mathscr{D}_2)$, *and* $\mathrm{cl}(\mathscr{D}_3)$.

Proof. *First step.* Inflation.

We will follow the same strategy as in the proof of Theorem 3.5. In particular, we will work with the group $H^1(\mathrm{Gal}(\mathbb{Q}(\sqrt{D})/\mathbb{Q}), \mathrm{Pic}(S_{\overline{\mathbb{Q}}})^{\mathrm{Gal}(\overline{\mathbb{Q}}/\mathbb{Q}(\sqrt{D}))})$.

Second step. Divisors.

The orbit structure of the 27 lines under the operation of $\mathrm{Gal}(\overline{\mathbb{Q}}/\mathbb{Q}(\sqrt{D}))$ is $[3,3,3,3,3,3,9]$. For the lines and double-sixes, we use the notation introduced in 4.2. Further, denote by

$$E^{(1)}, E^{(2)}, G^{(1)}, G^{(2)}, F^{(1)}, F^{(2)}, F^{(3)}$$

the divisors formed by summing over the orbits. These clearly define elements in $\mathrm{Pic}(S_{\overline{\mathbb{Q}}})^{\mathrm{Gal}(\overline{\mathbb{Q}}/\mathbb{Q}(\sqrt{D}))}$. We write P for the subgroup generated by these seven divisors.

Every element of $\mathrm{Pic}(S_{\overline{\mathbb{Q}}})$ is an integral linear combination of the divisors given by the 27 lines. Therefore, every element in $\mathrm{Pic}(S_{\overline{\mathbb{Q}}})^{\mathrm{Gal}(\overline{\mathbb{Q}}/\mathbb{Q}(\sqrt{D}))}$ is a $\mathbb{Q}$-linear combination of $E^{(1)}$, $E^{(2)}$, $G^{(1)}$, $G^{(2)}$, $F^{(1)}$, $F^{(2)}$, and $F^{(3)}$ and the denominators are divisors of nine. Consequently, P is of odd index in $\mathrm{Pic}(S_{\overline{\mathbb{Q}}})^{\mathrm{Gal}(\overline{\mathbb{Q}}/\mathbb{Q}(\sqrt{D}))}$. This implies that the natural homomorphism

$$H^1(\mathrm{Gal}(\mathbb{Q}(\sqrt{D})/\mathbb{Q}), P) \longrightarrow H^1(\mathrm{Gal}(\mathbb{Q}(\sqrt{D})/\mathbb{Q}), \mathrm{Pic}(S_{\overline{\mathbb{Q}}})^{\mathrm{Gal}(\overline{\mathbb{Q}}/\mathbb{Q}(\sqrt{D}))})$$

is a bijection.

Third step. The fundamental class.

Again, we write u for the non-zero element in $H^2(\mathrm{Gal}(\mathbb{Q}(\sqrt{D})/\mathbb{Q}), \mathbb{Z}) \cong \mathbb{Z}/2\mathbb{Z}$. Then, the periodicity isomorphism is given by

$$\cup u\colon \widehat{H}^{-1}(\mathrm{Gal}(\mathbb{Q}(\sqrt{D})/\mathbb{Q}), P) \longrightarrow H^1(\mathrm{Gal}(\mathbb{Q}(\sqrt{D})/\mathbb{Q}), P).$$

Fourth step. Computing $\widehat{H}^{-1}$.

We have $P = S/S_0$ for $S := \mathbb{Z}E^{(1)} \oplus \mathbb{Z}E^{(2)} \oplus \mathbb{Z}G^{(1)} \oplus \mathbb{Z}G^{(2)} \oplus \mathbb{Z}F^{(1)} \oplus \mathbb{Z}F^{(2)} \oplus \mathbb{Z}F^{(3)}$ and S_0 the group of the principal divisors contained in S. To simplify formulas, we will use the notation $D^{(1)} := E^{(1)} + G^{(2)}$, $D^{(2)} := E^{(2)} + G^{(1)}$, and $D^{(3)} := F^{(1)} + F^{(2)}$.

The relation $\widehat{H}^{-1}(\mathrm{Gal}(\mathbb{Q}(\sqrt{D})/\mathbb{Q}), S) = 0$ follows immediately from the definition. Hence, the short exact sequence $0 \to S_0 \to S \to P \to 0$ yields, as above,

$$\widehat{H}^{-1}(\mathrm{Gal}(\mathbb{Q}(\sqrt{D})/\mathbb{Q}), P) = (S_0^{\mathrm{Gal}(\mathbb{Q}(\sqrt{D})/\mathbb{Q})} \cap NS)/NS_0 .$$

Here, the norm map acts by the rule

$$\begin{aligned} N\colon a_1E^{(1)} + a_2E^{(2)} + b_1G^{(1)} + b_2G^{(2)} + c_1F^{(1)} + c_2F^{(2)} + c_3F^{(3)} \\ \mapsto (a_1 + b_2)D^{(1)} + (a_2 + b_1)D^{(2)} + (c_1 + c_2)D^{(3)} + 2c_3F^{(3)} . \end{aligned}$$

Hence, $NS = \langle D^{(1)}, D^{(2)}, D^{(3)}, 2F^{(3)} \rangle$. Principal divisors are characterized by the property that all intersection numbers are zero. A direct calculation shows

$$\begin{aligned} & S_0^{\mathrm{Gal}(\mathbb{Q}(\sqrt{D})/\mathbb{Q})} \cap NS \\ = \; & \{ d_1D^{(1)} + d_2D^{(2)} + d_3D^{(3)} + 2eF^{(3)} \mid d_1 + d_2 + d_3 + 3e = 0 \} \\ = \; & \langle D^{(1)} - D^{(2)}, D^{(1)} - D^{(3)}, D^{(1)} + D^{(2)} + D^{(3)} - 2F^{(3)} \rangle . \end{aligned}$$

It is easy to see that $E^{(1)} + G^{(1)} + F^{(1)} - F^{(3)}$ is a principal divisor. Hence,

$$D^{(1)} + D^{(2)} + D^{(3)} - 2F^{(3)} \in NS_0 .$$

Further, NS_0 contains all principal divisors which are divisible by 2. As it turns out that these two sorts of elements generate the whole of NS_0, assertion i) follows.

Fifth step. The representatives.

We actually constructed non-zero elements $c_1, c_2, c_3 \in H^1(\mathrm{Gal}(\overline{\mathbb{Q}}/\mathbb{Q}), \mathrm{Pic}(S_{\overline{\mathbb{Q}}}))$, represented by $D^{(1)} - D^{(2)}, D^{(1)} - D^{(3)}, D^{(2)} - D^{(3)} \in S_0^{\mathrm{Gal}(\mathbb{Q}(\sqrt{D})/\mathbb{Q})} \cap NS$. The first representative is equivalent to

$$3(D^{(1)} - D^{(2)}) + 2(D^{(1)} + D^{(2)} + D^{(3)} - 2F^{(3)}) + 6(D^{(2)} - D^{(3)}) = \mathrm{div}(F_{30}/F_{15}^2) .$$

Hence, $d(c_1) = (F_{30}/F_{15}^2) \cup u$. The corresponding Brauer class is $\mathrm{cl}\big(\big(\frac{E_0 \ldots E_5}{G_0 \ldots G_5}\big)\big)$. For the two other classes, the situation is analogous. □

4.5. Remark — In the $[6,6,6,9]$-case, the 45 tritangent planes decompose into five orbits.

- $[E_i, G_j, F_{ij}]$ for $i, j \in \{0,1,2\}$ or $i, j \in \{3,4,5\}$, $i \neq j$. (twelve planes)
- $[E_i, G_j, F_{ij}]$ for $i \in \{0,1,2\}$ and $j \in \{3,4,5\}$. (nine planes)
- $[E_i, G_j, F_{ij}]$ for $i \in \{3,4,5\}$ and $j \in \{0,1,2\}$. (nine planes)
- $[F_{i_0i_1}, F_{i_2i_3}, F_{i_4i_5}]$ for $\{i_0, i_1, i_2, i_3, i_4, i_5\} = \{0,1,2,3,4,5\}$,
  $i_0, i_1 \in \{0,1,2\}$, and $i_2, i_3 \in \{3,4,5\}$. (nine planes)
- $[F_{i_0i_1}, F_{i_2i_3}, F_{i_4i_5}]$ for $\{i_0, i_2, i_4\} = \{0,1,2\}$ and $\{i_1, i_3, i_5\} = \{3,4,5\}$. (six planes)

The three forms of type F_{30} are obtained by multiplying the linear forms defining the orbit of size twelve together with those for two of the orbits of size nine. Actually, the size twelve orbit is irrelevant. Up to a scalar factor, it is the square of a sextic form.

4.6. Remark — Triples of azygetic double-sixes have been studied by the classical algebraic geometers. See, for example, [Ko, §6]. A result from the 19th century states that there are exactly 120 triples of azygetic double-sixes on a smooth cubic surface. Actually, the automorphism group $W(E_6)$ acts transitively on them.

5. The General Case of a Galois Group Stabilizing a Double-six

5.1. — To explicitly compute $H^1(G, \text{Pic}(S_{\overline{\mathbb{Q}}}))$ as an abstract abelian group, one may use Manin's formula [Ma, Proposition 31.3]. This means the following.

$\text{Pic}(S_{\overline{\mathbb{Q}}})$ is generated by the 27 lines. The group of all permutations of the 27 lines respecting the intersection pairing is isomorphic to the Weyl group $W(E_6)$ of order 51 840. The group G operates on the 27 lines via a finite quotient G/H which is isomorphic to a subgroup of $W(E_6)$. Then,

$$H^1(G, \text{Pic}(S_{\overline{\mathbb{Q}}})) \cong \text{Hom}(ND \cap D_0/ND_0, \mathbb{Q}/\mathbb{Z}).$$

Here, D is the free abelian group generated by the 27 lines and D_0 is the subgroup of all principal divisors. $N \colon D \to D$ denotes the norm map as a G/H-module.

5.2. — Using Manin's formula, we computed $H^1(G, \text{Pic}(S_{\overline{\mathbb{Q}}}))$ for each of the 350 conjugacy classes of subgroups of $W(E_6)$. The computations in `GAP` took approximately 28 seconds of CPU time. Thereby, we recovered the following result of Sir Peter Swinnerton-Dyer [SD2]. (See also P. K. Corn [Co].)

5.3. Theorem (Swinnerton-Dyer). — *Let S be a non-singular cubic surface over $\mathbb{Q}$. Then, $H^1(\text{Gal}(\overline{\mathbb{Q}}/\mathbb{Q}), \text{Pic}(S_{\overline{\mathbb{Q}}}))$ may take only five values,* 0, $\mathbb{Z}/2\mathbb{Z}$, $\mathbb{Z}/2\mathbb{Z} \times \mathbb{Z}/2\mathbb{Z}$, $\mathbb{Z}/3\mathbb{Z}$, *and* $\mathbb{Z}/3\mathbb{Z} \times \mathbb{Z}/3\mathbb{Z}$. □

5.4. Remark — One has $H^1(G, \text{Pic}(S_{\overline{\mathbb{Q}}})) = 0$ in 257 of the 350 cases.

5.5. — More importantly, we make the following observation.

Proposition. *Let S be a non-singular cubic surface over $\mathbb{Q}$.*

i) *If* $H^1(\text{Gal}(\overline{\mathbb{Q}}/\mathbb{Q}), \text{Pic}(S_{\overline{\mathbb{Q}}})) = \mathbb{Z}/2\mathbb{Z}$ *then, on S, there is a Galois invariant double-six.*

ii) *If* $H^1(\text{Gal}(\overline{\mathbb{Q}}/\mathbb{Q}), \text{Pic}(S_{\overline{\mathbb{Q}}})) = \mathbb{Z}/2\mathbb{Z} \times \mathbb{Z}/2\mathbb{Z}$ *then, on S, there is a triple of azygetic double-sixes stabilized by* $\text{Gal}(\overline{\mathbb{Q}}/\mathbb{Q})$.

Proof. This is seen by a case-by-case study using `GAP`. □

5.6. Remarks — i) On the other hand, if there is a Galois invariant double-six on S then $H^1(\text{Gal}(\overline{\mathbb{Q}}/\mathbb{Q}), \text{Pic}(S_{\overline{\mathbb{Q}}}))$ is either 0, or $\mathbb{Z}/2\mathbb{Z}$ or $\mathbb{Z}/2\mathbb{Z} \times \mathbb{Z}/2\mathbb{Z}$.

ii) Proposition 5.5 immediately provokes the question whether the cohomology classes are always "the same" as in the $[12,15]$- and $[6,6,6,9]$-cases. I.e., of the type $\mathrm{cl}(\mathscr{D})$ for certain Galois invariant double-sixes. Somewhat surprisingly, this is indeed the case.

5.7. Lemma. — *Let $\mathscr{S}$ be a non-singular cubic surface over an algebraically closed field, H a group of automorphisms of the configuration of the 27 lines, and $H' \subseteq H$ any subgroup. Each of the criteria below is sufficient for*

$$\mathrm{res}\colon H^1(H,\mathrm{Pic}(\mathscr{S})) \to H^1(H',\mathrm{Pic}(\mathscr{S}))$$

being an injection.

i) *H and H' generate the same orbit structure.*

ii) *H' is of odd index in H and $H^1(H,\mathrm{Pic}(\mathscr{S}))$ is a 2-group.*

iii) *H' is a normal subgroup in H and $\mathrm{rk}\,\mathrm{Pic}(\mathscr{S})^H = \mathrm{rk}\,\mathrm{Pic}(\mathscr{S})^{H'}$.*

Proof. i) This follows immediately from the formula of Manin [Ma, Proposition 31.3].
ii) Here, $\mathrm{cores}\circ\mathrm{res}\colon H^1(H,\mathrm{Pic}(\mathscr{S})) \to H^1(H,\mathrm{Pic}(\mathscr{S}))$ is the multiplication by an odd number, hence the identity.
iii) The assumption ensures that H/H' operates trivially on $\mathrm{Pic}(\mathscr{S})^{H'}$. Hence, $H^1(H/H',\mathrm{Pic}(\mathscr{S})^{H'}) = 0$. The inflation-restriction sequence

$$0 \to H^1(H/H',\mathrm{Pic}(\mathscr{S})^{H'}) \to H^1(H,\mathrm{Pic}(\mathscr{S})) \to H^1(H',\mathrm{Pic}(\mathscr{S}))$$

yields the assertion. □

5.8. Proposition. — *Let $\mathscr{S}$ be a non-singular cubic surface over an algebraically closed field, U_1 the group of automorphisms of the configuration of the 27 lines stabilizing a double-six and U_3 the group stabilizing a triple of azygetic double-sixes.*

a) *Let $H \subseteq U_1$ be such that $H^1(H,\mathrm{Pic}(\mathscr{S})) = \mathbb{Z}/2\mathbb{Z}$. Then, the restriction*

$$\mathrm{res}\colon H^1(U_1,\mathrm{Pic}(\mathscr{S})) \longrightarrow H^1(H,\mathrm{Pic}(\mathscr{S}))$$

is a bijection.

b) *Let $H \subseteq U_3$ be a subgroup such that $H^1(H,\mathrm{Pic}(\mathscr{S})) = \mathbb{Z}/2\mathbb{Z}\times\mathbb{Z}/2\mathbb{Z}$. Then, the restriction*

$$\mathrm{res}\colon H^1(U_3,\mathrm{Pic}(\mathscr{S})) \longrightarrow H^1(H,\mathrm{Pic}(\mathscr{S}))$$

is a bijection.

Proof. The proof has a computer part. We use the machine to verify that the criteria formulated in Lemma 5.7 suffice to establish the result in all cases.

b) Here, the subgroup $(A_3\times A_3)\ltimes\mathbb{Z}/2\mathbb{Z}$ of order 18, as well as the two intermediate groups of order 36 produce the same orbit structure $[6,6,6,9]$. It turns out that every subgroup H which leads to $\mathbb{Z}/2\mathbb{Z}\times\mathbb{Z}/2\mathbb{Z}$ is a subgroup of odd (1, 3, or 9) index in one of those.

a) By Lemma 5.7.ii), we may test this on the 2-Sylow subgroups of H and U_1. $U_1^{(2)}$ is a group of order 32 such that the Picard rank is equal to two. It turns out that, for 2-groups H' such that $H^1(H',\mathrm{Pic}(\mathscr{S})) = \mathbb{Z}/2\mathbb{Z}$, the Picard rank may be only two or three.

There is a maximal 2-group such that the Picard rank is three. This is a group of order 16 generating the orbit structure $[1,1,1,4,4,4,4,4,4]$. To prove the assertion for this group, one first observes that it is of index three in a group of order 48 with orbit structure $[3,12,12]$. This group, in turn, is of index two in the maximal group with that orbit structure. That one, being of order 96, is the maximal group stabilizing a double-six and a tritangent plane containing three complementary lines. It is of index 15 in U_1. □

5.9. Corollary. — *Let $H' \subseteq H \subseteq U_1$ be arbitrary. Then, for the restriction map* $\text{res}\colon H^1(H,\text{Pic}(\mathscr{S})) \longrightarrow H^1(H',\text{Pic}(\mathscr{S}))$, *there are the following limitations.*

i) *If $H^1(H,\text{Pic}(\mathscr{S})) = 0$ then $H^1(H',\text{Pic}(\mathscr{S})) = 0$.*

ii) *If $H^1(H,\text{Pic}(\mathscr{S})) \cong \mathbb{Z}/2\mathbb{Z}$ and $H^1(H',\text{Pic}(\mathscr{S})) \neq 0$ then* res *is an injection.*

iii) *If $H^1(H,\text{Pic}(\mathscr{S})) \cong \mathbb{Z}/2\mathbb{Z} \times \mathbb{Z}/2\mathbb{Z}$ then $H^1(H',\text{Pic}(\mathscr{S})) \cong \mathbb{Z}/2\mathbb{Z} \times \mathbb{Z}/2\mathbb{Z}$ or 0. In the former case,* res *is a bijection. The latter is possible only when H' stabilizes a sixer.*

Proof. We know from Remark 5.6.i) that both groups may be only 0, $\mathbb{Z}/2\mathbb{Z}$,or $\mathbb{Z}/2\mathbb{Z} \times \mathbb{Z}/2\mathbb{Z}$.

i) If $H^1(H',\text{Pic}(\mathscr{S}))$ were isomorphic to $\mathbb{Z}/2\mathbb{Z}$ or $\mathbb{Z}/2\mathbb{Z} \times \mathbb{Z}/2\mathbb{Z}$ then the restriction from U_1, respectively U_3, to H' would be the zero map.

ii) is immediate from the computations above.

iii) If $H^1(H',\text{Pic}(\mathscr{S})) \cong \mathbb{Z}/2\mathbb{Z}$ then, by composition, we could produce the zero map on $\mathbb{Z}/2\mathbb{Z}$. The final assertion is an experimental observation. □

5.10. Remark (Pairs of syzygetic double-sixes). — U_2, the largest group stabilizing two syzygetic double-sixes is of order 96. In view of Proposition 5.5.ii), this ensures that $H^1(U_2,\text{Pic}(\mathscr{S})) \cong \mathbb{Z}/2\mathbb{Z}$ or 0. Actually, it is isomorphic to $\mathbb{Z}/2\mathbb{Z}$. Corollary 5.9.ii) implies that the Brauer classes associated with the two double-sixes coincide. Both are equal to the non-zero element.

Actually, the group U_2 leads to an orbit structure $[1,4,6,8,8]$. It is easy to compute $H^1(U_2,\text{Pic}(\mathscr{S}))$ directly using the same methods as in the proof of Theorem 4.4.

5.11. Theorem. — *Let $\pi\colon S \to \operatorname{Spec}\mathbb{Q}$ be an arbitrary smooth cubic surface. Then, consider the class map*

$$\text{cl}\colon \Phi_S^{\text{Gal}(\overline{\mathbb{Q}}/\mathbb{Q})} \to \text{Br}(S)/\pi^*\text{Br}(\mathbb{Q}),$$

introduced in Definition 4.1.ii), from the set of all Galois invariant double-sixes.

a) cl *has the properties below.*

i) *If $\mathscr{D}_1, \mathscr{D}_2$ are syzygetic double-sixes then $\text{cl}(\mathscr{D}_1) = \text{cl}(\mathscr{D}_2)$.*

ii) *If $\mathscr{D}_1, \mathscr{D}_2$ are azygetic then $\text{cl}(\mathscr{D}_1) + \text{cl}(\mathscr{D}_2) + \text{cl}(\mathscr{D}_3) = 0$ for $\mathscr{D}_3$ the third double-six of the corresponding triple.*

b) *If $\text{Br}(S)/\pi^*\text{Br}(\mathbb{Q}) \neq 0$ then*

i) *$\text{cl}(\mathscr{D}) \neq 0$ for every Galois invariant double-six. Further, $\text{cl}\big(\Phi_S^{\text{Gal}(\overline{\mathbb{Q}}/\mathbb{Q})}\big)$ contains exactly the elements of order two.*

ii) *Two double-sixes $\mathscr{D}_1, \mathscr{D}_2$ are syzygetic if and only if $\text{cl}(\mathscr{D}_1) = \text{cl}(\mathscr{D}_2)$ and azygetic if and only if $\text{cl}(\mathscr{D}_1) \neq \text{cl}(\mathscr{D}_2)$.* □

6. Explicit Brauer-Manin Obstruction

6.1. — Let S be a non-singular cubic surface with a Galois invariant double-six $\mathscr{D}$. This determines a class $c := \mathrm{cl}(\mathscr{D}) \in \mathrm{Br}(S)/\pi^*\mathrm{Br}(\mathbb{Q})$. Choose a representative $\underline{c} \in \mathrm{Br}(S)$ and the corresponding rational function $F_{30}/F_{15}^2 \in \mathbb{Q}(S)$. Finally, let $\mathbb{Q}(\sqrt{D})$ be the quadratic extension splitting the double-six into two sixers.

6.2. Fact — *The quaternion algebra over $\mathbb{Q}(S)$ corresponding to $\underline{c}$ is*

$$Q := \mathbb{Q}(S)\{X,Y\}/(XY+YX, X^2-D, Y^2-F_{30}/F_{15}^2)\,.$$

6.3. Remark — It is well known that a class in $\mathrm{Br}(S)$ is uniquely determined by its restriction to $\mathrm{Br}(\mathbb{Q}(S))$. The corresponding quaternion algebra over the whole of S may be described as follows.

Let $x \in S$. We know that $\mathrm{div}(F_{30}/F_{15}^2)$ is the norm of a divisor on $S_{\mathbb{Q}(\sqrt{D})}$. That one is necessarily locally principal. I.e., we have a rational function $f_x = a_x + b_x\sqrt{D}$ such that $\mathrm{div}(Nf_x) = \mathrm{div}(F_{30}/F_{15}^2)$ on a Zariski neighbourhood of x. Over the maximal such neighbourhood U_x, we define a quaternion algebra by

$$Q_x := \mathscr{O}_{U_x}\{X,Y_x\}/(XY_x+Y_xX, X^2-D, Y_x^2-\tfrac{F_{30}}{F_{15}^2 Nf_x})\,.$$

In particular, in a neighbourhood U_η of the generic point, we have the quaternion algebra $Q_\eta := \mathscr{O}_{U_\eta}\{X,Y\}/(XY+YX, X^2-D, Y^2-F_{30}/F_{15}^2)$.

Over $U_\eta \cap U_x$, there is the isomorphism $\iota_{\eta,x}\colon Q_\eta|_{U_\eta\cap U_x} \to Q_x|_{U_\eta\cap U_x}$, given by

$$X \mapsto X, \quad Y \mapsto (a_x + b_x X)Y_x\,.$$

For two points $x, y \in S$, the isomorphism $\iota_{\eta,y}\circ\iota_{\eta,x}^{-1}\colon Q_x|_{U_\eta\cap U_x\cap U_y} \to Q_y|_{U_\eta\cap U_x\cap U_y}$ extends to $U_x \cap U_y$.

Hence, the quaternion algebras Q_x may be glued together along these isomorphisms. This yields a quaternion algebra $\mathscr{Q}$ over S.

6.4. Corollary. — *Let $\pi\colon S \to \mathrm{Spec}\,\mathbb{Q}$ be a non-singular cubic surface with a Galois invariant double-six $\mathscr{D}$. Further, let p be a prime number and $\underline{c} \in \mathrm{Br}(S)$ a representative of the class $\mathrm{cl}(\mathscr{D}) \in \mathrm{Br}(S)/\pi^*\mathrm{Br}(\mathbb{Q})$.*

Then, the local evaluation map $\mathrm{ev}_p(\underline{c},\,.\,)\colon S(\mathbb{Q}_p) \to \mathbb{Q}/\mathbb{Z}$ is given as follows.

i) *Let $x \in S(\mathbb{Q}_p)$. Choose a rational function f_x such that $\mathrm{div}(Nf_x) = \mathrm{div}(F_{30}/F_{15}^2)$. Then,*

$$\mathrm{ev}_p(\underline{c},x) = \begin{cases} 0 & \text{if } \frac{F_{30}}{F_{15}^2 Nf_x}(x) \in \mathbb{Q}_p^* \text{ is in the image of } N\colon \mathbb{Q}_p(\sqrt{D}) \longrightarrow \mathbb{Q}_p\,, \\ \frac{1}{2} & \text{otherwise.} \end{cases}$$

Here, $F_{30}/F_{15}^2 \in \mathbb{Q}(S)$ is the rational function corresponding to the representative $\underline{c}$. $\mathbb{Q}(\sqrt{D})$ is the quadratic field splitting the double-six into two sixers.

ii) *If x is not contained in any of the 27 lines then $f_x \equiv 1$ is allowed.*

Proof. Assertion i) immediately follows from the above. For ii), recall that $\mathrm{div}(F_{30}/F_{15}^2)$ is a linear combination of the 27 lines. □

6.5. — As already noticed in Remark 2.6, the local evaluation map carries information only at finitely many primes. To exclude a particular prime, the following elementary criteria are highly practical.

6.6. Lemma (The local H^1-criterion). — *Let S be a non-singular cubic surface over $\mathbb{Q}$. Suppose that, for the decomposition group $G_p \cong \mathrm{Gal}(\overline{\mathbb{Q}}_p/\mathbb{Q}_p)$ at a prime number p,*

$$H^1(G_p, \mathrm{Pic}(S_{\overline{\mathbb{Q}}})) = 0 .$$

Then, for every $\alpha \in \mathrm{Br}(S)$, the value of $\mathrm{ev}_p(\alpha, x)$ is independent of $x \in S(\mathbb{Q}_p)$.

Proof. The local evaluation map ev_p factors via $\mathrm{Br}(S \times_{\mathrm{Spec}\,\mathbb{Q}} \mathrm{Spec}\,\mathbb{Q}_p)$.
By [Ma, Lemma 43.1.1], we have that

$$\mathrm{Br}(S \times_{\mathrm{Spec}\,\mathbb{Q}} \mathrm{Spec}\,\mathbb{Q}_p)/\mathrm{Br}(\mathbb{Q}_p) \cong H^1(\mathrm{Gal}(\overline{\mathbb{Q}}_p/\mathbb{Q}_p), \mathrm{Pic}(S_{\overline{\mathbb{Q}}_p})) .$$

Together with the assumption, this yields $\mathrm{Br}(S \times_{\mathrm{Spec}\,\mathbb{Q}} \mathrm{Spec}\,\mathbb{Q}_p) = \mathrm{Br}(\mathbb{Q}_p) = \mathbb{Q}/\mathbb{Z}$. The assertion follows. □

6.7. Remark — Recall from Remark 5.4 that we have $H^1(G, \mathrm{Pic}(S_{\overline{\mathbb{Q}}})) = 0$ for 257 of the 350 possible conjugacy classes of subgroups.

6.8. Corollary. — *Let $\pi\colon S \to \mathrm{Spec}\,\mathbb{Q}$ be a non-singular cubic surface with a Galois invariant double-six $\mathscr{D}$. Further, let $\underline{c} \in \mathrm{Br}(S)$ be a representative of the class $\mathrm{cl}(\mathscr{D}) \in \mathrm{Br}(S)/\pi^*\mathrm{Br}(\mathbb{Q})$. If a prime number p splits in the quadratic number field $\mathbb{Q}(\sqrt{D})$ splitting the two sixers then the value of $\mathrm{ev}_p(\underline{c}, x)$ is independent of $x \in S(\mathbb{Q}_p)$.*

Proof. This criterion is, of course, an immediate consequence of Corollary 6.4. In view of Fact 3.6, it is also a particular case of the local H^1-criterion. □

6.9. Proposition. — *Let $\pi\colon S \to \mathrm{Spec}\,\mathbb{Q}$ be a non-singular cubic surface with a Galois invariant double-six $\mathscr{D}$. Further, let $\underline{c} \in \mathrm{Br}(S)$ be a representative of the class $\mathrm{cl}(\mathscr{D}) \in \mathrm{Br}(S)/\pi^*\mathrm{Br}(\mathbb{Q})$. Then, for a prime number p such that*

- *the field extension $\mathbb{Q}(\sqrt{D})/\mathbb{Q}$ splitting the double-six is unramified at p,*
- *the reduction S_p is geometrically irreducible and no $\mathbb{Q}_p$-rational point on S reduces to a singularity of S_p,*

the value of $\mathrm{ev}_p(\underline{c}, x)$ is independent of $x \in S(\mathbb{Q}_p)$.

Proof. If p splits in the quadratic extension $\mathbb{Q}(\sqrt{D})$ then the assertion is true, trivially. Thus, we may assume that p remains prime in $\mathbb{Q}(\sqrt{D})$. The requirement that $z \in \mathbb{Q}_p^*$ is a norm from $\mathbb{Q}_p(\sqrt{D})$ then simply means that $\nu_p(z)$ is even.

Further, we may restrict our considerations to points $x \in S(\mathbb{Q}_p)$ which are not contained in any of the 27 lines on S. Indeed, the local evaluation map is p-adically continuous and the complement of the 27 lines is dense in $S(\mathbb{Q}_p)$ according to Hensel's lemma. In particular, we may work with F_{30}/F_{15}^2 itself.

By assumption, we have a model $\mathscr{S}$ of S over $\mathbb{Z}_p$ such that the special fiber of $\mathscr{S} \times_{\mathrm{Spec}\,\mathbb{Z}_p} \mathrm{Spec}\,\mathscr{O}_{\mathbb{Q}_p(\sqrt{D})}$ is irreducible. We delete its singularities to obtain a model $\underline{\mathscr{S}}$, smooth over $\mathscr{O}_{\mathbb{Q}_p(\sqrt{D})}$. According to the last assumption, every $x \in S(\mathbb{Q}_p)$ determines a unique extension $\underline{x} \in \mathscr{L}(\mathscr{O}_{\mathbb{Q}_p(\sqrt{D})})$.

It will suffice to construct a Zariski neighbourhood of $\underline{x}$ such that $\mathrm{ev}_p(\underline{c}, .)$ is constant. We have, on the geometric generic fiber,

$$\mathrm{div}(F_{30}/F_{15}^2) = 5E + 5G - 4F\,.$$

Here, the divisors $E := E_1 + \ldots + E_6$, $G := G_1 + \ldots + G_6$, and $F := F_{12} + \ldots + F_{56}$ are $\mathrm{Gal}(\overline{\mathbb{Q}}/\mathbb{Q}(\sqrt{D}))$-invariant, and, therefore, defined over $S \times_{\mathrm{Spec}\,\mathbb{Q}_p} \mathrm{Spec}\,\mathbb{Q}_p(\sqrt{D})$. $\mathscr{L}$ is a regular model of that variety. Hence, every divisor on $\mathscr{L}$ is locally principal. This yields, in a Zariski neighbourhood $\mathscr{U}_{\underline{x}}$,

$$F_{30}/F_{15}^2 = Cp^k e^5 g^5/f^4$$

for e, g, and f rational functions corresponding to the divisors E, G, and F, respectively, $k \in \mathbb{Z}$, and a certain $C \in \Gamma(\mathscr{U}_{\underline{x}}, \mathscr{O}^*_{\mathscr{U}_{\underline{x}}})$. Note that we get by with one power of p since the special fiber is irreducible.

The scheme $\mathscr{L}$ is acted upon by the conjugation $\sigma \in \mathrm{Gal}(\mathbb{Q}(\sqrt{D})/\mathbb{Q})$. Restricting to an open subscheme, if necessary, we may assume that $\mathscr{U}_{\underline{x}}$ is invariant under σ. The two sixers are interchanged by σ. Consequently,

$$\sigma(e) = cp^l g, \quad \sigma(f) = c'p^{l'} f$$

for $l, l' \in \mathbb{Z}$ and regular functions c, c', invertible on $\mathscr{U}_{\underline{x}}$. This yields

$$F_{30}/F_{15}^2 = Cc'^2 C^{-5} p^{k+2l'-5l} N(e^5/f^2)\,.$$

For $x_0 \in S(\mathbb{Q}_p)$ specializing to $\mathscr{U}_{\underline{x}}$, we therefore see

$$\nu_p\big((F_{30}/F_{15}^2)(x_0)\big) \equiv k + l \pmod 2\,.$$

In particular, the local evaluation map $\mathrm{ev}_p(\underline{c}, .)\colon S(\mathbb{Q}_p) \to \mathbb{Q}/\mathbb{Z}$ is constant on the set of all points specializing to $\mathscr{U}_{\underline{x}}$. As the point $x \in S(\mathbb{Q}_p)$ defining the open subset $\mathscr{U}_{\underline{x}}$ is arbitrary and the special fiber $\mathscr{S}_p$ is irreducible, this implies the assertion. □

6.10. Remarks — a) Assuming resolution of singularities in unequal characteristic, there is a proper model $\mathscr{S}$ of S being a regular scheme. Then, for p a prime unramified in $\mathbb{Q}(\sqrt{D})$, the evaluation $\mathrm{ev}_p(\underline{c}, x)$ depends only on the component of $\mathscr{S} \times_{\mathrm{Spec}\,\mathbb{Z}} \mathrm{Spec}\,\mathbb{F}_{p^2}$, the point x specializes to. If p is ramified and $p \neq 2$ then we have at least that $\mathrm{ev}_p(\underline{c}, x)$ is determined by the reduction of x modulo p.

b) As one might expect from the proof given, Proposition 6.9 is true in more generality. The reader might consult [B, Theorem 1].

7. Explicit Galois Descent

7.1. — Recall that in [EJ2], we described a method to construct non-singular cubic surfaces over $\mathbb{Q}$ with a Galois invariant double-six. The idea was to start with cubic surfaces in hexahedral form. For these, we developed an explicit version of Galois descent.

7.2. — More concretely, given a starting polynomial $f \in \mathbb{Q}[T]$ of degree six without multiple zeroes, we construct a cubic surface $S_{(a_0,\ldots,a_5)}$ over $\mathbb{Q}$ such that

$$S_{(a_0,\ldots,a_5)} \times_{\operatorname{Spec}\mathbb{Q}} \operatorname{Spec}\overline{\mathbb{Q}}$$

is isomorphic to the surface $S^{(a_0,\ldots,a_5)}$ in $\mathbf{P}^5$ given by

$$\begin{aligned} X_0^3 + X_1^3 + X_2^3 + X_3^3 + X_4^3 + X_5^3 &= 0\,, \\ X_0 + X_1 + X_2 + X_3 + X_4 + X_5 &= 0\,, \\ a_0X_0 + a_1X_1 + a_2X_2 + a_3X_3 + a_4X_4 + a_5X_5 &= 0\,. \end{aligned}$$

Here, $a_0,\ldots,a_5 \in \overline{\mathbb{Q}}$ are the zeroes of f.

The operation of an element $\sigma \in \operatorname{Gal}(\overline{\mathbb{Q}}/\mathbb{Q})$ on $S_{(a_0,\ldots,a_5)} \times_{\operatorname{Spec}\mathbb{Q}} \operatorname{Spec}\overline{\mathbb{Q}}$ goes over into the automorphism $\pi_\sigma \circ t_\sigma \colon S^{(a_0,\ldots,a_5)} \to S^{(a_0,\ldots,a_5)}$. Here, π_σ permutes the coordinates according to the rule $a_{\pi_\sigma(i)} = \sigma(a_i)$ while t_σ is the naive operation of σ on $S_{(a_0,\ldots,a_5)}$ as a morphism of schemes twisted by σ.

7.3. Remarks — i) More details on the theory are given in [EJ2, Theorem 6.6].

ii) On $S_{(a_0,\ldots,a_5)}$, there are the 15 obvious lines given by

$$X_{i_0} + X_{i_1} = X_{i_2} + X_{i_3} = X_{i_4} + X_{i_5} = 0$$

for $\{i_0,i_1,i_2,i_3,i_4,i_5\} = \{0,1,2,3,4,5\}$. They clearly form a Galois invariant set. The complement is a double-six. Correspondingly, there are the 15 obvious tritangent planes given by $X_i + X_j = 0$ for $i \neq j$.

There are formulas for the 30 non-obvious tritangent planes, too [EJ2, Proposition 7.1.ii)]. What is important is that they are defined over $\mathbb{Q}(a_0,\ldots,a_5,\sqrt{d_4})$ for

$$d_4(a_0,\ldots,a_5) := \sigma_2^2 - 4\sigma_4 + \sigma_1(2\sigma_3 - \tfrac{3}{2}\sigma_1\sigma_2 + \tfrac{5}{16}\sigma_1^3)$$

the Coble quartic. Here, σ_i is the i-th elementary symmetric function in $a_0,\ldots,a_5$.

Further, an element $\sigma \in \operatorname{Gal}(\overline{\mathbb{Q}}/\mathbb{Q})$ flips the double-six if and only if it defines the conjugation of $\mathbb{Q}(\sqrt{D})$ for $D := d_4 {\cdot} \Delta$, the second factor denoting the discriminant of $a_0,\ldots,a_5$ [EJ2, Proposition 7.4].

iii) The smooth manifold $S(\mathbb{R})$ has two connected components if and only if exactly four of the $a_0,\ldots,a_5$ are real and $d_4(a_0,\ldots,a_5) > 0$. Otherwise, $S(\mathbb{R})$ is connected [EJ2, Corollary 8.4].

iv) The descent variety $S_{(a_0,\ldots,a_5)}$ may easily be computed completely explicitly. In fact, [EJ2, Algorithm 6.7] yields a quaternary cubic form with 20 rational coefficients.

7.4. — Using the criteria provided in section 6, we have the following strategy to compute the Brauer-Manin obstruction on $S_{(a_0,\ldots,a_5)}$.

Strategy (to explicitly compute the Brauer-Manin obstruction on $S_{(a_0,\dots,a_5)}$).

i) Compute $D := d_4(a_0,\dots,a_5)\cdot\Delta(a_0,\dots,a_5)$. Determine the list L_1 of all primes at which $\mathbb{Q}(\sqrt{D})$ is ramified.

ii) By a Gröbner basis calculation, determine all the primes outside L_1 at which $S_{(a_0,\dots,a_5)}$ has bad reduction. Write them into a list L_2.

iii) From L_2, delete all primes which split in $\mathbb{Q}(\sqrt{D})$. Further, erase all those primes from L_2 for which the singular points on the reduction modulo p are not defined over $\mathbb{F}_p$ or do not lift to $S_{(a_0,\dots,a_5)} \times_{\operatorname{Spec}\mathbb{Z}} \operatorname{Spec}\mathbb{Z}/p^k\mathbb{Z}$ for k large.

iv) Put $L := L_1 \cup L_2$. If $D < 0$ and $S_{(a_0,\dots,a_5)}(\mathbb{R})$ is not connected then the infinite place has to be added to this list of critical primes.

v) Delete all the primes from L for which the local H^1-criterion works successfully.

vi) Scale the form F_{30} by a constant factor such that the local evaluation maps are zero for all the primes outside L.

vii) For the primes p that remained in L, the form F_{30} has to be evaluated. For that, cover $S(\mathbb{Q}_p)$ by finitely many open subsets which are sufficiently small to ensure that the first p-adic digit of F_{30} does not change. If $p = 2$ then the first three digits have to be taken into account.

In the case that we have a Galois invariant triple of azygetic double-sixes, the last step has to be executed three times, once for each of the corresponding forms of type F_{30}.

8. Application: Manin's Conjecture

8.1. — Recall that a conjecture, due to Yu. I. Manin, asserts that the number of $\mathbb{Q}$-rational points of anticanonical height $\leq B$ on a Fano variety S is asymptotically equal to $\tau B \log^{\operatorname{rk}\operatorname{Pic}(S)-1} B$, for $B \to \infty$. Further, the coefficient $\tau \in \mathbb{R}$ is conjectured to be the Tamagawa-type number $\tau(S)$ introduced by E. Peyre in [Pe]. In the particular case of a cubic surface, the anticanonical height is the same as the naive height.

8.2. — E. Peyre's Tamagawa-type number is defined in [PT, Definition 2.4] as

$$\tau(S) := \alpha(S)\cdot\beta(S)\cdot\lim_{s\to 1}(s-1)^t L(s,\chi_{\operatorname{Pic}(S_{\overline{\mathbb{Q}}})})\cdot\tau_H\big(S(\mathbb{A}_\mathbb{Q})^{\operatorname{Br}}\big)$$

for $t = \operatorname{rk}\operatorname{Pic}(S)$.

Here, the factor $\beta(S)$ is simply defined as $\beta(S) := \#H^1\big(\operatorname{Gal}(\overline{\mathbb{Q}}/\mathbb{Q}),\operatorname{Pic}(S_{\overline{\mathbb{Q}}})\big)$. $\alpha(S)$ is given as follows [Pe, Définition 2.4]. Let $\Lambda_{\text{eff}}(S) \subset \operatorname{Pic}(S)\otimes_\mathbb{Z}\mathbb{R}$ be the cone generated by the effective divisors. Identify $\operatorname{Pic}(S)\otimes_\mathbb{Z}\mathbb{R}$ with $\mathbb{R}^t$ via a mapping induced by an isomorphism $\operatorname{Pic}(S) \xrightarrow{\cong} \mathbb{Z}^t$. Consider the dual cone $\Lambda^\vee_{\text{eff}}(S) \subset (\mathbb{R}^t)^\vee$. Then, $\alpha(S) := t\cdot\operatorname{vol}\{x \in \Lambda^\vee_{\text{eff}} \mid \langle x, -K\rangle \leq 1\}$.

$L(\,\cdot\,,\chi_{\operatorname{Pic}(S_{\overline{\mathbb{Q}}})})$ denotes the Artin L-function of the $\operatorname{Gal}(\overline{\mathbb{Q}}/\mathbb{Q})$-representation $\operatorname{Pic}(S_{\overline{\mathbb{Q}}})\otimes_\mathbb{Z}\mathbb{C}$ which contains the trivial representation t times as a direct summand. Therefore, $L(s,\chi_{\operatorname{Pic}(S_{\overline{\mathbb{Q}}})}) = \zeta(s)^t\cdot L(s,\chi_P)$ and

$$\lim_{s\to 1}(s-1)^t L(s,\chi_{\operatorname{Pic}(S_{\overline{\mathbb{Q}}})}) = L(1,\chi_P)$$

where ζ denotes the Riemann zeta function and P is a representation which does not contain trivial components. [Mu, Corollary 11.5 and Corollary 11.4] show that $L(s,\chi_P)$ has neither a pole nor a zero at $s=1$.

Finally, τ_H is the *Tamagawa measure* on the set $S(\mathbb{A}_\mathbb{Q})$ of adelic points on S and $S(\mathbb{A}_\mathbb{Q})^{\mathrm{Br}} \subseteq S(\mathbb{A}_\mathbb{Q})$ denotes the part which is not affected by the Brauer-Manin obstruction.

8.3. — As S is projective, we have $S(\mathbb{A}_\mathbb{Q}) = \prod_{\nu\in\mathrm{Val}(\mathbb{Q})} S(\mathbb{Q}_\nu)$. Then, the Tamagawa measure τ_H is defined to be the product measure $\tau_H := \prod_{\nu\in\mathrm{Val}(\mathbb{Q})} \tau_\nu$.

Here, for a prime number p, the local measure τ_p on $S(\mathbb{Q}_p)$ is given as follows. Let $a \in S(\mathbb{Z}/p^k\mathbb{Z})$ and put $\mathfrak{U}_a^{(k)} := \{x \in S(\mathbb{Q}_p) \mid x \equiv a \pmod{p^k}\}$.Then,

$$\tau_p(\mathfrak{U}_a^{(k)}) := \det(1-p^{-1}\mathrm{Frob}_p \mid \mathrm{Pic}(S_{\overline{\mathbb{Q}}})^{I_p}) \cdot \lim_{m\to\infty} \frac{\#\{y \in S(\mathbb{Z}/p^m\mathbb{Z}) \mid y \equiv a \pmod{p^k}\}}{p^{m\dim S}}.$$

$\mathrm{Pic}(S_{\overline{\mathbb{Q}}})^{I_p}$ denotes the fixed module under the inertia group.

τ_∞ is described in [Pe, Lemme 5.4.7]. In the case of a cubic surface, defined by the equation $f=0$, this yields

$$\tau_\infty(U) = \frac{1}{2} \int\limits_{\substack{CU \\ |x_0|,\dots,|x_3|\leq 1}} \omega_{\mathrm{Leray}}$$

for $U \subset S(\mathbb{R})$. Here, ω_{Leray} is the *Leray measure* on the cone $CS(\mathbb{R})$. It is related to the usual hypersurface measure by the formula $\omega_{\mathrm{Leray}} = \frac{1}{\|\mathrm{grad} f\|}\,\omega_{\mathrm{hyp}}$.

8.4. — Using [EJ2, Algorithm 6.7], we constructed many examples of smooth cubic surfaces over $\mathbb{Q}$ with a Galois invariant double-six. For each of them, one may apply Strategy 7.4 to compute the effect of the Brauer-Manin obstruction. Then, the method described in [EJ1] applies for the computation of Peyre's constant.

From the ample supply, the examples below were chosen in the hope that they indicate the main phenomena. The Brauer-Manin obstruction may work at many primes simultaneously but examples where few primes are involved are the most interesting. We show that the fraction of the Tamagawa measure excluded by the obstruction can vary greatly. We also show that there may be an obstruction at the infinite prime.

8.5. Example — The polynomial

$$f := T^6 - 390T^4 - 10180T^3 + 10800T^2 + 2164296T + 13361180 \in \mathbb{Q}[T]$$

yields the cubic surface S given by the equation

$$-x^2z - x^2w - 3xy^2 + xz^2 + 14xzw + 8xw^2 - 2y^3 - 11y^2z$$
$$+ y^2w + 4yz^2 + 4yzw + 10yw^2 + 4z^3 - 11z^2w + 9zw^2 - 6w^3 = 0.$$

S has bad reduction at 2, 3, 5, 11, and 9265613761. The Galois group operating on the 27 lines is S_6 acting in such a way that the orbit structure is $[12,15]$. Therefore, we have $H^1(\mathrm{Gal}(\overline{\mathbb{Q}}/\mathbb{Q}), \mathrm{Pic}(S_{\overline{\mathbb{Q}}})) \cong \mathbb{Z}/2\mathbb{Z}$. The quadratic field splitting the double-six is $\mathbb{Q}(\sqrt{10})$.

The primes 3 and 9265613761 split in $\mathbb{Q}(\sqrt{10})$. The local H^1-criterion excludes the prime 5. Further, it turns out that the local evaluation map at 11 is constant. Hence, the

Brauer-Manin obstruction works only at the prime 2. From the whole of $S(\mathbb{Q}_2)$ which is of measure 4 only a subset of measure $\frac{9}{4}$ is allowed.

Using this, for Peyre's constant, we find $\tau(S) \approx 1.7005$. There are actually 6641 $\mathbb{Q}$-rational points of height at most 4000 in comparison with a prediction of 6802.

8.6. Example — The polynomial

$$f := T^6 + 60T^4 - 40T^3 - 900T^2 + 15\,072T - 27\,860 \in \mathbb{Q}[T]$$

yields the cubic surface S given by the equation

$$5x^3 - 9x^2y + x^2z + 6x^2w + 3xy^2 + xyz + 6xyw - 2xz^2$$
$$- 4xzw - y^3 - 3y^2z + 2yz^2 + 2yzw + 4z^3 + 2z^2w + 2zw^2 = 0\,.$$

S has bad reduction at 2, 3, 5, and 73. The Galois group operating on the 27 lines is $A_6 \times \mathbb{Z}/2\mathbb{Z}$ and the orbit structure is $[12, 15]$. The quadratic field $\mathbb{Q}(\sqrt{2})$ splits the double-six.

The prime 73 splits in $\mathbb{Q}(\sqrt{2})$. Further, the local H^1-criterion excludes the prime 5. Hence, the Brauer-Manin obstruction works only at the primes 2 and 3. At 2, the local evaluation map decomposes $S(\mathbb{Q}_2)$ into two sets of measures 1 and $\frac{1}{2}$, respectively. At 3, the corresponding measures are $\frac{70}{81}$ and $\frac{28}{81}$. An easy calculation shows $\tau_H(S(\mathbb{A}_\mathbb{Q})^{\mathrm{Br}}) = \frac{4}{7}\tau_H(S(\mathbb{A}_\mathbb{Q}))$.

Using this, for Peyre's constant, we find $\tau(S) \approx 5.0879$. Up to a search bound of 4000, there are actually 19 302 $\mathbb{Q}$-rational points in comparison with a prediction of 20 352.

8.7. Example — The polynomial $f := T(T^5 - 5T - 2) \in \mathbb{Q}[T]$ yields the cubic surface S given by the equation

$$2x^3 + x^2y - 4x^2z - x^2w + 2xy^2 + 2xyz + 2xyw - 2xz^2 - 4xzw$$
$$- 2xw^2 + 2y^2z - y^2w + yz^2 + 2yzw - 5yw^2 - 3z^2w + 6zw^2 + 9w^3 = 0\,.$$

S has bad reduction at 2, 3, and 5. Further, $S(\mathbb{R})$ consists of two connected components. The Galois group operating on the 27 lines is isomorphic to $S_5 \times \mathbb{Z}/2\mathbb{Z}$ and the orbit structure is $[12, 15]$. $\mathbb{Q}(\sqrt{-15})$ is the field splitting the double-six.

The prime 2 splits in $\mathbb{Q}(\sqrt{-15})$. Further, the local H^1-criterion excludes the prime 5. Hence, the Brauer-Manin obstruction works only at 3 and the infinite prime. At 3, the local evaluation map decomposes $S(\mathbb{Q}_2)$ into two sets of measures $\frac{2}{3}$ and $\frac{4}{9}$, respectively. At the infinite prime, the corresponding measures are approximately 1.9179 and 1.1673. An easy calculation shows $\tau_H(S(\mathbb{A}_\mathbb{Q})^{\mathrm{Br}}) \approx 0.5243 \cdot \tau_H(S(\mathbb{A}_\mathbb{Q}))$.

Using this, for Peyre's constant, we find $\tau(S) \approx 3.7217$. Up to a search bound of 4000, there are actually 14 249 $\mathbb{Q}$-rational points in comparison with a prediction of 14 887.

8.8. Example — The polynomial

$$f := T(T^5 - 60T^3 - 90T^2 + 675T + 810) \in \mathbb{Q}[T]$$

yields the cubic surface S given by the equation

$$3x^3 + 2x^2z + xy^2 - 2xyz - 2xyw - xzw + 2xw^2 - yzw - yw^2 - z^3 + z^2w = 0\,.$$

S has bad reduction at 2, 3, and 5. The Galois group operating on the 27 lines is isomorphic to S_5 and the orbit structure is $[12,15]$. The quadratic field $\mathbb{Q}(\sqrt{-3})$ splits the double-six.

The local H^1-criterion excludes the prime 5. Further, the local evaluation maps turn out to be constant on $S(\mathbb{Q}_2)$ and $S(\mathbb{R})$. At the real prime, the reason is simply that $S(\mathbb{R})$ is connected. Consequently, the Brauer-Manin obstruction works only at the prime 3. From the whole of $S(\mathbb{Q}_3)$ measuring $\frac{2}{3}$ a subset of measure $\frac{4}{9}$ is allowed.

Using this, for Peyre's constant, we find $\tau(S) \approx 2.2647$. Up to a search bound of 4000, there are actually 8886 $\mathbb{Q}$-rational points in comparison with a prediction of 9059.

8.9. Example — The polynomial $f := T(T^5 + 20T + 16) \in \mathbb{Q}[T]$ yields the cubic surface S given by the equation

$$-3x^3 - 7x^2y - 4x^2z + 5x^2w + 4xy^2 + 10xyz - 4xyw - 2xz^2 \\ + 2xzw + xw^2 - 4y^2z + yz^2 - 4yzw - 16yw^2 + z^2w - 5zw^2 = 0.$$

S has bad reduction at 2 and 5. The Galois group operating on the 27 lines is isomorphic to $A_5 \times \mathbb{Z}/2\mathbb{Z}$. The orbit structure is $[12,15]$. The quadratic field $\mathbb{Q}(\sqrt{-5})$ splits the double-six.

The local H^1-criterion excludes the prime 5. At the infinite prime, the local evaluation map is constant since $S(\mathbb{R})$ is connected. Hence, the Brauer-Manin obstruction works only at the prime 2. It allows a subset of measure $\frac{17}{16}$ out of $S(\mathbb{Q}_2)$ measuring $\frac{5}{4}$.

Using this, for Peyre's constant, we find $\tau(S) \approx 2.4545$. Up to a search bound of 4000, there are actually 9736 $\mathbb{Q}$-rational points in comparison with a prediction of 9818.

8.10. Example — The polynomial

$$f := T^6 - 456T^4 - 904T^3 + 102\,609T^2 + 1\,041\,060T + 2\,935\,300 \in \mathbb{Q}[T]$$

yields the cubic surface S given by the equation

$$-2x^3 + 3x^2z + 9x^2w - 4xy^2 - 8xyz - 10xzw + 4xw^2 - 4y^3 - 3y^2z \\ - 4y^2w - 2yz^2 - 2yzw + 8yw^2 - z^3 + z^2w - 6zw^2 - 2w^3 = 0.$$

S has bad reduction at 2, 3, 5, 31, and 11 071. The Galois group operating on the 27 lines is isomorphic to $(S_3 \times S_3) \ltimes \mathbb{Z}/2\mathbb{Z}$ of order 72. The orbit structure is $[6,6,6,9]$. There is a triple of Galois invariant double-sixes. Therefore, we have that $H^1(\mathrm{Gal}(\overline{\mathbb{Q}}/\mathbb{Q}), \mathrm{Pic}(S_{\overline{\mathbb{Q}}})) \cong$ $\cong \mathbb{Z}/2\mathbb{Z} \times \mathbb{Z}/2\mathbb{Z}$. The quadratic field splitting the double-sixes is $\mathbb{Q}(\sqrt{2})$.

The primes 31 and 11 071 split in $\mathbb{Q}(\sqrt{2})$. Further, the local evaluation maps turn out to be constant on $S(\mathbb{Q}_3)$. Consequently, the Brauer-Manin obstruction works only at the primes 2 and 5.

The local evaluation maps decompose $S(\mathbb{Q}_2)$ into four sets of measures $\frac{7}{16}$, $\frac{5}{16}$, $\frac{1}{4}$, and $\frac{1}{4}$, respectively. At the prime 5, the corresponding measures are $\frac{516}{625}$, 0, $\frac{96}{625}$, and 0. Observe, for one of the three non-zero Brauer classes, the local evaluation map is constant on $S(\mathbb{Q}_5)$.

A simple calculation shows that $\tau_H(S(\mathbb{A}_\mathbb{Q})^{\mathrm{Br}}) = \frac{111}{340}\tau_H(S(\mathbb{A}_\mathbb{Q}))$. Using this, for Peyre's constant, we find $\tau(S) \approx 1.8532$. Up to a search bound of 4000, there are actually 6994 $\mathbb{Q}$-rational points in comparison with a prediction of 7413.

References

[B] M. Bright, Efficient evaluation of the Brauer-Manin obstruction, *Math. Proc. Cambridge Philos. Soc.* 142 (2007), 13–23

[CE] H. Cartan and S. Eilenberg, Homological Algebra, *Princeton Univ. Press, Princeton.* 1956

[CG] J. W. S. Cassels and M. J. T. Guy, On the Hasse principle for cubic surfaces, *Mathematika* 13 (1966), 111–120

[CKS] J.-L. Colliot-Thélène, D. Kanevsky, and J.-J. Sansuc, Arithmétique des surfaces cubiques diagonales, in: Diophantine approximation and transcendence theory (Bonn 1985), *Lect. Notes in Math.,* Vol. 1290, Springer, Berlin 1987, 1–108

[Co] P.K. Corn, Del Pezzo surfaces and the Brauer-Manin obstruction, *Ph.D. thesis, Harvard* 2005

[EJ1] A. S. Elsenhans and J. Jahnel, Experiments with general cubic surfaces, in: Algebra, Arithmetic, and Geometry (in honor of Yu. I. Manin), Y. Tschinkel and Y. Zarhin (eds.), *Volume I, Progress in Mathematics* 269, Birkhäuser, Boston 2007, 637–654

[EJ2] A. S. Elsenhans and J. Jahnel, Cubic surfaces with a Galois invariant double-six, *Central European J. Math.* 8 (2010), 646–661

[Ha] R. Hartshorne, Algebraic Geometry, *Graduate Texts in Mathematics* 52, Springer, New York 1977

[Ko] G. Kohn, Über einige Eigenschaften der allgemeinen Fläche dritter Ordnung, *Sitzungsberichte der Kaiserlichen Akademie der Wissenschaften Wien* 117 (1908), 53–73

[Ma] Yu. I. Manin, *Cubic forms, algebra, geometry, arithmetic,* North-Holland Publishing Co. and American Elsevier Publishing Co., Amsterdam, London, and New York 1974

[Mo] L. J. Mordell, On the conjecture for the rational points on a cubic surface, *J. London Math. Soc.* 40 (1965), 149–158

[Mu] M. R. Murty, Applications of symmetric power L-functions, in: Lectures on automorphic L-functions, Fields Inst. Monogr. 20, *Amer. Math. Soc.,* Providence 2004, 203–283

[Pe] E. Peyre: Hauteurs et mesures de Tamagawa sur les variétés de Fano, *Duke Math. J.* 79 (1995), 101–218

[PT] E. Peyre and Y. Tschinkel, Tamagawa numbers of diagonal cubic surfaces, numerical evidence, *Math. Comp.* 70 (2001) 367–387

[SD1] S. P. Swinnerton-Dyer, Two special cubic surfaces, *Mathematika* 9 (1962) 54–56

[SD2] S. P. Swinnerton-Dyer, The Brauer group of cubic surfaces, *Math. Proc. Cambridge Philos. Soc.* 113 (1993) 449–460

[Ta] J. Tate, Global class field theory, in: Algebraic number theory, Edited by J. W. S. Cassels and A. Fröhlich, *Academic Press and Thompson Book Co.*, London and Washington (1967)

In: Frontiers of Combinatorics and Number Theory, Volume 2 ISBN: 978-1-61942-918-5
Editor: Zhi-Wei Sun, pp. 135-166

Chapter 9

CONGRUENCES OF ALTERNATING MULTIPLE HARMONIC SUMS

Roberto Tauraso[1,*] ***and Jianqiang Zhao*** [2,†]
[1]Dipartimento di Matematica,
Università di Roma "Tor Vergata",
Roma, Italy
[2]Department of Mathematics, Eckerd College,
St. Petersburg, FL, US
Max-Planck Institut für Mathematik,
Vivatsgasse, Bonn, Germany

Abstract

In this sequel to [16], we continue to study the congruence properties of the alternating version of multiple harmonic sums. As contrast to the study of multiple harmonic sums where Bernoulli numbers and Bernoulli polynomials play the key roles, in the alternating setting the Euler numbers and the Euler polynomials are also essential.

2010 Mathematics Subject Classification: Primary 11A07, 11B50; Secondary 11M41.

1. Introduction

In proving a Van Hamme type congruence the first author was led to consider some congruences involving alternating multiple harmonic sums (AMHS for short) which are defined as follows. Let $d > 0$ and let $\mathbf{s} := (s_1, \ldots, s_d) \in (\mathbb{Z}^*)^d$. We define the *alternating multiple harmonic sum* as

$$H(\mathbf{s}; n) := \sum_{1 \le k_1 < k_2 < \cdots < k_d \le n} \prod_{i=1}^{d} \frac{\operatorname{sgn}(s_i)^{k_i}}{k_i^{|s_i|}}.$$

By convention we set $H(\mathbf{s}; n) = 0$ if $n < d$. We call $\ell(\mathbf{s}) := d$ and $|\mathbf{s}| := \sum_{i=1}^{d} |s_i|$ its *depth* and *weight*, respectively. We point out that $\ell(\mathbf{s})$ is sometimes called the length

*E-mail addresses: tauraso@mat.uniroma2.it
†E-mail addresses: zhaoj@eckerd.edu

in the literature. When every s_i is positive we recover the multiple harmonic sums (MHS for short) whose congruence properties are studied in [10, 11, 18, 19]. There is another "non-strict" version of the AMHS defined as follows:

$$S(\mathbf{s};n) := \sum_{1\le k_1\le k_2\le\cdots\le k_d\le n} \prod_{i=1}^{d} \frac{\operatorname{sgn}(s_i)^{k_i}}{k_i^{|s_i|}}.$$

By the Inclusion-Exclusion Principle it is easy to see that

$$S(\mathbf{s};n) = \sum_{\mathbf{r}\preceq\mathbf{s}} H(\mathbf{r};n), \tag{1.1}$$

$$H(\mathbf{s};n) = \sum_{\mathbf{r}\preceq\mathbf{s}} (-1)^{\ell(\mathbf{s})-\ell(\mathbf{r})} S(\mathbf{r};n), \tag{1.2}$$

where $\mathbf{r} \prec \mathbf{s}$ means $\mathbf{r}$ can be obtained from $\mathbf{s}$ by combining some of its parts.

The main goal of this paper is to provide a systematic study of the congruence property of $H(\mathbf{s};p-1)$ (and $S(\mathbf{s};p-1)$) for primes $p > |\mathbf{s}|+2$ by using intimate relations between Bernoulli polynomials, Bernoulli numbers, Euler polynomials, and Euler numbers. Throughout the paper, we often use the abbreviation $S(-) = S(-;p-1)$ and $H(-) = H(-;p-1)$ if no confusion arises. The following congruences concerning harmonic sums will be crucial for us.

Theorem 1.1. *Let $k\in\mathbb{N}$ and p be a prime. Set $X_p(k) := \frac{B_{p-k}}{p-k} - \frac{B_{2p-1-k}}{2(2p-1-k)}$.*
(a) [15, Theorem 5.1] *If $p\ge k+3$ then*

$$H(k) \equiv \begin{cases} k(k+1)B_{p-2-k}p^2/2(p-2-k) & \pmod{p^3}, \quad \text{if } k \text{ is odd;} \\ -2kX_p(k+1)p & \pmod{p^3}, \quad \text{if } k \text{ is even.} \end{cases} \tag{1.3}$$

(b) [15, Theorem 5.2] *If $p\ge k+4$ then*

$$H(k;(p-1)/2) \equiv \begin{cases} 2(2^k-2)X_p(k) & \pmod{p^2}, \quad \text{if } k>1 \text{ is odd;} \\ -k(2^{k+1}-1)X_p(k+1)p & \pmod{p^3}, \quad \text{if } k \text{ is even;} \\ -2q_p + pq_p^2 - \frac{2}{3}p^2q_p^3 - \frac{7}{12}p^2B_{p-3} & \pmod{p^3}, \quad \text{if } k=1. \end{cases} \tag{1.4}$$

Here $q_p = (2^{p-1}-1)/p$ is the Fermat quotient.

We now sketch the outline of the paper. We start section 2 by recalling some important relations among AMHS such as the stuffle and reversal relations. We then present some basic properties of Euler polynomials which provide one of the fundamental tools for us in the alternating setting. In section 2.3 we describe two reduction procedures for $H(\mathbf{s}) \pmod p$ for general $\mathbf{s}$, Theorem 2.1 and Theorem 2.2, which are used to derive congruences in depth two and depth three cases in section 3 and section 4, respectively.

In section 5 we deal with the homogeneous AMHS of arbitrary depth and provide an explicit formula using the relation between the power sum and elementary symmetric functions and the partition functions. Section 6 is devoted to a comprehensive study of the weight four AMHS in which identities involving Bernoulli numbers such as those proved in [18] play the leading roles. For example, by writing $H(-) = H(-; p-1)$ we find the following interesting relations (see Proposition 6.1, Proposition 6.2 and Proposition 6.4):

$$H(1,-3) \equiv \frac{1}{2}H(-2,2) \equiv \sum_{k=0}^{p-3} 2^k B_k B_{p-3-k} \pmod p,$$

$$H(-1,3) \equiv -\frac{1}{2}q_p B_{p-3} \pmod p,$$

$$H(1,-2,-1) \equiv H(1,-3) - \frac{5}{4}q_p B_{p-3} \pmod p,$$

$$H(-1,1,-1,1) \equiv H(1,-1,1,-1) \equiv -\frac{1}{12}\left(q_p B_{p-3} + 2q_p^4\right) \pmod p,$$

$$H(-1,1,1,-1) \equiv \frac{1}{12}\left(6H(1,-3) + 7q_p B_{p-3} + 2q_p^4\right) \pmod p,$$

for all primes $p \geq 7$. None of the above congruences can be obtained simply by the stuffle and reversal relations.

After studying some special types of AMHS of weight four in section 5, we turn to congruence relations involving lower weight AMHS modulo higher powers of primes in the last section. One of the main ideas in these sections is to relate AMHS to sums $U(\mathbf{s}; n)$ and $V(\mathbf{s}; n)$ defined by (5.9) and (5.10), respectively. These sums appeared previously in congruences involving powers of Fermat quotient (see [2, 3, 5, 8]).

Most of results of this paper were obtained while the second author was visiting the Max-Planck-Institut für Mathematik whose support is gratefully acknowledged.

2. Properties of AMHS

2.1. Stuffle Relations

The most important relation between AMHS is the so called stuffle relation. It is possible to formalize this using words as in [20, §2] or [14, §2.2] which is a generalization of the MHS case (see [10, §2]). Unfortunately, for AMHS we don't have the integral representations which provide another product structure for the alternating multiple zeta values which are the infinite sum version of AMHS.

Fix a positive integer n. Let $\mathfrak{A}$ be the algebra generated by letters y_s for $s \in \mathbb{Z}^*$. Define a multiplication $*$ on $\mathfrak{A}$ by requiring that $*$ distribute over addition, that $\mathbf{1} * w = w * \mathbf{1} = w$ for the empty word $\mathbf{1}$ and any word w, and that, for any two words w_1, w_2 and two letters y_s, y_t $(s, t \in \mathbb{Z}^*)$

$$y_s w_1 * y_t w_2 = y_s(w_1 * y_t w_2) + y_t(y_s w_1 * w_2) + y_{s\oplus t}(w_1 * w_2) \tag{2.1}$$

where $s \oplus t = \mathrm{sgn}(st)(|s| + |t|)$. Then we get an algebra homomorphism

$$\begin{aligned} H:\quad (\mathfrak{A}, *) &\longrightarrow \{H(\mathbf{s}; n) : \mathbf{s} \in \mathbb{Z}^r, r \in \mathbb{N}\} \\ \mathbf{1} &\longmapsto 1 \\ y_{s_1}\cdots y_{s_r} &\longmapsto H(s_1, \dots, s_r; n). \end{aligned}$$

For example,

$$H(-2;n)H(-3,2;n) = H(-2,-3,2;n) + H(-3,-2,2;n) + H(-3,2,-2;n) + H(5,2;n) + H(-3,-4;n).$$

There is another kind of relation caused by the reversal of the arguments which we call the *reversal relations*. For any $\mathbf{s} = (s_1, \ldots, s_r)$ they have the form

$$\begin{aligned} H(\mathbf{s};p-1) &\equiv \operatorname{sgn}\Big(\prod_{j=1}^{r} s_j\Big)(-1)^r H(\overleftarrow{\mathbf{s}};p-1) \pmod p, \\ S(\mathbf{s};p-1) &\equiv \operatorname{sgn}\Big(\prod_{j=1}^{r} s_j\Big)(-1)^r S(\overleftarrow{\mathbf{s}};p-1) \pmod p, \end{aligned} \tag{2.2}$$

for any odd prime $p > |\mathbf{s}|$, where $\overleftarrow{\mathbf{s}} = (s_r, \ldots, s_1)$ is the reversal of s.

2.2. Euler Polynomials

In the study of congruences of MHS [10, 18, 19] we have seen that Bernoulli numbers play the key roles by virtue of the following identity: ([1, p. 804, 23.1.4-7])

$$\sum_{j=1}^{n-1} j^d = \sum_{r=0}^{d} \binom{d+1}{r} \frac{B_r}{d+1} n^{d+1-r}, \quad \forall n, d \geq 1. \tag{2.3}$$

In the case of AMHS, however, the Euler polynomials and the Euler numbers are indispensable, too. Recall that the Euler polynomials $E_n(x)$ are defined by the generating function

$$\frac{2e^{tx}}{e^t+1} = \sum_{n=0}^{\infty} E_n(x) \frac{t^n}{n!}.$$

Lemma 2.1. *Let $n \in \mathbb{Z}_{\geq 0}$. Then we have*

$$\sum_{i=1}^{d-1} (-1)^i i^n = \frac{1}{2}\Big((-1)^{d-1}E_n(d) + E_n(0)\Big) = \sum_{a=0}^{n} \binom{n}{a} F_{n,d,a} d^{n-a}, \tag{2.4}$$

where

$$F_{n,d,a} = \begin{cases} (-1)^{d-1}E_a(0)/2, & \text{if } a < n; \\ (1-(-1)^d)E_n(0)/2, & \text{if } a = n > 0; \\ -(1+(-1)^d)/2, & \text{if } a = n = 0. \end{cases}$$

Moreover, $E_0(0) = 1$ and for all $a \in \mathbb{N}$

$$E_a(0) = \frac{2^{a+1}}{a+1}\Big(B_{a+1}\Big(\frac{1}{2}\Big) - B_{a+1}\Big) = \frac{2}{a+1}(1-2^{a+1})B_{a+1}. \tag{2.5}$$

Proof. Consider the generating function

$$\begin{aligned}\sum_{n=0}^{\infty}\left(\sum_{i=1}^{d-1}(-1)^i i^n\right)\frac{t^n}{n!}=&\sum_{i=1}^{d-1}(-1)^i e^{ti}\\ =&\frac{(-e^t)^d-1}{-e^t-1}-1\\ =&\frac{(-1)^{d-1}e^{dt}+1}{e^t+1}-1\\ =&\frac{1}{2}\sum_{n=0}^{\infty}\Big((-1)^{d-1}E_n(d)+E_n(0)\Big)\frac{t^n}{n!}-1. \end{aligned}\tag{2.6}$$

Now (2.4) follows from the notorious equation (see, for e.g., [1, p. 805, 23.1.7])

$$E_n(x)=\sum_{a=0}^{n}\binom{n}{a}E_a(0)x^{n-a}\tag{2.7}$$

for all $n>0$. Equation (2.5) is also well-known (see, for example, item 23.1.20 on p. 805 of loc. cit.). □

Remark 2.1. The classical Euler numbers E_k is defined by

$$\frac{2}{e^t+e^{-t}}=\sum_{k=0}^{\infty}E_k\frac{t^k}{k!}.$$

They are related to $E_k(0)$ by the formula (see [1, p. 805, 23.1.7])

$$E_m(0)=\sum_{k=0}^{m}\binom{m}{k}\frac{E_k}{2^k}\left(-\frac{1}{2}\right)^{m-k}.$$

Corollary 2.1. *Let $a\in\mathbb{Z}_{\geq 0}$ and p be a prime such that $p\geq a+2$. Then*

$$H(-a;p-1)\equiv\begin{cases}-\dfrac{2(1-2^{p-a})}{a}B_{p-a} & \pmod p, \quad \textit{if } a \textit{ is odd;}\\[2mm] \dfrac{a(1-2^{p-1-a})}{a+1}pB_{p-1-a} & \pmod{p^2}, \quad \textit{if } a \textit{ is even.}\end{cases}\tag{2.8}$$

Proof. Taking $d=p$ and $n=p(p-1)-a$ in Lemma 2.1 we see that

$$\begin{aligned}H(-a;p-1)\equiv&F_{p(p-1)-a,p,p(p-1)}+p(p(p-1)-a)F_{p(p-1)-a,p,p(p-1)-1-a}\\ \equiv&E_{p(p-1)-a}(0)-\frac{1}{2}paE_{p(p-1)-1-a}(0) \pmod{p^2},\end{aligned}$$

since all the coefficients in (2.4) are p-integral by (2.5) and the property of Bernoulli numbers: B_m is not p-integral if and only if $p-1$ divides $m>0$. Then the corollary directly follows from (2.5) and Kummer congruences

$$\begin{aligned}\frac{B_{p(p-1)-a}}{p(p-1)-a}\equiv&\frac{B_{p-1-a}}{p-1-a} \pmod p,\\ \frac{B_{p(p-1)-a+1}}{p(p-1)-a+1}\equiv&\frac{B_{p-a}}{p-a} \pmod p.\end{aligned}$$

□

Remark 2.2. (a). The corollary can also be obtained from [16, Theorem 2.1] combined with (1.4). Notice that both terms in [16, Theorem 2.1] contribute nontrivially when a is even since the modulus is p^2. (b). Notice that if a is odd we allow $p = a+2$ to be a prime modulus unlike [4, Lemma 5.1]. If a is even then the corollary is given also by [4, (6.2)].

2.3. Two Reduction Formulae

We now prove two reduction formulae of $H(\mathbf{s})$ for arbitrary composition $\mathbf{s}$, corresponding to the two cases where $\mathbf{s}$ begins with a positive or a negative number.

Theorem 2.1. *Let $a, \ell \in \mathbb{N}$, $\mathbf{s} = (s_1, \dots, s_\ell) \in (\mathbb{Z}^*)^\ell$ and $\mathbf{s}' = (s_2, \dots, s_\ell)$. For any prime $p \ge a+2$ write $H(-) = H(-; p-1)$. Then*

$$H(a,\mathbf{s}) \equiv -\frac{1}{a} H\big((a-1)\oplus s_1, \mathbf{s}'\big) + \sum_{k=1}^{p-1-a} \binom{p-a}{k} \frac{B_k}{p-a} H\big((k+a-1)\oplus s_1, \mathbf{s}'\big) \pmod p. \tag{2.9}$$

Proof. By definition

$$H(a,\mathbf{s}) \equiv \sum_{j_\ell=1}^{p-1} \frac{\operatorname{sgn}(s_\ell)^{j_\ell}}{j_\ell^{|s_\ell|}} \sum_{j_{\ell-1}=1}^{j_\ell-1} \frac{\operatorname{sgn}(s_{\ell-1})^{j_{\ell-1}}}{j_{\ell-1}^{|s_{\ell-1}|}} \cdots \sum_{j_1=1}^{j_2-1} \frac{\operatorname{sgn}(s_1)^{j_1}}{j_1^{|s_1|}} \sum_{j=1}^{j_1-1} j^{p-1-a}$$

$$\equiv \sum_{k=0}^{p-1-a} \binom{p-a}{k} \frac{B_k}{p-a} \sum_{j_\ell=1}^{p-1} \frac{\operatorname{sgn}(s_\ell)^{j_\ell}}{j_\ell^{|s_\ell|}} \sum_{j_{\ell-1}=1}^{j_\ell-1} \frac{\operatorname{sgn}(s_{\ell-1})^{j_{\ell-1}}}{j_{\ell-1}^{|s_{\ell-1}|}} \cdots \sum_{j_1=1}^{j_2-1} \frac{\operatorname{sgn}(s_1)^{j_1}}{j_1^{|s_1|}} j_1^{p-a-k}$$

which is exactly the right hand side of (2.9). □

Theorem 2.2. *Let $a, \ell \in \mathbb{N}$, $\mathbf{s} = (s_1, \dots, s_\ell) \in (\mathbb{Z}^*)^\ell$ and $\mathbf{s}' = (s_2, \dots, s_\ell)$. For any prime $p \ge a+2$ write $H(-) = H(-; p-1)$. Then*

$$H(-a,\mathbf{s}) \equiv \frac{(1-2^{p-a})B_{p-a}}{p-a}\Big(H(\mathbf{s}) - H(-s_1, \mathbf{s}')\Big) - \sum_{k=0}^{p-2-a} \binom{p-1-a}{k} \frac{(1-2^{k+1})B_{k+1}}{k+1} H\big((k+a)\oplus(-s_1), \mathbf{s}'\big) \pmod p.$$

Proof. By definition and Lemma 2.1

$$H(a,\mathbf{s}) \equiv \sum_{j_\ell=1}^{p-1} \frac{\operatorname{sgn}(s_\ell)^{j_\ell}}{j_\ell^{|s_\ell|}} \sum_{j_{\ell-1}=1}^{j_\ell-1} \frac{\operatorname{sgn}(s_{\ell-1})^{j_{\ell-1}}}{j_{\ell-1}^{|s_{\ell-1}|}} \cdots \sum_{j_1=1}^{j_2-1} \frac{\operatorname{sgn}(s_1)^{j_1}}{j_1^{|s_1|}} \sum_{j=1}^{j_1-1} (-1)^j j^{p-1-a}$$

$$\equiv \sum_{k=0}^{p-2-a} \binom{p-1-a}{k} \frac{E_k(0)}{2} \sum_{j_\ell=1}^{p-1} \frac{\operatorname{sgn}(s_\ell)^{j_\ell}}{j_\ell^{|s_\ell|}} \sum_{j_{\ell-1}=1}^{j_\ell-1} \frac{\operatorname{sgn}(s_{\ell-1})^{j_{\ell-1}}}{j_{\ell-1}^{|s_{\ell-1}|}} \cdots \sum_{j_1=1}^{j_2-1} \frac{\operatorname{sgn}(s_1)^{j_1}}{j_1^{|s_1|+k+a}} (-1)^{j_1-1}$$

$$+ \frac{E_{p-1-a}(0)}{2} \sum_{j_\ell=1}^{p-1} \frac{\operatorname{sgn}(s_\ell)^{j_\ell}}{j_\ell^{|s_\ell|}} \sum_{j_{\ell-1}=1}^{j_\ell-1} \frac{\operatorname{sgn}(s_{\ell-1})^{j_{\ell-1}}}{j_{\ell-1}^{|s_{\ell-1}|}} \cdots \sum_{j_1=1}^{j_2-1} \frac{\operatorname{sgn}(s_1)^{j_1}}{j_1^{|s_1|}} \Big(1 - (-1)^{j_1}\Big)$$

$$\equiv \frac{E_{p-1-a}(0)}{2}\Big(H(\mathbf{s}) - H(-s_1, \mathbf{s}')\Big) - \sum_{k=0}^{p-2-a} \binom{p-1-a}{k} \frac{E_k(0)}{2} H\big((k+a)\oplus(-s_1), \mathbf{s}'\big).$$

The theorem follows from (2.5) easily. □

3. AMHS of Depth Two

In this section we will provide congruence formulae for all depth two AMHS. All but one case are given by very concise values involving Bernoulli numbers or Euler numbers (which are closely related by identity (2.5)).

Theorem 3.1. *Let $a, b \in \mathbb{N}$ and a prime $p \geq a+b+2$. Write $S(-) = S(-;p-1)$ and $H(-) = H(-;p-1)$. If $a+b$ is odd then we have*

$$S(a,b) \equiv H(a,b) \equiv \frac{(-1)^b}{a+b}\binom{a+b}{a} B_{p-a-b} \pmod p, \quad (3.1)$$

$$H(-a,-b) \equiv \frac{2^{p-a-b}-1}{a+b}(-1)^b\binom{a+b}{a} B_{p-a-b} \pmod p, \quad (3.2)$$

$$S(-a,-b) \equiv \frac{2^{p-a-b}-1}{a+b}\left(2+(-1)^b\binom{a+b}{a}\right) B_{p-a-b} \pmod p, \quad (3.3)$$

$$H(-a,b) \equiv H(a,-b) \equiv \frac{1-2^{p-a-b}}{a+b} B_{p-a-b} \pmod p, \quad (3.4)$$

$$S(-a,b) \equiv S(a,-b) \equiv \frac{2^{p-a-b}-1}{a+b} B_{p-a-b} \pmod p. \quad (3.5)$$

If $a+b$ is even then we have

$$S(a,b) \equiv H(a,b) \equiv 0 \pmod p, \quad (3.6)$$

$$S(-a,-b) \equiv H(-a,-b) \equiv \frac{2(1-2^{p-a})(1-2^{p-a})}{ab} B_{p-a}B_{p-b} \pmod p. \quad (3.7)$$

Proof. Congruences (3.1) and (3.6) follow from [18, Theorem 3.1] (see [10, Theorem 6.1] for a different proof). Congruences (3.2), (3.4), and (3.7) are given by [4, Lemma 6.2] (notice that $(1-2^{p-1})B_{p-1} \equiv q_p \pmod p$). Finally, congruences for S version of AMHS are all easy consequences of the H version by the relation $S(\alpha,\beta) = H(\alpha,\beta) + H(\alpha \oplus \beta)$. □

Even though we don't have compact congruence formulae for $H(-a,b)$ and $H(a,-b)$ when $a+b$ is even we can prove two general statements using the two reduction procedures provided by Theorem 2.1 and Theorem 2.2.

Proposition 3.1. *Let $a, b \in \mathbb{N}$ and a prime $p \geq a+b+2$. Write $S(-) = S(-;p-1)$ and $H(-) = H(-;p-1)$. If $a+b$ is even then we have*

$$S(a,-b) \equiv H(a,-b) \equiv -S(-b,a) \equiv -H(-b,a)$$
$$\equiv \sum_{k=0}^{p-a-1}\binom{p-a}{k}\frac{2(1-2^{2p-a-b-k})B_k B_{2p-a-b-k}}{(p-a)(2p-a-b-k)} \pmod p. \quad (3.8)$$

Proof. By Theorem 2.1 we have

$$H(a,-b) \equiv \sum_{k=0}^{p-a-1} \binom{p-a}{k} \frac{B_k}{p-a} \cdot H(-(a+b+k-1)) \pmod p.$$

To use Corollary 2.1 we need to break the sum into two parts, i.e., when $a+b+k<p$ and when $a+b+k \geq p$. In the first case we can replace k by $k+p-1$ and then to get the correct term in (3.8) we only need to use Fermat's Little Theorem $2^{p+1-a-b-k} \equiv 2^{2p-a-b-k} \pmod p$ and Kummer congruence $B_{p+1-a-b-k}/(p+1-a-b-k) \equiv B_{2p-a-b-k}/(2p-a-b-k) \pmod p$. This finishes the proof of the proposition. □

The second reduction procedure, Theorem 2.2, provides us another useful result on AMHS of even weight and depth two.

Proposition 3.2. *Let $a,b \in \mathbb{N}$ and a prime $p \geq a+b+2$. Write $S(-)=S(-;p-1)$ and $H(-)=H(-;p-1)$. If $a+b$ is even then we have*

$$\begin{aligned}
&S(-a,b) \equiv H(-a,b) \equiv -S(b,-a) \equiv -H(b,-a)\\
\equiv &\sum_{k=1}^{p-2-a-b} \binom{p-1-a}{k} \frac{2(1-2^{k+1})(1-2^{p-a-b-k})B_{k+1}B_{p-a-b-k}}{(k+1)(a+b+k)}\\
+ &\sum_{k=p-1-a-b}^{p-1-a} \binom{p-1-a}{k} \frac{2(1-2^{k+1})(1-2^{2p-1-a-b-k})B_{k+1}B_{2p-1-a-b-k}}{(k+1)(1+a+b+k)} \pmod p.
\end{aligned}$$

Proof. By Theorem 2.2 we have

$$H(-a,b) \equiv \sum_{k=0}^{p-1-a} \binom{p-1-a}{k} \frac{(1-2^{k+1})B_{k+1}}{k+1} H\big(-(k+a+b)\big) \pmod p$$

since $H(b) \equiv 0 \pmod p$. The rest follows from Corollary 2.1 quickly, similar to the proof of Proposition 3.1. □

The two propositions above will be used in section 6 to compute some AMHS congruences explicitly.

4. AMHS of Depth Three

All the congruences in this section are modulo a prime p. Write $H(-)=H(-;p-1)$. Recall that for depth three MHS modulo p can be determined [18, Theorem 3.5], [10, Theorem 6.2] and [18, (3.13)]. For AMHS, we first observe that for any $\alpha,\beta,\gamma \in \mathbb{Z}$

$$\begin{aligned}
H(\alpha,\beta,\gamma) =& H(\alpha)H(\beta)H(\gamma) - H(\gamma)H(\beta,\alpha) - H(\gamma)H(\beta\oplus\alpha)\\
&-H(\gamma,\beta)H(\alpha) - H(\gamma\oplus\beta)H(\alpha) + H(\gamma,\beta,\alpha)\\
&+H(\gamma\oplus\beta,\alpha) + H(\gamma,\beta\oplus\alpha) + H(\gamma\oplus\beta\oplus\alpha).
\end{aligned}$$

This can be easily checked by stuffle relations but the idea is hidden in the general framework set up by Hoffman [9]. Combining with the reversal relations we can obtained the following results without much difficulty. We leave its proof to the interested reader.

Theorem 4.1. *Let p be a prime and $a, b, c \in \mathbb{N}$ such that $p > w$ where $w := a + b + c$. Write $H(-) = H(-; p-1)$. Then we have*

(1). If w is even then

$$2H(a,-b,c) \equiv H(-c-b,a) + H(c,-b-a) \pmod{p}, \quad (4.1)$$
$$2H(a,b,-c) \equiv H(-c-b,a) + H(-c,b+a) - H(-c)H(b,a) \pmod{p}, \quad (4.2)$$
$$2H(-a,-b,-c) \equiv H(c+b,-a) + H(-c,a+b) - H(-c)H(-b,-a) - H(-c,-b)H(-a) \pmod{p}. \quad (4.3)$$

(2). If w is odd then

$$2H(a,-b,-c) \equiv H(c+b,a) + H(-c,-b-a) - H(-c)H(-b,a) \pmod{p}, \quad (4.4)$$
$$2H(-a,b,-c) \equiv - H(-c)H(b,-a) - H(-c,b)H(-a) \quad (4.5)$$
$$+ H(-c-b,-a) + H(-c,-b-a) \pmod{p}. \quad (4.6)$$

Because of the reversal relations when the weight is even there remains essentially only one more case to consider in depth three. This is given by the next result which will be used in section 6.

Theorem 4.2. *Let a, b, c be positive integers such that $w := a+b+c$ is even. Then for any prime $p \geq w+3$ we have*

$$H(a,-b,-c) \equiv -\sum_{k=2}^{p-w+1} \binom{p-a}{p-w-k+1}\binom{k+c-1}{c}\frac{(1-2^k)B_kB_{p-w-k+1}}{ak}$$
$$-\sum_{k=p+1-b-c}^{p-c} \binom{p-a}{2p-w-k}\binom{k+c-1}{c}\frac{(1-2^k)B_kB_{2p-w-k}}{ak}$$
$$-\frac{(1-2^{p-c})(1-2^{p-a-b})B_{p-a-b}B_{p-c}}{(a+b)c} \pmod{p}.$$

Proof. The proof is essentially a repeated application of Theorem 2.1. But we spell out all the details below because there are some subtle details that we need to attend to.

By (2.3) and Fermat's Little Theorem we have modulo p

$$H(a,-b,-c) \equiv \sum_{i=1}^{p-1} \frac{(-1)^i}{i^c} \sum_{j=1}^{i-1} (-1)^j j^{p-1-b} \sum_{k=1}^{j-1} k^{p-1-a}$$
$$\equiv \sum_{i=1}^{p-1} \frac{(-1)^i}{i^c} \sum_{k=0}^{p-1-a} \binom{p-a}{k} \frac{B_k}{p-a} \sum_{j=1}^{i-1} (-1)^j j^{\rho(k)+p-a-b-k},$$

where $\rho(k) = 0$ if $k < p-a-b$ and $\rho(k) = p-1$ if $k \geq p-a-b$ (to make sure all exponents are positive in the sum of the second line above). By Lemma 2.1

$$\begin{aligned}
H(a,-b,-c) &\equiv \sum_{k=0}^{p-1-a}\binom{p-a}{k}\frac{B_k}{p-a}\sum_{r=0}^{n}\binom{n}{r}\sum_{i=1}^{p-1}(-1)^i i^{n-r-c}F_{n,i,r}\\
&\equiv -\sum_{k=0}^{p-1-a}\binom{p-a}{k}\sum_{r=1}^{n-1}\binom{n}{r}\frac{(1-2^{r+1})B_kB_{r+1}}{(p-a)(r+1)}\sum_{i=1}^{p-1} i^{n-c-r}\\
&\quad+\sum_{k=0}^{p-1-a}\binom{p-a}{k}\frac{(1-2^{n+1})B_kB_{n+1}}{(p-a)(n+1)}\sum_{i=1}^{p-1}((-1)^i-1)i^{-c},
\end{aligned}$$

where $n=\rho(k)+p-a-b-k$. Here we have used the fact that when $r=0$ we have $F_{n,i,r}=(-1)^{i-1}$ and thus the inner sum is $\sum_{i=1}^{p-1} i^{n-c}\equiv 0 \pmod p$ except when $p-1|n-c$, i.e., except when $n=c$ and $k=p-w$. But then $B_k=0$ since w is even by assumption. Thus

$$\begin{aligned}
H(a,-b,-c) \equiv &\sum_{\substack{0\le k<p-a\\ c<n}}\binom{p-a}{k}\binom{n}{n-c}\frac{(1-2^{n-c+1})B_kB_{n-c+1}}{(p-a)(n-c+1)}\\
&+H(-c)\sum_{k=0}^{p-1-a}\binom{p-a}{k}\frac{(1-2^{n+1})B_kB_{n+1}}{(p-a)(n+1)}.
\end{aligned}$$

Now if c is even then $H(-c)\equiv 0 \pmod p$. So we may assume c is odd in the last line above. Then $k+n+1$ is always odd so that $B_kB_{n+1}\neq 0$ if and only if $k=1$ and $n=p-a-b-1$. Hence

$$\begin{aligned}
H(a,-b,-c)\equiv &\sum_{k=0}^{p-w-1}\binom{p-a}{k}\binom{p-a-b-k}{p-w-k}\frac{(1-2^{p-w-k+1})B_kB_{p-w-k+1}}{(p-a)(p-w-k+1)}\\
&+\sum_{k=p-a-b}^{p-1-a}\binom{p-a}{k}\binom{2p-1-a-b-k}{2p-1-w-k}\frac{(1-2^{2p-w-k})B_kB_{2p-w-k}}{(p-a)(2p-w-k)}\\
&-\frac{(1-2^{p-c})(1-2^{p-a-b})B_{p-a-b}B_{p-c}}{(a+b)c}.
\end{aligned}$$

After substitutions $k\to p-w+1-k$ in the first sum and $k\to 2p-w-k$ in the second sum the theorem follows immediately from Corollary 2.1, □

Remark 4.1. The condition $p\ge w+3$ in Theorem 4.2 can not be weakened since

$$H(1,-2,-3)\equiv \text{RHS}+5\not\equiv \text{RHS} \pmod 7.$$

5. AMHS of Arbitrary Depth

In this section we provide some general results on AMHS without restrictions on the depth. We first consider the homogeneous case for which the key idea comes from [18, Lemma 2.12] and [10, Theorem 2.3].

Let $p_i = \sum_{j\geq 1} x_j^i$ be the power-sum symmetric functions and $e_i = \sum_{j_1<\cdots<j_i} x_{j_i}\cdots x_{j_i}$ be the elementary symmetric functions of degree i. Let $P(\ell)$ be the set of unordered partitions of ℓ. For $\lambda = (\lambda_1,\dots,\lambda_r) \in P(\ell)$ we set $p_\lambda = \prod_{i=1}^r p_{\lambda_i}$. Recall that the expression of e_l in terms of p_i is given by the following formula (see [13, p.28]):

$$\ell! e_\ell = \begin{vmatrix} p_1 & 1 & 0 & \cdots & 0 \\ p_2 & p_1 & 2 & \cdots & 0 \\ & & & \ddots & 0 \\ p_{\ell-1} & p_{\ell-2} & p_{\ell-3} & \cdots & \ell-1 \\ p_\ell & p_{\ell-1} & p_{\ell-2} & \cdots & p_1 \end{vmatrix} = \sum_{\lambda\in P(\ell)} c_\lambda p_\lambda. \tag{5.1}$$

Denote by $O(\ell) \subset P(\ell)$ the subset of odd partitions $\lambda = (\lambda_1,\dots,\lambda_r)$ (i.e., λ_i is odd for every part).

Lemma 5.1. *Let $a,\ell \in \mathbb{N}$ and p a prime such that $a\ell < p-1$. Set $H(-) = H(-;p-1)$. For an odd partition $\lambda = (\lambda_1,\dots,\lambda_r) \in O(\ell)$ we put $H_\lambda(-a) = \prod_{i=1}^r H(-\lambda_i a)$. Then*

$$\ell! H(\{-a\}^\ell) \equiv \sum_{\lambda\in O(\ell)} c_\lambda H_\lambda(-a) \pmod p, \tag{5.2}$$

where c_λ are given by (5.1). *In particular, if a is even then* (5.2) $\equiv 0 \pmod p$. *If $a > 1$ is odd then* (5.2) *is congruent to a $\mathbb{Q}$-linear combination of $B_{p-\lambda_1 a}\cdots B_{p-\lambda_r a}$ for odd partitions $\lambda = (\lambda_1,\dots,\lambda_r)$. If $a = 1$ then* (5.2) *is congruent to a $\mathbb{Q}$-linear combination of $q_p^s B_{p-\lambda_{s+1}}\cdots B_{p-\lambda_r}$ for odd partitions $\lambda = (\{1\}^s,\lambda_{s+1},\dots,\lambda_r)$ ($\lambda_i > 1$ for all $i > s$), where $q_p = (2^{p-1}-1)/p$ is the Fermat quotient.*

Proof. Congruence (5.2) follows from [18, Lemma 2.12] and [10, Theorem 2.3]. The last part follows from [16, Theorem 2.1] and (1.4). □

For example, by [16, Theorem 2.1] and (1.4) we have

$$H(-1) = -H(1) + H(1;\frac{p-1}{2}) \equiv -2q_p + pq_p^2 - \frac{2}{3}p^2q_p^3 - \frac{1}{4}p^2 B_{p-3} \pmod{p^3}. \tag{5.3}$$

Observe that $O(2) = \{(1,1)\}, O(3) = \{(1,1,1),(3)\}, O(4) = \{(1,1,1,1),(1,3)\}$. It is obvious that $c_{(1,\cdots,1)} = 1$, $c_{(3)} = 2$, $c_{(1,3)} = 8$, $c_{(1,1,3)} = 20$, and $c_{(1,1,1,3)} = c_{(3,3)} = 40$, which implies that modulo prime p

$$2H(\{-1\}^2) \equiv 4q_p^2 \quad \Longrightarrow H(\{-1\}^2) \equiv 2q_p^2, \tag{5.4}$$

$$6H(\{-1\}^3) \equiv -8q_p^3 + 2H(-3) \quad \Longrightarrow H(\{-1\}^3) \equiv -\frac{4}{3}q_p^3 - \frac{1}{6}B_{p-3}, \tag{5.5}$$

$$24H(\{-1\}^4) \equiv 16q_p^4 + 8H(-1)H(-3) \quad \Longrightarrow H(\{-1\}^4) \equiv \frac{2}{3}q_p^4 + \frac{1}{3}q_pB_{p-3}, \tag{5.6}$$

$$5!H(\{-1\}^5) \equiv 20H(-1)^2H(-3) - 32q_p^5 \Longrightarrow H(\{-1\}^5) \equiv -\frac{4}{15}q_p^5 - \frac{1}{3}q_p^2B_{p-3} \tag{5.7}$$

and

$$\begin{aligned} 6!H(\{-1\}^6) &\equiv 64q_p^6 + 40H(-1)^3H(-3) + 40H(-3)^2 \\ \Longrightarrow H(\{-1\}^6) &\equiv \frac{4}{45}q_p^6 + \frac{2}{9}q_p^3B_{p-3} + \frac{1}{72}B_{p-3}^2. \end{aligned} \tag{5.8}$$

Remark 5.1. Congruences (5.4) and (5.5) are not new. See the Remarks on [4, p. 365].

For non-homogeneous s we don't know too much except for those of very special forms. For example, we have the follow easy statement.

Proposition 5.1. *Suppose* $\mathbf{s} = \overleftarrow{\mathbf{s}}$ *is palindromic. If the number of negative components in* s *and the weight* $|\mathbf{s}|$ *have different parity then for every odd prime* p

$$S(\mathbf{s}; p-1) \equiv H(\mathbf{s}; p-1) \equiv 0 \pmod{p}.$$

Proof. This is obvious by the reversal relations (2.2) □

In order to state and prove Proposition 5.2, we need to investigate the following two types of sums. Define

$$U(\mathbf{s}; n) := \sum_{1 \le k_1 < k_2 < \cdots < k_\ell \le n} \prod_{i=1}^{\ell} \frac{(-\operatorname{sgn}(s_i)/2 + 3/2)^{k_i}}{k_i^{|s_i|}}, \tag{5.9}$$

$$V(\mathbf{s}; n) := \sum_{1 \le k_1 < k_2 < \cdots < k_\ell \le n} \prod_{i=1}^{\ell} \frac{(\operatorname{sgn}(s_i)/4 + 3/4)^{k_i}}{k_i^{|s_i|}}. \tag{5.10}$$

For example, $U(-m; n) = \sum_{k=1}^{n} 2^k/k^m$, $V(-m; n) = \sum_{k=1}^{n} 1/(2^k k^m)$, and for s with only positive components $U(\mathbf{s}; n) = V(\mathbf{s}; n) = H(\mathbf{s}; n)$. The sums $U(-m; p-1)$ appeared previously in congruences involving powers of Fermat quotient (see [2, 3, 5, 8]). To compute the congruence involving these sums we need the following preliminary results which will also be needed in section 7.

Lemma 5.2. *Assume* $\mathbf{s} = (s_1, \ldots, s_\ell) \in (\mathbb{Z}^*)^\ell$. *Let* $\mathbf{e}_j$ *be the standard* j*-th unit vector. Then for every odd prime* p *we have*

$$U(\mathbf{s}; n) \equiv 2^{p\sharp\{j: s_j<0\}} (-1)^{|\mathbf{s}|} \Big(V(\overleftarrow{\mathbf{s}}; n) + p \sum_{j=1}^{\ell} |s_j| V(\overleftarrow{\mathbf{s} \oplus \mathbf{e}_j}; n) \Big) \pmod{p^2}, \tag{5.11}$$

$$V(\mathbf{s}; n) \equiv 2^{-p\sharp\{j: s_j<0\}} (-1)^{|\mathbf{s}|} \Big(U(\overleftarrow{\mathbf{s}}; n) + p \sum_{j=1}^{\ell} |s_j| U(\overleftarrow{\mathbf{s} \oplus \mathbf{e}_j}; n) \Big) \pmod{p^2}. \tag{5.12}$$

Further,

$$V(-1) \equiv -\frac{1}{2^p} \Big(U(-1) + pU(-2) + p^2 U(-3) \Big) \pmod{p^3}. \tag{5.13}$$

Proof. By substitution $k_i \to p - k_i$ we get

$$\begin{aligned} U(\mathbf{s}; n) &= \sum_{1 \le k_1 < k_2 < \cdots < k_\ell \le n} \prod_{i=1}^{\ell} \frac{(-\operatorname{sgn}(s_i)/2 + 3/2)^{p-k_i}}{k_i^{|s_i|} (p/k_i - 1)^{|s_i|}} \\ &\equiv \sum_{1 \le k_1 < k_2 < \cdots < k_\ell \le n} \prod_{i=1}^{\ell} \frac{(-\operatorname{sgn}(s_i)/2 + 3/2)^{p-k_i}}{k_i^{|s_i|}} (-1)^{|s_i|} \left(1 + \frac{|s_i| p}{k_i}\right) \pmod{p^2}. \end{aligned}$$

Thus (5.11) quickly follows from

$$-\frac{\operatorname{sgn}(s_i)}{2}+\frac{3}{2}=\begin{cases}1, & \text{if } s_i>0;\\ 2, & \text{if } s_i<0,\end{cases} \quad \text{and} \quad \frac{\operatorname{sgn}(s_i)}{4}+\frac{3}{4}=\begin{cases}1, & \text{if } s_i>0;\\ 1/2, & \text{if } s_i<0.\end{cases}$$

Congruence (5.12) can be proved similarly. By the same idea

$$V(-1)=\sum_{k=1}^{p-1}\frac{1}{2^k k}=\sum_{k=1}^{p-1}\frac{1}{2^{p-k}k(p/k-1)}\equiv -\frac{1}{2^p}\sum_{k=1}^{p-1}\frac{2^k}{k}\left(1+\frac{p}{k}+\frac{p^2}{k^2}\right) \pmod{p^3}.$$

The congruence (5.13) follows immediately. □

Lemma 5.3. *Let p be an odd prime. For all positive integers d and m we have*

$$\sum_{1\le n_1\le\cdots\le n_d\le m}\frac{(1-x)^{n_1}-1}{n_1\cdots n_d}=\sum_{j=1}^{m}\frac{(-x)^j}{j^d}\binom{m}{j}, \tag{5.14}$$

$$\sum_{1\le n_1\le\cdots\le n_d\le p-1}\frac{(1-x)^{n_1}-1}{n_1\cdots n_d}\equiv\sum_{j=1}^{p-1}\frac{x^j}{j^d}\Big(1-pH(1;j)+p^2H(1,1;j)\Big) \pmod{p^3}. \tag{5.15}$$

Proof. Since for all positive integer $j<p$

$$(-1)^j\binom{p-1}{j}\equiv 1-pH(1;j)+p^2H(1,1;j) \pmod{p^3}, \tag{5.16}$$

congruence (5.15) follows easily from (5.14) which we now prove by mimicking the argument in [12].

Let $\mathbb{Q}^*=\mathbb{Q}\setminus\{0\}$. Define the injective operator

$$\begin{aligned}\Delta:\mathbb{Q}[x]\setminus\mathbb{Q}^* &\longrightarrow \mathbb{Q}[x]\setminus\mathbb{Q}^*\\ P(x) &\longmapsto xP'(x).\end{aligned}$$

It suffices to show that

$$\Delta^d\left(\sum_{1\le n_1\le\cdots\le n_d\le m}\frac{(1-x)^{n_1}-1}{n_1\cdots n_d}\right)=\Delta^d\left(\sum_{j=1}^{m}\frac{(-x)^j}{j^d}\binom{m}{j}\right).$$

Clearly

$$\Delta^d\left(\sum_{j=1}^{m}\frac{(-x)^j}{j^d}\binom{m}{j}\right)=\sum_{j=1}^{m}(-x)^j\binom{m}{j}=(1-x)^m-1.$$

On the other hand,

$$
\begin{aligned}
&\Delta^d \left(\sum_{1\le n_1\le n_2\le\cdots\le n_d\le m} \frac{(1-x)^{n_1}-1}{n_1 n_2\cdots n_d} \right) \\
=&\Delta^{d-1} \left(\sum_{1\le n_2\le\cdots\le n_d\le m} \frac{(1-x)^{n_2}-1}{n_2\cdots n_d} \right) \\
&\vdots \\
=&\Delta \left(\sum_{1\le n_d\le m} \frac{(1-x)^{n_d}-1}{n_d} \right) \\
=&-x \sum_{1\le n_d\le m} (1-x)^{n_d-1} \\
=&(1-x)^m-1.
\end{aligned}
$$

By injectivity of Δ this completes the proof of the lemma. □

Lemma 5.4. *Let d and m be two positive integers. Then*

$$
\sum_{1\le n_1\le\cdots\le n_d\le m} \frac{(-1)^{n_d}(1-x)^{n_1}}{n_1\cdots n_d}\binom{m}{n_d} = \sum_{k=1}^{m} \frac{x^k}{k^d} - \sum_{k=1}^{m} \frac{1}{k^d}. \tag{5.17}
$$

Proof. The proof, which is completely similar to [12], is left to the interested reader. Note that there is a misprint in [12] where in the definition of $f(x)$ the range of i should be from 1 to j. Namely

$$
f(x) = \sum_{k=1}^{n} \frac{(-1)^{k-1}}{k}\binom{n}{k} \sum_{j=1}^{k} \frac{1}{j} \sum_{i=1}^{j} \frac{(1-x)^i}{i}.
$$

□

Proposition 5.2. *Let $n \in \mathbb{N}$. For any prime $p > n+2$ write $H(-) = H(-;p-1)$ and similarly for S, U and V. Then*

$$
\begin{aligned}
H(-1,\{1\}^n) \equiv S(-1,\{1\}^n) \equiv (-1)^n H(\{1\}^n,-1) \equiv (-1)^n S(\{1\}^n,-1) \\
\equiv U(-n-1) \equiv (-1)^{n+1} 2V(-n-1) \pmod p.
\end{aligned} \tag{5.18}
$$

Proof. The congruence $(-1)^n H(\{1\}^n,-1) \equiv U(-n-1) \pmod p$ is the content of [16, Theorem 2.3]. By taking $m = p-1$, $d = n+1$ and $x = 2$ in Lemma 5.3 or Lemma 5.4 we get $S(-1,\{1\}^n) \equiv U(-n-1) \pmod p$. The congruence for V follows from (5.11). The other two congruences follows from the reversal relations (2.2). □

We now generalize this to the following

Proposition 5.3. *Let m, n be two nonnegative integers and a a positive integer. For any prime $p > a(m+n)+2$ write $H(-) = H(-; p-1)$. Then*

$$H(\{a\}^m, -a, \{a\}^n) \equiv (-1)^{m+n} S(\{a\}^n, -a, \{a\}^m) \pmod p, \tag{5.19}$$
$$H(\{a\}^m, -a, \{a\}^n) \equiv (-1)^{(m+n+1)(a+1)} S(\{a\}^m, -a, \{a\}^n) \pmod p. \tag{5.20}$$

Proof. The first congruence (5.19) follows from (5.21) in the next Lemma by taking $x = -1$, $k = m+1$ and $d = m+n+1$. The second congruence (5.20) follows (5.19) by the reversal relation. □

Let $a, d \in \mathbb{N}$ and a prime $p \geq da+3$. We identify the finite field $\mathbb{F}_p$ of p elements with $\mathbb{Z}/p\mathbb{Z}$. For $1 \leq k \leq d$ define

$$H_{d,k}^{(a)}(x) = \sum_{0<i_1<\cdots<i_d<p} \frac{x^{i_k}}{(i_1 \cdots i_d)^a} \in \mathbb{F}_p[x],$$
$$S_{d,k}^{(a)}(x) = \sum_{1\leq i_1\leq\cdots\leq i_d<p} \frac{x^{i_k}}{(i_1 \cdots i_d)^a} \in \mathbb{F}_p[x]$$
$$h_{d,k}^{(a)}(x) = \sum_{0<i_1<\cdots<i_d<p} \frac{(-1)^{\sum i_j} x^{i_k}}{(i_1 \cdots i_d)^a} \in \mathbb{F}_p[x],$$
$$s_{d,k}^{(a)}(x) = \sum_{1\leq i_1\leq\cdots\leq i_d<p} \frac{(-1)^{\sum i_j} x^{i_k}}{(i_1 \cdots i_d)^a} \in \mathbb{F}_p[x],$$

where $\sum i_j = i_1 + \cdots + i_d$. For convenience we set $H_{d,0}^{(a)}(x) = H(\{a\}^d) = 0$, $S_{d,d+1}^{(a)}(x) = x^{p-1}S(\{a\}^d) = 0$, $S_{d,0}^{(a)}(x) = xS(\{a\}^d) = 0$ and $H_{d,d+1}^{(a)}(x) = x^p H(\{a\}^d) = 0$ by [18, Theorem 2.13]. Moreover for even a we set $h_{d,0}^{(a)}(x) = H(\{-a\}^d) = 0$, $s_{d,d+1}^{(a)}(x) = (-x)^{p-1}S(\{-a\}^d) = 0$, $s_{d,0}^{(a)}(x) = -xS(\{-a\}^d) = 0$ and $h_{d,d+1}^{(a)}(x) = (-x)^p H(\{-a\}^d) = 0$ by Lemma 5.1 and (2.8).

Lemma 5.5. *For $1 \leq k \leq d$ we have the identity in $\mathbb{F}_p[x]$*

$$H_{d,k}^{(a)}(x) + (-1)^d S_{d,d+1-k}^{(a)}(x) = 0, \tag{5.21}$$
$$h_{d,k}^{(a)}(x) + (-1)^d s_{d,d+1-k}^{(a)}(x) = 0. \tag{5.22}$$

Proof. To prove (5.21) we proceed by induction on d. For $d = 1$ it holds since

$$H_{1,1}^{(a)}(x) = \sum_{k=1}^{p-1} \frac{x^k}{k^a} = S_{1,1}^{(a)}(x), \quad H_{1,0}^{(a)}(x) = S_{1,0}^{(a)}(x) = H_{1,2}^{(a)}(x) = S_{1,2}^{(a)}(x) = 0.$$

Now we will follow the idea of the proof of [16, Theorem 2.3]. For $1 \le k \le d$ we have that

$$\begin{aligned}\left(x\frac{d}{dx}\right)^a H^{(a)}_{d,k}(x) &= \sum_{0<i_1<\cdots<i_{k-1}<i_{k+1}<\cdots<i_d<p} \frac{1}{(i_1\cdots i_{k-1}i_{k+1}\cdots i_d)^a} \sum_{i_k=i_{k-1}+1}^{i_{k+1}-1} x^{i_k} \\ &= \sum_{0<i_1<\cdots<i_{k-1}<i_{k+1}<\cdots<i_d<p} \frac{1}{(i_1\cdots i_{k-1}i_{k+1}\cdots i_d)^a} \cdot \frac{x^{i_{k+1}} - x^{i_{k-1}+1}}{x-1} \\ &= \frac{1}{x-1} H^{(a)}_{d-1,k}(x) - \frac{x}{x-1} H^{(a)}_{d-1,k-1}(x),\end{aligned}$$

that is

$$(x-1)\left(x\frac{d}{dx}\right)^a H^{(a)}_{d,k}(x) = H^{(a)}_{d-1,k}(x) - xH^{(a)}_{d-1,k-1}(x).$$

In a similar way we have that

$$(x-1)\left(x\frac{d}{dx}\right)^a S^{(a)}_{d,k}(x) = xS^{(a)}_{d-1,k}(x) - S^{(a)}_{d-1,k-1}(x).$$

Hence

$$\begin{aligned}&(x-1)\left(x\frac{d}{dx}\right)^a \left(H^{(a)}_{d,k}(x) + (-1)^d S^{(a)}_{d,d+1-k}(x)\right) \\ =&H^{(a)}_{d-1,k}(x) - xH^{(a)}_{d-1,k-1}(x) + (-1)^d xS^{(a)}_{d-1,d+1-k}(x) - (-1)^d S^{(a)}_{d-1,d-k}(x) \\ =&H^{(a)}_{d-1,k}(x) + (-1)^{d-1} S^{(a)}_{d-1,d-k}(x) - x\left(H^{(a)}_{d-1,k-1}(x) + (-1)^{d-1} S^{(a)}_{d-1,d-(k-1)}(x)\right) \\ =&0\end{aligned}$$

by inductive assumption. Thus $(x\frac{d}{dx})^{a-1}\left(H^{(a)}_{d,k}(x) + (-1)^d S^{(a)}_{d,d+1-k}(x)\right) = c$ for some constant $c \in \mathbb{F}_p$ since this polynomial has degree less than p. By letting $x = 0$ we see that $c = 0$. Repeating this process a times yields (5.21).

The congruence (5.22) can be proved in a similar way. In particular, we can show easily that

$$\begin{aligned}-(x+1)\left(x\frac{d}{dx}\right)^a h^{(a)}_{d,k}(x) =&h^{(a)}_{d-1,k}(-x) + xh^{(a)}_{d-1,k-1}(-x), \\ (x+1)\left(x\frac{d}{dx}\right)^a s^{(a)}_{d,k}(x) =&xs^{(a)}_{d-1,k}(-x) + s^{(a)}_{d-1,k-1}(-x).\end{aligned}$$

So by induction we get

$$(x+1)\left(x\frac{d}{dx}\right)^a \left(H^{(a)}_{d,k}(x) + (-1)^d S^{(a)}_{d,d+1-k}(x)\right) = 0,$$

which quickly leads to (5.22). This concludes the proof of the lemma. □

Similar to Proposition 5.3 we also have the following

Proposition 5.4. *Let m, n be two nonnegative integers. Let a be a positive even integer. For any prime $p > a(m+n)+2$ write $H(-) = H(-;p-1)$. Then*

$$H(\{-a\}^m, a, \{-a\}^n) \equiv S(\{-a\}^m, a, \{-a\}^n) \pmod{p}, \tag{5.23}$$

$$H(\{-a\}^m, a, \{-a\}^n) \equiv (-1)^{(m+n)} S(\{-a\}^n, a, \{-a\}^m) \pmod{p}. \tag{5.24}$$

Proof. The second congruence (5.23) follows from (5.22) by taking $x = -1$, $k = m+1$ and $d = m+n+1$. The first congruence (5.24) follows (5.23) by the reversal relation since a is even. □

Proposition 5.5. *Let r and s be two even negative or two odd positive numbers. Then for every positive integer n and every prime $p \geq n|r+s|+2$*

$$H(\{r,s\}^n; p-1) \equiv S(\{r,s\}^n; p-1) \equiv 0 \pmod{p}.$$

Proof. The idea of the proof is exactly the same as in that of [18, Theorem 3.18(ii)]. Since there is a slight inaccuracy in that proof we want to correct it here.

Write $H(-) = H(-;p-1)$ and $S(-) = S(-;p-1)$. We now prove the proposition by induction on n. When $n = 1$ it follows quickly from Theorem 3.1. Suppose $n > 1$ and the proposition is true up to $n-1$. Then by the shuffle relations and the Inclusion-Exclusion Principle

$$S(\{r,s\}^n) = \sum_{\mathbf{t} \preceq \{r,s\}^n} H(\mathbf{t}) = H(\{r,s\}^n) + \sum_{k=1}^{n} (-1)^{k-1} H(\{r+s\}^k) \cdot S(\{r,s\}^{n-k}).$$

But $H(\{r+s\}^k) \equiv 0$ by Lemma 5.1 and Corollary 2.1 when r, s are both even negative numbers and by [18, Theorem 1.5] when r, s are both odd positive numbers. Thus

$$S(\{r,s\}^n) \equiv H(\{r,s\}^n) \pmod{p}. \tag{5.25}$$

On the other hand, from [18, (2.11)] we get

$$S(\{r,s\}^n) \equiv \sum_{\sqcup_{i=1}^{l} \mathbf{s}_i = \{r,s\}^n} (-1)^l H(\mathbf{s}_1) \cdots H(\mathbf{s}_l) \pmod{p}.$$

Now if $l > 1$ then either $\mathbf{s}_1 = \{r,s\}^d$ for some $d < n$ in which case $H(\mathbf{s}_1) \equiv 0 \pmod{p}$ by induction assumption, or else $\mathbf{s}_1 = \{r,s\}^d \sqcup \{r\}$, in which case $\mathbf{s}_1$ is a palindrome hence $H(\mathbf{s}_1) \equiv 0 \pmod{p}$ by Proposition 5.1. Consequently

$$S(\{r,s\}^n) \equiv -H(\{r,s\}^n) \pmod{p}.$$

Together with (5.25) it shows that $H(\{r,s\}^n) \equiv S(\{r,s\}^n) \equiv 0 \pmod{p}$ and the proposition is proved. □

We conclude this section with the following three conjectures.

Conjecture 5.1. *Let n, r and s be positive integers with r odd and s even. Then for every prime p*

$$\begin{aligned} H(\{r+s\}^n;(p-1)/2) &\equiv (-2)^{n(r+s)} H(\{r,-s\}^n;p-1) \pmod p, \\ S(\{r+s\}^n;(p-1)/2) &\equiv 2^{n(r+s)}\, S(\{r,-s\}^n;p-1) \pmod p. \end{aligned}$$

Even in the case of $(r,s)=(1,2)$ these are not obvious except for the case $n=1$. In fact, by Theorem 1.1(b) and (7.30) to be proved in Proposition 7.2 we can even get

$$H(3;(p-1)/2) \equiv -8H(1,-2;p-1) \pmod{p^2}.$$

Further, for $(r,s)=(1,2)$, we have a corresponding family of identities involving alternating Euler sums proved in [20]. Of course, the two congruences in Conjecture 5.1 are equivalent for large primes p if we can prove the following

Conjecture 5.2. *Let n, r and s be positive integers with r odd and s even. Let p be a prime such that $p \geq n(r+s)+2$. Then*

$$S(\{r,-s\}^n;p-1) \equiv (-1)^n H(\{r,-s\}^n;p-1) \pmod p.$$

Conjecture 5.3. *Let n and s be integers with n positive . Let p be a prime such that $p \geq n|s|+2$. Then*

$$S(\{s\}^n;(p-1)/2) \equiv H(\{s\}^n;(p-1)/2) \pmod p.$$

6. Weight Four AMHS Congruences Modulo Prime p

In [16] the first author studied the congruence properties of AMHS of weight less than four. In this section, applying the results obtained in the previous sections we can analyze the weight four AMHS in some detail. First we treat some special congruences which can not be obtained by just using the stuffle relations and the reversal relations. Let $p \geq 7$ be a prime and set $A = A_p, \cdots, K = K_p$ as follows:

$$\begin{aligned} A &:= \sum_{k=2}^{p-3} B_k B_{p-3-k}, & B &:= \sum_{k=2}^{p-3} 2^k B_k B_{p-3-k}, & C &:= \sum_{k=2}^{p-3} 2^{p-3-k} B_k B_{p-3-k}, \\ D &:= \sum_{k=2}^{p-3} \frac{B_k B_{p-3-k}}{k}, & E &:= \sum_{k=2}^{p-3} \frac{2^k B_k B_{p-3-k}}{k}, & F &:= \sum_{k=2}^{p-3} \frac{2^{p-3-k} B_k B_{p-3-k}}{k}, \\ G &:= \sum_{k=2}^{p-3} k B_k B_{p-3-k}, & J &:= \sum_{k=2}^{p-3} 2^k k B_k B_{p-3-k}, & K &:= \sum_{k=2}^{p-3} 2^{p-3-k} k B_k B_{p-3-k}. \end{aligned}$$

Then by [18, Corollary 3.6] and simple computation

$$A \equiv -B_{p-3}, \quad G \equiv 0, \quad C \equiv B - \frac{3}{4}A, \quad K \equiv -3B - J + 3A \pmod p. \tag{6.1}$$

Proposition 6.1. *For all prime* $p \geq 7$ *write* $H(-) = H(-;p-1)$. *Then we have*

$$H(1,-3) \equiv \frac{1}{2}H(-2,2) \equiv B - A \equiv \sum_{k=0}^{p-3} 2^k B_k B_{p-3-k} \pmod p, \tag{6.2}$$

$$H(-1,3) \equiv -\frac{1}{2}q_p B_{p-3} \pmod p. \tag{6.3}$$

Proof. We take congruence modulo p throughout this proof. By Proposition 3.2 we have

$$\begin{aligned} H(-3,1) \equiv & -\frac{1}{3}\sum_{k=1}^{p-6}(k+2)(k+3)(1-2^{k+1})(1-2^{p-4-k})\frac{B_{k+1}B_{p-4-k}}{k+4} - \frac{1}{2}q_p B_{p-3} \\ \equiv & \frac{1}{3}\sum_{k=2}^{p-3}(k+1)(k+2)(1-2^k)(1-2^{p-3-k})\frac{B_k B_{p-3-k}}{k} - \frac{1}{2}q_p B_{p-3}. \end{aligned}$$

by the substitution $k \to p-4-k$. Similarly we can get

$$\begin{aligned} H(-2,2) \equiv & -\sum_{k=2}^{p-3}(k+2)(1-2^k)(1-2^{p-3-k})\frac{B_k B_{p-3-k}}{k} + \frac{3}{2}q_p B_{p-3}, \\ H(-1,3) \equiv & 2\sum_{k=2}^{p-3}(1-2^k)(1-2^{p-3-k})\frac{B_k B_{p-3-k}}{k} - 2q_p B_{p-3}. \end{aligned}$$

Using (6.1) we reduce the above to

$$3H(-3,1) \equiv 3A - 3B + \frac{5}{2}D - 2E - 2F - \frac{3}{2}q_p B_{p-3}, \tag{6.4}$$

$$H(-2,2) \equiv 2B - 2A - \frac{5}{2}D + 2E + 2F + \frac{3}{2}q_p B_{p-3}, \tag{6.5}$$

$$H(-1,3) \equiv \quad + \frac{5}{2}D - 2E - 2F - 2q_p B_{p-3}. \tag{6.6}$$

On the other hand, by Proposition 3.1 we get

$$\begin{aligned} H(1,-3) \equiv & \frac{2}{p-1}\sum_{k=0}^{p-2}\binom{p-1}{k}\frac{1-2^{2p-4-k}}{2p-4-k}B_k B_{2p-4-k} \\ \equiv & -2\sum_{k=0}^{p-5}\frac{1-2^{p-3-k}}{p-3-k}B_k B_{p-3-k} + 2q_p B_{p-3} \\ \equiv & -2\sum_{k=2}^{p-3}\frac{1-2^k}{k}B_k B_{p-3-k} + 2q_p B_{p-3} \\ \equiv & 2E - 2D + 2q_p B_{p-3}. \end{aligned}$$

by the substitution $k \to p-3-k$. Thus by the reversal relation

$$H(-3,1) \equiv -H(1,-3) \equiv 2D - 2E - 2q_p B_{p-3}. \tag{6.7}$$

Similarly we find

$$H(-2,2) \equiv -H(2,-2) \equiv B - A + 2E - 2D + 2q_pB_{p-3}, \tag{6.8}$$

$$H(-1,3) \equiv -H(3,-1) \equiv \frac{1}{3}\Big(-J + 3A - 3B + 2D - 2E\Big). \tag{6.9}$$

Then by adding (6.4), (6.5), (6.7) and (6.8) altogether we have

$$4H(-3,1) + 2H(-2,2) \equiv 0$$

which implies the first congruence in (6.2). Now adding (6.7) and (6.8) yields

$$-H(-3,1) \equiv H(-3,1) + H(-2,2) \equiv B - A \tag{6.10}$$

which is the second congruence in (6.2). Plugging this into (6.4) we see that

$$\frac{5}{2}D - 2E - 2F - \frac{3}{2}q_pB_{p-3} \equiv 0$$

which combined with (6.6) produces (6.3). This finishes the proof of the proposition. □

Proposition 6.2. *For all prime $p \geq 7$ write $H(-) = H(-;p-1)$. Then we have*

$$H(1,-1,-2) \equiv \frac{1}{2}H(1,-3) + \frac{1}{2}J \pmod p, \tag{6.11}$$

$$H(1,-2,-1) \equiv H(1,-3) - \frac{5}{4}q_pB_{p-3} \pmod p, \tag{6.12}$$

$$H(2,-1,-1) \equiv -H(1,-3) - \frac{1}{2}J + \frac{3}{4}q_pB_{p-3} \pmod p. \tag{6.13}$$

Proof. By Theorem 4.2, Proposition 6.1 and (6.1) we get that modulo p

$$\begin{aligned} 2H(1,-1,-2) &\equiv -\sum_{k=2}^{p-3}(k+1)(1-2^k)B_kB_{p-3-k} \\ &\equiv B - A + J - G = H(1,-3) + J, \end{aligned} \tag{6.14}$$

$$H(1,-2,-1) \equiv -\sum_{k=2}^{p-3}(1-2^k)B_kB_{p-3-k} - \frac{5}{4}q_pB_{p-3} \equiv H(1,-3) - \frac{5}{4}q_pB_{p-3},$$

$$\begin{aligned} H(2,-1,-1) &\equiv \sum_{k=2}^{p-3}\frac{(k+2)(1-2^k)B_kB_{p-3-k}}{-2} + \frac{3}{4}q_pB_{p-3} \\ &\equiv -H(1,-3) - \frac{1}{2}J + \frac{3}{4}q_pB_{p-3}, \end{aligned}$$

as claimed. □

By (4.2) and (6.10) it is readily seen that

$$2H(1,1,-2) \equiv H(-3,1) + H(-2,2) \equiv -H(-3,1) \pmod p.$$

In fact, by the stuffle and the reversal relations we can find congruences of all weight four AMHS of depth up to three. By the reversal relations we only need to list about half of the values.

Proposition 6.3. *For all prime $p \geq 7$ write $H(-) = H(-;p-1)$ and set $H_{\bar{2}11} := H(-2,1,1)$. Then $H_{\bar{2}11} = (A-B)/2$ and modulo p we have*

$$\begin{array}{ll} H(4) \equiv H(-4) \equiv H(2,2) \equiv H(-2,-2) \equiv H(1,3) \equiv H(1,-2,1) \equiv 0, & \\ H(-1,-2,-1) \equiv 0, & H(1,-2,-1) \equiv -2H_{\bar{2}11} - \frac{5}{4}q_p, \\ H(2,-2) \equiv 4H_{\bar{2}11}, & H(-1,1,2) \equiv 2H_{\bar{2}11} - \frac{3}{4}q_pB_{p-3}, \\ H(1,-3) \equiv -2H_{\bar{2}11}, & H(-1,2,1) \equiv -H_{\bar{2}11} + \frac{5}{4}q_pB_{p-3}, \\ H(1,-1,2) \equiv 3H_{\bar{2}11}, & H(-2,-1,-1) \equiv 2H_{\bar{2}11} - q_pB_{p-3}, \\ H(3,-1) \equiv \frac{1}{2}q_pB_{p-3}, & H(-2,1,-1) \equiv 3H_{\bar{2}11} - \frac{1}{2}J + \frac{3}{4}q_pB_{p-3}, \\ H(-1,-3) \equiv \frac{1}{2}q_pB_{p-3}, & H(2,-1,-1) \equiv 2H_{\bar{2}11} - \frac{1}{2}J + \frac{3}{4}q_pB_{p-3}, \\ H(1,-1,-2) \equiv -H_{\bar{2}11} + \frac{1}{2}J, & H(-1,2,-1) \equiv -4H_{\bar{2}11} + J - \frac{5}{2}q_pB_{p-3}B_{p-3}. \end{array}$$

We now turn to the depth four cases.

Proposition 6.4. *Let $p \geq 7$ be a prime and write $H(-) = H(-;p-1)$. Then we have*

$$H(-1,-1,1,1) \equiv H(1,1,-1,-1), \qquad H(-1,1,-1,1) \equiv H(1,-1,1,-1) \pmod p,$$

and

$$H(1,-1,-1,1) \equiv -\frac{1}{2}\Big(H(1,-3) + J + q_p^4\Big) \pmod p, \tag{6.15}$$

$$H(1,1,-1,-1) \equiv \frac{1}{24}\Big(6J + 7q_pB_{p-3} + 8q_p^4\Big) \pmod p, \tag{6.16}$$

$$H(1,-1,1,-1) \equiv -\frac{1}{12}\Big(q_pB_{p-3} + 2q_p^4\Big) \pmod p, \tag{6.17}$$

$$H(-1,1,1,-1) \equiv \frac{1}{12}\Big(6H(1,-3) + 7q_pB_{p-3} + 2q_p^4\Big) \pmod p. \tag{6.18}$$

Proof. By Theorem 2.1

$$H(1,-1,-1,1) \equiv -H(-1,-1,1) - \frac{1}{2}H(-2,-1,1) - \sum_{k=2}^{p-3} B_k H(-(k+1),-1,1),$$

$$H(1,1,-1,-1) \equiv -H(1,-1,-1) - \frac{1}{2}H(2,-1,-1) - \sum_{k=2}^{p-3} B_k H(k+1,-1,-1).$$

Using reversal relations and (4.4) we see that

$$H(1,-1,-1,1) \equiv -H(-1,-1,1) - \frac{1}{2}H(-2,-1,1)$$
$$+\frac{1}{2}\sum_{k=2}^{p-3} B_k\Big(H(k+2,1) + H(-(k+1),-2) - H(-(k+1))H(-1,1)\Big), \tag{6.19}$$

$$H(1,1,-1,-1) \equiv -H(1,-1,-1) - \frac{1}{2}H(2,-1,-1)$$
$$-\frac{1}{2}\sum_{k=2}^{p-3} B_k\Big(H(2,k+1) + H(-1,-(k+2)) - H(-1)H(-1,k+1)\Big). \tag{6.20}$$

Note that by Theorem 2.1

$$
\begin{aligned}
-H(-1,1) \equiv& H(1,-1) \equiv -\sum_{k=0}^{p-3}(-1)^k B_k H(-(k+1)),\\
H(1,1,-1) \equiv& -H(1,-1) - \frac{1}{2}H(2,-1) + \sum_{k=2}^{p-3} B_k H(-1,k+1).
\end{aligned}
$$

We may use (3.1), and (3.2) to simplify (6.19) and (6.20) further. For all $k = 2, \ldots, p-5$ we have

$$
H(k+2,1) \equiv -B_{p-3-k},\ H(-(k+1),-2) \equiv \frac{1}{2}(2^{p-3-k}-1)(k+2)B_{p-3-k}, \tag{6.21}
$$

$$
H(2,k+1) \equiv -\frac{(k+2)B_{p-3-k}}{2},\ H(-1,-(k+2)) \equiv (2^{p-3-k}-1)B_{p-3-k}. \tag{6.22}
$$

However, one has to be very careful in applying these formulae because the formulae might fail when $k = p-3$. We need to compute these separately as follows:

$$
\begin{aligned}
H(p-1,1) =& \sum_{i=1}^{p-1}\frac{1}{i}\sum_{j=1}^{i-1}\frac{1}{j^{p-1}} \equiv \sum_{i=1}^{p-1}\frac{i-1}{i} \equiv -1 \equiv -B_0,\\
H(-(p-2),-2) =& \sum_{i=1}^{p-1}\frac{(-1)^i}{i^2}\sum_{j=1}^{i-1}\frac{(-1)^j}{j^{p-2}} \equiv \sum_{i=1}^{p-1}\frac{(-1)^i}{i^2}\left((-1)^i\frac{1-2i}{4}-\frac{1}{4}\right) \equiv 0,\\
H(2,p-2) \equiv& H(p-2,2) = -\sum_{i=1}^{p-1}\frac{1}{i^2}\sum_{j=1}^{i-1}\frac{1}{j^{p-2}} \equiv -\sum_{i=1}^{p-1}\frac{i-1}{2i} \equiv \frac{1}{2} \equiv 0,\\
H(-1,-(p-1)) \equiv& -H(-(p-1),-1) = -\sum_{i=1}^{p-1}\frac{(-1)^i}{i}\sum_{j=1}^{i-1}\frac{(-1)^j}{j^{p-1}}\\
\equiv& \sum_{i=1}^{p-1}\frac{1+(-1)^i}{2i} \equiv \frac{1}{2}H(-1) \equiv -q_p
\end{aligned}
$$

by (5.3). We see that only $H(-1,-(p-1))$ fails the formula in (6.22) and therefore we get

$$
\begin{aligned}
H(1,-1,-1,1) \equiv& -H(-1,-1,1) - \frac{1}{2}H(-2,-1,1) + \frac{1}{2}H(-1)H(-1,1)\\
&-\frac{1}{2}H(-1,1)^2 + \frac{1}{2}\sum_{k=2}^{p-3} B_k\left(-B_{p-3-k} + \frac{1}{2}(2^{p-3-k}-1)(k+2)B_{p-3-k}\right)
\end{aligned}
$$

and

$$
\begin{aligned}
H(1,1,-1,-1) \equiv& H(-1,-1,1) + \frac{1}{2}H(-1)\Big(H(1,1,-1) + H(1,-1) + \frac{1}{2}H(2,-1)\Big)\\
&-\frac{1}{2}H(2,-1,-1) - \frac{1}{2}\sum_{k=2}^{p-3} B_k\left(-\frac{(k+2)B_{p-3-k}}{2} + (2^{p-3-k}-1)B_{p-3-k}\right) + \frac{1}{2}q_p B_{p-3}.
\end{aligned}
$$

Now by [16, Corollary 2.4, Corollary 2.5] we know $H(-1,1) \equiv -q_p^2$, $H(-1,-1,1) \equiv q_p^3 + \frac{7}{8}B_{p-3}$, and $H(1,1,-1) \equiv -\frac{1}{3}q_p^3 - \frac{7}{24}B_{p-3}$. Together with (5.3) these yield

$$\begin{aligned} H(1,-1,-1,1) \equiv& -\frac{1}{2}B_{p-3} - \frac{1}{2}H(-2,-1,1) - \frac{1}{2}q_p^4 - \frac{1}{2}A \\ &+ \frac{1}{4}\sum_{k=0}^{p-3}(2^{p-3-k}-1)(k+2)B_kB_{p-3-k}, \\ H(1,1,-1,-1) \equiv& B_{p-3} - \frac{1}{2}H(2,-1,-1) + \frac{1}{3}q_p^4 + \frac{2}{3}q_pB_{p-3} + A + \frac{1}{4}G \\ &- \frac{1}{2}\sum_{k=0}^{p-3}2^{p-3-k}B_kB_{p-3-k}. \end{aligned}$$

It now follows from (6.1) and the substitution $k \to p-3-k$ that

$$\begin{aligned} H(1,-1,-1,1) \equiv& -\frac{1}{2}H(-2,-1,1) - \frac{1}{2}q_p^4 + \frac{1}{4}\sum_{k=0}^{p-3}(1-2^k)(k+1)B_kB_{p-3-k}, \\ H(1,1,-1,-1) \equiv& -\frac{1}{2}H(2,-1,-1) - \frac{1}{2}H(1,-3) + \frac{2}{3}q_pB_{p-3-k} + \frac{1}{3}q_p^4. \end{aligned}$$

Hence (6.15) and (6.16) quickly follow from (6.14) and (6.13).

Finally, (6.17) follows from stuffle relations applied to $H(-1)H(1,-1,1)$ and then (6.18) from stuffle relations applied to $H(-1)H(1,1,-1)$. This finishes the proof of the proposition. □

For other depth four and weight four AMHS we have the following relations derived from the stuffle relations and the congruences obtained above:

$$\begin{aligned} H(1,1,-1,1) \equiv& 2H_{\bar{2}11} + 3H(-1,1,1,1) + \frac{1}{2}q_pB_{p-3} && \pmod{p}, \\ H(-1,-1,1,-1) \equiv& 6H_{\bar{2}11} + 3H(1,-1,-1,-1) - 4q_pB_{p-3} - 2q_p^4 && \pmod{p}, \\ H(-1,-1,-1,-1) \equiv& \frac{1}{3}q_pB_{p-3} + \frac{2}{3}q_p^4 && \pmod{p}. \end{aligned}$$

On the other hand, we can only deduce from Theorem 2.1 and Theorem 2.2 that modulo p

$$\begin{aligned} H(1,1,1,-1) \equiv& -H(1,1,-1) - \frac{1}{2}H(2,1,-1) - \sum_{k=2}^{p-3}B_kH(k+1,1,-1), \\ H(-1,-1,-1,1) \equiv& -q\Big(H(-1,-1,1) - H(1,-1,1)\Big) - \frac{1}{2}H(2,-1,1) \\ &+ \sum_{k=2}^{p-3}(1-2^k)B_kH(k+1,-1,1), \end{aligned}$$

where, by the reduction theorems again,

$$\begin{aligned}H(k+1,1,-1) \equiv & -\frac{1}{k+1}H(k+1,-1)-\frac{1}{2}H(k+2,-1)\\ & +\sum_{j=2}^{p-k-2}\binom{p-k-1}{j}\frac{B_j}{p-k-1}H(j+k+1,-1),\\ H(k+1,-1,1) \equiv & -\frac{1}{k+1}H(-(k+1),1)-\frac{1}{2}H(-(k+2),1)\\ & +\sum_{j=2}^{p-k-2}\binom{p-k-1}{j}\frac{B_j}{p-k-1}H(-(j+k+1),1).\end{aligned}$$

Observe that the indices k and j in the above sums can be both taken to be even numbers. Thus by Proposition 3.1 and Proposition 3.2 (setting $s=i+j+k$)

$$\begin{aligned}H(j+k+1,-1) \equiv & \sum_{i=0}^{p-j-k-2}\binom{p-j-k-1}{i}\frac{2(1-2^{2p-s-2})B_iB_{2p-s-2}}{(p-j-k-1)(2p-s-2)}\\ \equiv & \sum_{i=0}^{p-j-k-2}\binom{p-j-k-1}{i}\frac{2(1-2^{p-s-1})B_iB_{p-s-1}}{(j+k+1)(s+1)},\end{aligned}$$

$$\begin{aligned}H(-(j+k+1),1) \equiv & \sum_{i=1}^{p-j-k-4}\binom{p-2-j-k}{i}\frac{2(1-2^{i+1})(1-2^{p-s-2})B_{i+1}B_{p-s-2}}{(i+1)(s+2)}\\ & +\frac{2(1-2^{p-j-k-1})(1-2^{p-1})B_{p-j-k-1}B_{p-1}}{p-j-k-1}\\ \equiv & \sum_{i=2}^{p-j-k-3}\binom{p-2-j-k}{i}\frac{2(1-2^{i})(1-2^{p-s-1})B_iB_{p-s-1}}{i(s+1)}\\ & -q_p\frac{2(1-2^{p-j-k-1})B_{p-j-k-1}}{j+k+1}.\end{aligned}$$

Consequently, both $H(1,1,1,-1)$ and $H(-1,-1,-1,1)$ can be written as a triple sum with most of the terms involving products $B_iB_jB_kB_{p-i-j-k-2}$. It is very likely that modulo p we cannot reduce $H(1,1,1,-1)$ and $H(-1,-1,-1,1)$ to a linear combination of AMHS of depths up to three. At least in theory one possible way to check this hypothesis is to find six infinite sets of primes $S_1=\{p_1^{(k)}:k\geq 1\},\ldots,S_6=\{p_6^{(k)}:k\geq 1\}$ for each of the following six elements:

$$\begin{aligned}& b_1(p)=J_p, \quad b_2(p)=H(1,-3), \quad b_3(p)=H(1,-1,-1,-1),\\ & b_4(p)=q_p^4, \quad b_5(p)=q_pB_{p-3}, \quad b_6(p)=H(-1,1,1,1),\end{aligned}$$

such that for each choice $(p_1^{(k)},\ldots,p_6^{(k)})$ we always have $b_j(p_i^{(k)})\equiv 0 \pmod{p_i^{(k)}}$ for all $i\neq j$ and $b_j(p_j^{(k)})\not\equiv 0 \pmod{p_j^{(k)}}$ for all $j=1,\ldots,6$. In practice this is extremely difficult to carry out. For example, if $b_4(p) \equiv 0 \pmod p$ then the prime p is called a Wieferich prime. The only known Wieferich primes are 1093 and 3511 and if any other Wieferich primes exist, they must be greater than 6.7×10^{15} according to [6]. It turns out that

$$[J_p, H(1,-3), H(1,-1,-1,-1), H(-1,1,1,1)] \equiv [1023, 529, 670, 952] \pmod{1093},$$
$$[J_p, H(1,-3), H(1,-1,-1,-1), H(-1,1,1,1)] \equiv [1618, 2160, 1620, 540] \pmod{3511}.$$

In order to understand the general mod p structure of AMHS we need to consider some infinite algebras similar to the adeles (see [21]).

7. Congruences Modulo Prime Powers

In this last section we shall study the congruence properties of AMHS of small weights with prime power moduli. We first consider the sums $U(\mathbf{s}; p-1)$ defined by (5.9).

Proposition 7.1. *Let p be a prime ≥ 7. Let A and B be defined as in section 6. Write $U(-) = U(-; p-1)$. Then we have*

$$U(-1) \equiv -2q_p - \frac{7}{12}p^2 B_{p-3} \pmod{p^3}, \tag{7.1}$$
$$U(-2) \equiv -q_p^2 + \frac{2}{3}pq_p^3 + \frac{7}{6}pB_{p-3} \pmod{p^2}, \tag{7.2}$$
$$U(-1,1) \equiv q_p^2 - \frac{2}{3}pq_p^3 - \frac{1}{12}pB_{p-3} \pmod{p^2}, \tag{7.3}$$
$$U(1,-1) \equiv -\frac{13}{12}pB_{p-3} \pmod{p^2}, \tag{7.4}$$
$$U(-3) \equiv -\frac{1}{3}q_p^3 - \frac{7}{24}B_{p-3} \pmod{p}, \tag{7.5}$$
$$U(-2,1) \equiv \frac{1}{3}q_p^3 - \frac{23}{24}B_{p-3} \pmod{p}, \tag{7.6}$$
$$U(1,-2) \equiv \frac{5}{4}B_{p-3} \pmod{p}, \tag{7.7}$$
$$U(2,-1) \equiv -\frac{3}{4}B_{p-3} \pmod{p}, \tag{7.8}$$
$$U(-1,2) \equiv \frac{1}{3}q_p^3 + \frac{25}{24}B_{p-3} \pmod{p}, \tag{7.9}$$
$$U(1,1,-1) \equiv -\frac{1}{2}B_{p-3} \pmod{p}, \tag{7.10}$$
$$U(1,-1,1) \equiv \frac{1}{2}B_{p-3} \pmod{p}, \tag{7.11}$$
$$U(-1,1,1) \equiv -\frac{1}{3}q_p^3 - \frac{7}{24}B_{p-3} \pmod{p}, \tag{7.12}$$
$$U(-4) \equiv H(-1,1,1,1) \equiv -H(1,1,1,-1) \pmod{p}, \tag{7.13}$$
$$U(1,-3) \equiv A - B + \frac{5}{4}q_pB_{p-3} \pmod{p}, \tag{7.14}$$
$$U(-3,1) \equiv H(1,1,1,-1) + B - A - \frac{5}{4}q_pB_{p-3} \pmod{p}. \tag{7.15}$$

Proof. Throughout the proof we write $H(-) = H(-;p-1)$ and similarly for U and V. We will prove the congruences in the following order: (7.5), (7.1), (7.6), (7.7), (7.2), (7.3), (7.4), (7.10), (7.8), (7.9), and (7.11) to (7.15).

First, (7.5) follows from [5, (4)] and $H(-3) \equiv -\frac{1}{2}B_{p-3}$ (take $a = 3$ in (2.8)). Taking $d = 1$ and $m = p-1$ in (5.14) we have

$$\begin{aligned}\sum_{k=1}^{p-1} \frac{(1-x)^k - 1}{k} &= \sum_{j=1}^{p-1} \frac{(-x)^j}{j}\binom{p-1}{j} = \sum_{j=1}^{p-1} \frac{(-x)^j}{j}\left(\binom{p}{j} - \binom{p-1}{j-1}\right) \\ &= -p\sum_{j=1}^{p-1} \frac{x^j}{j^2}(-1)^{j-1}\binom{p-1}{j-1} - \frac{1}{p}\sum_{j=1}^{p-1}(-x)^j\binom{p}{j} \\ &= -p\sum_{j=1}^{p-1} \frac{x^j}{j^2}(-1)^{j-1}\binom{p-1}{j-1} + \frac{(x-1)^p - x^p + 1}{p}. \qquad (7.16)\end{aligned}$$

Since

$$(-1)^{j-1}\binom{p-1}{j-1} \equiv 1 - pH(1;j-1) \pmod{p^2}$$

letting $x = -1$ in (7.16) we get

$$U(-1) - H(1) \equiv -pH(-2) + p^2H(1,-2) - 2q_p \pmod{p^3}. \qquad (7.17)$$

So it is not hard to see that (7.1) can be obtained from the following: $H(1,-2) \equiv \frac{1}{4}B_{p-3} \pmod{p}$ by [16, Corollary 2.4], $H(-2) \equiv \frac{1}{2}pB_{p-3} \pmod{p^2}$ by (2.8), and the well-known fact $H(1) \equiv -\frac{1}{3}p^2B_{p-3} \pmod{p^3}$ (see, for e.g., (1.3)). Letting $x = 1/2$ in (7.16) we have that

$$V(-1) - H(1) \equiv -pV(-2) + p^2V(1,-2) + \frac{q_p}{2^{p-1}} \pmod{p^3}.$$

Multiplied by -2^p this yields by Lemma 5.2

$$U(-1) + 2^pH(1) \equiv p^2U(-3) + p^2U(-2,1) - 2q_p \pmod{p^3}. \qquad (7.18)$$

Thus (7.6) follows from (7.1), (7.5) and $H(1) \equiv -\frac{1}{3}p^2B_{p-3} \pmod{p^3}$. Further, by the stuffle relation

$$U(1,-2) \equiv H(1)U(-2) - U(-2,1) - U(-3) \equiv \frac{5}{4}B_{p-3} \pmod{p}$$

we get (7.7). Now letting $x = 2$ in (7.16) we see that

$$H(-1) - H(1) \equiv -pU(-2) + p^2U(1,-2) - 2q_p \pmod{p^3}. \qquad (7.19)$$

Hence (7.2) follows from (7.7) and (5.3). Then taking $d = 2$ and $x = -1$ in (5.15) we get

$$\begin{aligned}U(-1,1) + U(-2) \equiv& H(1,1) + H(2) + H(-2) - pH(1,-2) - pH(-3) \\ \equiv& \frac{13}{12}pB_{p-3} \pmod{p^2}. \qquad (7.20)\end{aligned}$$

Thus $U(1,-1) = U(-1)H(1) - U(-1,1) - U(-2) \equiv -\frac{13}{12}pB_{p-3} \pmod{p^2}$ which is (7.3). Then (7.4) follows from the stuffle relation $U(-1,1) = H(1)U(-1) - U(1,-1) - U(-2)$. Moreover, taking $d = 1$ and $x = 2$ in (5.15) we get

$$H(-1)-H(1) \equiv U(-1)-p\Big(U(-2)+U(1,-1)\Big)+p^2\Big(U(1,-2)+U(1,1,-1)\Big). \pmod{p^3}$$

This implies (7.10) because of (7.1), (7.2), (7.3) and (7.7).

Next, taking $d = 3$ and $x = -1$ in (5.15) we get

$$\begin{aligned} U(-1,1,1) + U(-1,2) + U(-2,1) \equiv& H(-3) + S(1,1,1) - U(-3) \\ \equiv& \frac{1}{3}q_p^3 - \frac{5}{24}B_{p-3} \pmod{p} \end{aligned} \tag{7.21}$$

since $S(1,1,1) = H(1,1,1) + H(1)H(2) \equiv 0 \pmod{p}$. Taking $d = 3$ and $x = 1/2$ in (5.15) we get

$$V(-1,1,1) + V(-2,1) + V(-1,2) \equiv 0 \pmod{p}$$

which implies by Lemma 5.2 that

$$U(1,1,-1) + U(1,-2) + U(2,-1) \equiv 0 \pmod{p}. \tag{7.22}$$

Thus (7.8) follows from (7.7) and (7.10) immediately. Then (7.9) follows easily from the stuffle relation of $H(2)U(-1)$. Also, (7.11) and (7.12) follow from the stuffle relations:

$$\begin{aligned} H(1)U(1,-1) =& U(2,-1) + U(1,-2) + 2U(1,1,-1) + U(1,-1,1) \equiv 0 \pmod{p}, \\ H(1)U(-1,1) =& U(-2,1) + U(-1,2) + 2U(-1,1,1) + U(1,-1,1) \equiv 0 \pmod{p}. \end{aligned}$$

We now turn to the last three congruences of weight four. By [16, Theorem 2.3] we know (7.13) holds. Hence taking $d = 4$, $m = p - 1$ and $x = 2$ in Lemma 5.4 and using (5.16) we get

$$\begin{aligned} S(-1,1,1) + p\Big(U(-4) + H(-1,2,1) + H(-2,1,1) + H(-3,1) - H(1)S(-1,1,1)\Big) \\ \equiv U(-3) - H(1) \pmod{p^2}. \end{aligned} \tag{7.23}$$

On the other hand, taking $d = 2$ and $x = 2$ in (5.15) we get

$$S(-1,1,1) - S(1,1,1) \equiv U(-3) - pU(-4) - pU(1,-3) \pmod{p^2}. \tag{7.24}$$

Comparing (7.23) and (7.24), using $S(1,1,1) = H(1,1,1) + H(1)H(2) \equiv 0 \pmod{p^2}$ by [18, Theorem 2.13], and $H(-1,2,1) + H(-2,1,1) \equiv \frac{5}{4}q_pB_{p-3} \pmod{p}$ by Proposition 6.3, we can deduce (7.14). Finally, (7.15) follows from (7.13) and the stuffle relation $U(-3,1) = H(1)U(-3) - U(1,-3) - U(-4)$. This finishes the proof of the proposition. □

Lemma 7.1. *Let $a, b \in \mathbb{Z}^*$ and write $H(-) = H(-;p-1)$. Then*

$$H(a,b) \equiv (-1)^{a+b}\operatorname{sgn}(ab)\Big(H(b,a) + p|b|H(1\oplus b,a) + p|a|H(b,1\oplus a)\Big) \pmod{p^2}.$$

Proof. By definition

$$H(a,b) = \sum_{1\le m<n<p} \frac{\operatorname{sgn}(a)^m \operatorname{sgn}(b)^n}{m^{|a|}n^{|b|}} = \sum_{1\le n<m<p} \frac{\operatorname{sgn}(a)^{p-m} \operatorname{sgn}(b)^{p-n}}{(p-m)^{|a|}(p-n)^{|b|}}$$
$$=(-1)^{a+b}\operatorname{sgn}(ab) \sum_{1\le n<m<p} \frac{\operatorname{sgn}(b)^n \operatorname{sgn}(a)^m}{n^{|b|}m^{|a|}} \frac{1}{(1-p/n)^{|b|}(1-p/m)^{|a|}}.$$

The lemma follows easily. □

Proposition 7.2. *Let A and B be defined as in section 6. For all prime $p \ge 7$ write $H(-) = H(-;p-1)$ and set (see Theorem 1.1(b))*

$$X = X_p(3) := \frac{B_{p-3}}{p-3} - \frac{B_{2p-4}}{4p-8}.$$

Then we have

$$H(-1,-1) \equiv 2q_p^2 + p\left(2X - 2q_p^3\right) + p^2\left(\frac{11}{6}q_p^4 + \frac{1}{2}q_pB_{p-3}\right) \pmod{p^3} \tag{7.25}$$
$$\equiv 2q_p^2 - 2pq_p^3 - \frac{1}{3}pB_{p-3} \pmod{p^2}, \tag{7.26}$$
$$H(1,-1) \equiv q_p^2 - pq_p^3 - \frac{13}{24}pB_{p-3} \pmod{p^2}, \tag{7.27}$$
$$H(-1,1) \equiv -q_p^2 + pq_p^3 + \frac{1}{24}pB_{p-3} \pmod{p^2}, \tag{7.28}$$
$$H(-3) \equiv 3X \pmod{p^2}, \tag{7.29}$$
$$H(-2,1) \equiv H(1,-2) \equiv -\frac{3}{2}X \pmod{p^2}, \tag{7.30}$$
$$H(2,-1) \equiv -\frac{3}{2}X - \frac{7}{6}pq_pB_{p-3} + p(B-A) \pmod{p^2}, \tag{7.31}$$
$$H(-1,2) \equiv -\frac{3}{2}X - \frac{1}{6}pq_pB_{p-3} + p(A-B) \pmod{p^2}. \tag{7.32}$$

Proof. First, (7.25) follows readily from the shuffle relation $H(-1)^2 = 2H(-1,-1) + H(2)$, the congruence (5.3), and $H(2) \equiv -4pX \pmod{p^3}$ by (1.3). This clearly implies (7.26) since by Kummer congruence $X \equiv -B_{p-3}/6 \pmod{p}$.

Next, taking $d=1$, and $x=-1$ in (5.15) we get

$$U(-1) - H(1) \equiv H(-1) - p\Big(H(-2) + H(1,-1)\Big) \pmod{p^2}.$$

Combining this with (7.1) and using (5.3) we get

$$H(1,-1) \equiv q_p^2 - \frac{2}{3}pq_p^3 - \frac{1}{4}pB_{p-3} + pH(1,1,-1) \equiv q_p^2 - pq_p^3 - \frac{13}{24}pB_{p-3} \pmod{p^2}$$

which is (7.27). Then (7.28) is deduced from the stuffle relation

$$H(-1,1) \equiv H(-1)H(1) - H(1,-1) - H(-2) \equiv -q_p^2 + pq_p^3 + \frac{1}{24}pB_{p-3} \pmod{p^2}.$$

Turning to weight three we get by [16, Theorem 2.1]

$$H(-3) \equiv \frac{1}{4} H(3; (p-1)/2) \pmod{p^2}.$$

Hence (7.29) follows from (1.4). Now by Lemma 7.1 we have the reversal relation

$$H(-2,1) \equiv H(1,-2) + pH(2,-2) + 2pH(1,-3) \equiv H(1,-2) \pmod{p^2} \quad (7.33)$$

by Proposition 6.1. On the other hand, by stuffle relation

$$H(-2,1) + H(1,-2) = H(1)H(-2) - H(-3) \equiv -3X \pmod{p^2}.$$

Together with (7.33) this clearly yields (7.30).

Finally, by stuffle relation and (7.29)

$$H(2,-1)+H(-1,2) = H(2)H(-1)-H(-3) \equiv -\frac{4}{3} pq_p B_{p-3} - 3X \pmod{p^2}. \quad (7.34)$$

By Lemma 7.1 we have the reversal relation

$$\begin{aligned} H(2,-1) \equiv& H(-1,2) + pH(-2,2) + 2pH(-1,3) \\ \equiv& H(1,-2) + 2p(B-A) - pq_p B_{p-3} \pmod{p^2} \end{aligned}$$

by Proposition 6.1. Combining with (7.34) this implies (7.31) and (7.32).

We have finished the proof of the proposition. □

Corollary 7.1. *For every prime $p \geq 7$ we have*

$$U(-1; p-1) \equiv -2q_p + \frac{7}{2} X p^2 + \frac{1}{2} p^3 H(-3,1; p-1) \pmod{p^4}. \quad (7.35)$$

Proof. Write $U(-) = U(-; p-1)$ and $H(-) = H(-; p-1)$. Taking $x = -1$ in (7.16) we get

$$U(-1) - H(1) \equiv -pH(-2) + p^2 H(1,-2) - p^3 H(1,1,-2) - 2q_p \pmod{p^4}.$$

Thus the corollary follows from (7.30), Proposition 6.3, the following congruences

$$\begin{aligned} H(1) &\equiv 2p^2 X && \pmod{p^4} && \text{(by [15, Remark 5.1])}, \\ H(2) &\equiv -4pX && \pmod{p^3} && \text{(by (1.3))}, \\ H(2,(p-1)/2) &\equiv -14pX && \pmod{p^3} && \text{(by (1.4))}, \end{aligned}$$

and $H(-2) = \frac{1}{2} H(2,(p-1)/2) - H(2)$. □

Proposition 7.3. *For all prime $p \geq 7$ write $H(-) = H(-; p-1)$, $h_{31} := H(3,1; (p-1)/2)$, and set X as above. Then*

$$H(-1,-2) \equiv \frac{9}{2} X - \frac{5}{6} pq_p B_{p-3} + \frac{1}{4} ph_{31} \pmod{p^2}, \quad (7.36)$$

$$H(-2,-1) \equiv -\frac{9}{2} X - \frac{1}{6} pq_p B_{p-3} - \frac{1}{4} ph_{31} \pmod{p^2}. \quad (7.37)$$

Proof. By expanding $H(1,2;p-1)$ we get

$$\begin{aligned}H(1,2;(p-1)/2)-H(2,1;(p-1)/2)\equiv pH(2,2;(p-1)/2)+2ph_{31}\\ -H(1;(p-1)/2)\Big(H(2;(p-1)/2)+2pH(3;(p-1)/2)\Big)+H(1,2;p-1)\\ \equiv -6X-\frac{10}{3}pq_pB_{p-3}+2ph_{31} \pmod{p^2}\end{aligned}$$

since $2H(2,2;(p-1)/2)=H(2;(p-1)/2)^2-H(4;(p-1)/2)\equiv 0 \pmod p$ and

$$H(1,2)\equiv -H(2,1)\equiv -6X \pmod{p^2} \tag{7.38}$$

by [17, Theorem 2.3]. Hence

$$\begin{aligned}H(-1,-2)\equiv&\frac{1}{4}\Big(H(1,2;(p-1)/2)-H(2,1;(p-1)/2)-ph_{31}\Big)-H(1,2)\\ \equiv&\frac{9}{2}X-\frac{5}{6}pq_pB_{p-3}+\frac{1}{4}ph_{31} \pmod{p^2}\end{aligned}$$

which is (7.36). Finally, (7.37) follows from

$$H(-1,-2)+H(-2,-1)=H(-1)H(-2)-H(3)\equiv -pq_pB_{p-3} \pmod{p^2}.$$

This completes the proof of the proposition. □

Finally, we consider the depth three cases.

Proposition 7.4. *Let A and B be defined as in section 6. For all prime $p\geq 7$ set X as above, write $H(-)=H(-;p-1)$ and $h_{31}:=H(3,1;(p-1)/2)$ as above. Then we have (setting $C=A-B$)*

$$H(-1,-1,-1)\equiv -\frac{4}{3}q_p^3+X+p\Big(2q_p^4+\frac{2}{3}q_pB_{p-3}\Big) \pmod{p^2}, \tag{7.39}$$

$$H(-1,1,-1)\equiv \frac{p}{2}\Big(q_pB_{p-3}-C\Big) \pmod{p^2}, \tag{7.40}$$

$$H(1,-1,-1)\equiv -q_p^3+\frac{21}{4}X+p\Big(\frac{3}{2}q_p^4+\frac{3}{8}q_pB_{p-3}+\frac{C}{4}+\frac{h_{31}}{8}\Big) \pmod{p^2}, \tag{7.41}$$

$$H(-1,-1,1)\equiv q_p^3-\frac{21}{4}X+p\Big(-\frac{3}{2}q_p^4+\frac{1}{8}q_pB_{p-3}+\frac{C}{4}-\frac{h_{31}}{8}\Big) \pmod{p^2}, \tag{7.42}$$

$$H(1,-1,1)\equiv -2H(1,1,-1)+3X+p\Big(\frac{7}{6}q_pB_{p-3}+C\Big), \pmod{p^2}, \tag{7.43}$$

$$H(-1,1,1)\equiv H(1,1,-1)-p\Big(\frac{1}{2}q_pB_{p-3}+C\Big) \pmod{p^2}. \tag{7.44}$$

Proof. First, (7.39) follows from the stuffle relation

$$3H(-1,-1,-1)=H(-1)H(-1,-1)-H(-1,2)-H(2,-1).$$

By reversal relation (similar to the proof of Lemma 7.1) we can show easily that

$$\begin{aligned}&H(-1,1,-1)\equiv\\ &-H(-1,1,-1)-p\Big(H(-2,1,-1)+H(-1,2,-1)+H(-1,1,-2)\Big) \pmod{p^2}.\end{aligned}$$

Notice that $H(-1,1,-2) \equiv H(-2,1,-1) \pmod p$. Hence (7.40) follows from Proposition 6.3. It then implies (7.41) and (7.42) by the two stuffle relations:

$$2H(1,-1,-1) = H(-1)H(1,-1) - H(-1,1,-1) - H(-2,-1) - H(1,2),$$
$$2H(-1,-1,1) = H(-1)H(-1,1) - H(-1,1,-1) - H(-1,-2) - H(2,1),$$

Proposition 7.2, Proposition 7.3 and (7.38). Similarly, the last two congruences follow immediately from the stuffle relations:

$$H(1,-1,1) = H(1)H(1,-1) - 2H(1,1,-1) - H(2,-1) - H(1,-2),$$
$$2H(-1,1,1) = H(1)H(-1,1) - H(1,-1,1) - H(-2,1) - H(-1,2),$$

and Proposition 7.2. This completes the proof of the proposition. □

Remark 7.1. Currently, we are not able to express $H(1,1,-1) \pmod{p^2}$ explicitly. In fact, by (7.23) it is not hard to find that

$$H(1,1,-1) \equiv U(-3) - p\Big(U(-4) + \frac{7}{12}q_pB_{p-3} + A - B\Big) \pmod{p^2},$$

hence it is equivalent to determining $U(-3) \pmod{p^2}$.

In conclusion, we remark that to study weight w AMHS modulo p^2 we need information of weight $w+1$ AMHS modulo p, and if we change modulus to p^3 then we would need to know AMHS of weight $w+2$ modulo p. Hence, it is possible that results in [2] might provide some help in determining the congruence of weight three AMHS modulo prime squares.

References

[1] M. Abramowitz and I.A. Stegun, *Handbook of Mathematical Functions with Formulas, Graphs, and Mathematical Tables*, Dover, New York, 1972.

[2] T. Agoh and L. Skula, The fourth power of the Fermat quotient, *J. Number Theory* 128 (2008), 2865–2873.

[3] P. Bachmann, Über den Rest von $\frac{2^{p-1}-1}{p} \pmod p$, *J. Reine Angew. Math.* 142 (1913), 41–50.

[4] M. Chamberland and K. Dilcher, Divisibility properties of a class of binomial sums, *J. Number Theory* 120 (2006), 349–371.

[5] K. Dilcher and L. Skula, The cube of the Fermat quotient, Integers 6 (2006), #A24.

[6] F. Dorais and D. Klyve, Near Wieferich primes up to 6.7×10^{15}, preprint, 2008.

[7] A. Gardiner, Four problems on prime power divisibility, *Amer. Math. Monthly* 95 (1988), 926–931.

[8] A. Granville, The square of the Fermat quotient, Integers 4 (2004), #A22.

[9] M.E. Hoffman, The Hopf algebra structure of multiple harmonic sums, *Nuclear Phys. B Proc. Suppl.* 135 (2004), 214–219.

[10] M.E. Hoffman, Quasi-symmetric functions and mod p multiple harmonic sums, preprint, 2004, arXiv: math/0401319.

[11] M.E. Hoffman, Algebraic aspects of multiple zeta values, in: Zeta Functions, Topology and Quantum Physics, Developments in Math. 14, T. Aoki et. al. (eds.), Springer, New York, 2005, pp. 51–74.

[12] O.P. Lossers, Solution III of problem 10490: a reciprocal summation identity, *Amer. Math. Monthly* 106 (1999), 589–590.

[13] I.G. Macdonald, Symmetric Functions and Hall Polynomials, 2nd ed., *Oxford Univ. Press,* New York, 1995.

[14] G. Racinet, Doubles mélanges des polylogarithmes multiples aux racines de l'unité, *Publ. Math. Inst. Hautes 'Etudes Sci.* 95 (2002), 185–231.

[15] Z.-H. Sun, Congruences concerning Bernoulli numbers and Bernoulli polynomials, *Discrete Appl. Math.* 105 (2000), 193–223.

[16] R. Tauraso, Congruences involving alternating multiple harmonic sum, *Electron. J. Combin.* 17 (2010), #R16.

[17] R. Tauraso, More congruences for central binomial coefficients, *J. Number Theory* 130(2010), 2639–2649.

[18] J. Zhao, Wolstenholme-type theorem for multiple harmonic sums, *Int. J. Number Theory* 4(2008), 73–106.

[19] J. Zhao, Multiple harmonic sums II: finiteness of p-divisible sets, preprint, 2003, arXiv: math/0303043.

[20] J. Zhao, On a conjecture of Borwein, Bradley and Broadhurst, *J. Reine Angew. Math.* 639 (2010), 223–233.

[21] J. Zhao, Mod p structure of alternating and non-alternating multiple harmonic sums, preprint, 2010.

In: Frontiers of Combinatorics and Number Theory, Volume 2 ISBN: 978-1-61942-918-5
Editor: Zhi-Wei Sun, pp. 167-176

Chapter 10

More Generating Functions for Values of Certain L-Functions

Alexander E. Patkowski*
University of Regina, Regina,
Saskatchewan, Canada

Abstract

We prove some identities on generating functions for values of L-functions related to the work of two papers [2, 5].

2010 Mathematics Subject Classification: 11P81, 11E16, 05A17

Keywords: q-series, L-functions, generating functions

1. Introduction and Main Results

In a paper by Zagier [10], generating functions for values of L-functions and sums of tails of q-series were investigated. Zagier showed that

$$-2e^{-t/24}\sum_{n=0}^{\infty}(1-e^{-t})(1-e^{-2t})\cdots(1-e^{-nt}) = \sum_{n=0}^{\infty}\left(-\frac{1}{24}\right)^{n} L(\chi_{12},-2n-1)\frac{t^n}{n!}, \quad (1.1)$$

where

$$\chi_{12}(n) := \begin{cases} 1, & \text{if } n \equiv 1,\, 11 \pmod{12}, \\ -1, & \text{if } n \equiv 5,\, 7 \pmod{12}, \\ 0, & \text{otherwise.} \end{cases}$$

Set $\chi(a)$ to be the order 2 character of conductor $4(3+\sqrt{6})$ on ideals in $\mathbb{Z}[\sqrt{6}]$ where

$$\chi(a) := \begin{cases} i^{yx}\left(\frac{12}{x}\right), & \text{if } y \equiv 0 \pmod{2}, \\ i^{yx+1}\left(\frac{12}{x}\right), & \text{if } y \equiv 1 \pmod{2}, \end{cases}$$

*E-mail address: alexpatk@hotmail.com

and $a = (x + y\sqrt{6})$. Consider the Hecke L-function

$$\Upsilon(\rho, s) = \sum_{n=1}^{\infty} \frac{a(n)}{n^s} := \sum_{a \subseteq \mathbb{Z}[\sqrt{6}]} \chi(a) N(a)^{-s}.$$

For integers $1 \leq r < 48$ let the partial L-function be given by

$$\Upsilon_r(\rho, s) := \sum_{n \equiv r \pmod{48}}^{\infty} \frac{a(n)}{n^s}.$$

In [2], Andrews, Jiménez-Urroz and Ono (using the work of [1]) gave the identity

$$-2e^{t/24} \sum_{n=0}^{\infty} (1 - e^{-t})(1 - e^{-3t}) \cdots (1 - e^{-(2n-1)t}) = \sum_{n=0}^{\infty} \left(-\frac{1}{24}\right)^n \bar{\Upsilon}(\rho, -n) \frac{t^n}{n!}, \quad (1.2)$$

(where $\bar{\Upsilon}(\rho, s) = \Upsilon_{23}(\rho, s) + \Upsilon_{47}(\rho, s)$) as a formal power series in t.

Other examples of generating functions for values of L-functions are given in [2, 4, 5, 8]. In this paper we offer some results related to the work of [2] as well as [5]. Our first result is related to the right side of (1.2).

Theorem 1.1. *As a formal power series in* t, *we have*

$$\begin{aligned} &- e^{t/24} \sum_{n=0}^{\infty} \frac{(1 - e^{-2t})(1 - e^{-6t}) \cdots (1 - e^{-(4n-2)t})}{(1 + e^{-t})(1 + e^{-3t}) \cdots (1 + e^{-(4n+1)t})} \\ &= \frac{1}{4} \sum_{n=0}^{\infty} \left(-\frac{1}{24}\right)^n \bar{\Upsilon}(\rho, -n) \frac{t^n}{n!}, \end{aligned} \quad (1.3)$$

where $\bar{\Upsilon}(\rho, s) = \Upsilon_{23}(\rho, s) + \Upsilon_{47}(\rho, s)$.

The next result is obtained using the work of Corson et. al [5]. We consider the Hecke L-function with $K = \mathbb{Q}(\sqrt{2})$, given by

$$\zeta_K(s) = \sum_{\substack{a \subseteq O_K \\ N(a) \equiv -1 \pmod{8}}} \chi(a) N(a)^{-s}, \quad (1.4)$$

where $\chi(a) = (-1)^{\frac{N(a)+1}{8}}$ and $N(a) \equiv -1 \pmod{8}$, and where O_K is the ring of integers of K.

Theorem 1.2. *As a formal power series in* t, *we have*

$$-e^{t/8} \sum_{n=0}^{\infty} \frac{(1 - e^{-2t})^2 (1 - e^{-6t})^2 \cdots (1 - e^{-(4n-2)t})^2}{(1 + e^{-t})(1 + e^{-3t}) \cdots (1 + e^{-(4n+1)t})} = \frac{1}{2} \sum_{n=0}^{\infty} \left(-\frac{1}{8}\right)^n \zeta_K(-n) \frac{t^n}{n!}. \quad (1.5)$$

We can also obtain the following equally deep generating function for the $\Upsilon_r(\rho, s)$ functions in our next theorem.

Theorem 1.3. *As a formal power series in t, we have*

$$e^{t/24}\sum_{n=0}^{\infty}\frac{(1-e^{-t})(1+e^{-2t})\cdots(1+(-1)^{n}e^{-nt})}{(1+e^{-2t})(1+e^{-4t})\cdots(1+e^{-2nt})}=\frac{1}{2}\sum_{n=0}^{\infty}\left(-\frac{1}{24}\right)^{n}\bar{\Upsilon}(\rho,-n)\frac{t^{n}}{n!}, \tag{1.6}$$

where, again, $\bar{\Upsilon}(\rho,s)=\Upsilon_{23}(\rho,s)+\Upsilon_{47}(\rho,s)$.

Next we consider a result that is related to both Theorem 5 of [2] and D. Corson et. al [5]. First we give the sum of tails identity. We shall invoke the notation [6] $(x;q)_n:=\prod_{j=0}^{n-1}(1-xq^j)$.

Theorem 1.4. *We have*

$$\begin{aligned}
&q^{1/8}\sum_{n=0}^{\infty}\left(\frac{(q^2;q^2)_\infty}{(q;q^2)_\infty}-\frac{(q^2;q^2)_{2n}}{(q;q^2)_{2n+1}}\right)\\
&=-\frac{1}{4}\sum_{\substack{n\geq 0\\|j|\leq n}}^{\infty}q^{(2n+1)^2/8-2(2j)^2/8}(1+q^{(16n+8)/8})\\
&+\frac{1}{2}q^{1/8}\frac{(q^2;q^2)_\infty}{(q;q^2)_\infty}\left(-\frac{1}{2}+\sum_{n=1}^{\infty}\frac{(-1)^n q^n}{1-q^n}\right)+\frac{1}{4}\sum_{n=0}^{\infty}(2n+1)q^{(2n+1)^2/8}.
\end{aligned} \tag{1.7}$$

Following [5], define the Hecke L-function for $K=\mathbb{Q}(\sqrt{2})$:

$$\lambda_K(s)=\sum_{\substack{a\subseteq O_K\\ N(a)\equiv 1 \pmod 8}}\chi(a)N(a)^{-s},$$

where $\chi(a)=(-1)^{(N(a)-1)/8}$ when $N(a)\equiv 1 \pmod 8$.

Theorem 1.5. *As a formal power series in t, we have*

$$\begin{aligned}
&e^{-t/8}\sum_{n=0}^{\infty}(-1)^n\frac{(1-e^{-2t})(1-e^{-4t})\cdots(1-e^{-(2n)t})}{(1+e^{-t})(1+e^{-3t})\cdots(1+e^{-(2n+1)t})}\\
&=\frac{1}{2}\sum_{n=0}^{\infty}\left(-\frac{1}{8}\right)^n\lambda_K(-n)\frac{t^n}{n!}.
\end{aligned} \tag{1.8}$$

2. Proofs of Main Theorems

We prove our main results using similar methods to those found in [2]. We require Andrews' transformation [7].

Lemma 2.1. *(Andrews' transformation [7])*

$$\sum_{n=0}^{\infty}\frac{(a;q^2)_n(b;q)_{2n}}{(q^2;q^2)_n(c;q)_{2n}}t^n=\frac{(at;q^2)_\infty(b;q)_\infty}{(t;q^2)_\infty(c;q)_\infty}\sum_{n=0}^{\infty}\frac{(c/b)_n(t;q^2)_n b^n}{(q)_n(at;q^2)_n}. \tag{2.1}$$

We also require a sum of tails identity.

Proposition 2.1. *We have*

$$\sum_{n=0}^{\infty}\left(\frac{(a;q^2)_\infty(b;q)_\infty}{(q^2;q^2)_\infty(c;q)_\infty}-\frac{(a;q^2)_n(b;q)_{2n}}{(q^2;q^2)_n(c;q)_{2n}}\right)$$

$$=\frac{(a;q^2)_\infty(b;q)_\infty}{(q^2;q^2)_\infty(c;q)_\infty}\left(\sum_{n=1}^{\infty}\frac{q^{2n}}{1-q^{2n}}-a\sum_{n=0}^{\infty}\frac{q^{2n}}{1-aq^{2n}}-\sum_{n=1}^{\infty}\frac{(c/b)_n(q^2;q^2)_{n-1}b^n}{(q)_n(a;q^2)_n}\right). \tag{2.2}$$

Proof. Let, as in [2], ϵ be the differential operator $\lim_{t\to 1^-}\frac{d}{dt}$. Applying Andrews' transformation to Proposition 2.1 of [2] we find

$$\sum_{n=0}^{\infty}\left(\frac{(a;q^2)_\infty(b;q)_\infty}{(q^2;q^2)_\infty(c;q)_\infty}-\frac{(a;q^2)_n(b;q)_{2n}}{(q^2;q^2)_n(c;q)_{2n}}\right)$$

$$=\epsilon(1-t)\frac{(at;q^2)_\infty(b;q)_\infty}{(t;q^2)_\infty(c;q)_\infty}\sum_{n=0}^{\infty}\frac{(c/b)_n(t;q^2)_nb^n}{(q)_n(at;q^2)_n}$$

$$=\frac{(a;q^2)_\infty(b;q)_\infty}{(q^2;q^2)_\infty(c;q)_\infty}\left(\sum_{n=1}^{\infty}\frac{q^{2n}}{1-q^{2n}}-a\sum_{n=0}^{\infty}\frac{q^{2n}}{1-aq^{2n}}+\epsilon\sum_{n=0}^{\infty}\frac{(c/b)_n(t;q^2)_nb^n}{(q)_n(at;q^2)_n}\right)$$

$$=\frac{(a;q^2)_\infty(b;q)_\infty}{(q^2;q^2)_\infty(c;q)_\infty}\left(\sum_{n=1}^{\infty}\frac{q^{2n}}{1-q^{2n}}-a\sum_{n=0}^{\infty}\frac{q^{2n}}{1-aq^{2n}}-\sum_{n=1}^{\infty}\frac{(c/b)_n(q^2;q^2)_{n-1}b^n}{(q)_n(a;q^2)_n}\right),$$

where we have used $\epsilon(t;q)_n=-(q;q)_{n-1}$ in the last line. □

Theorem 2.6. *We have the following sums of tails:*

$$\sum_{n=0}^{\infty}\left(\frac{(q^2;q^4)_\infty}{(-q;q^2)_\infty}-\frac{(q^2;q^4)_n}{(-q;q^2)_{2n+1}}\right)=\frac{(q^2;q^4)_\infty}{(-q;q^2)_\infty}D_6(q)+\frac{1}{2}\sum_{n=1}^{\infty}\frac{(-1)^{n-1}q^{n^2}}{(q;q^2)_n}, \tag{2.3}$$

and

$$\sum_{n=0}^{\infty}\left(\frac{(q^2;q^4)_\infty^2}{(-q;q^2)_\infty}-\frac{(q^2;q^4)_n^2}{(-q;q^2)_{2n+1}}\right)=\frac{(q^2;q^4)_\infty^2}{(-q;q^2)_\infty}D_2(q)+\frac{1}{2}\sum_{n=1}^{\infty}\frac{(-1;q^2)_n(-q)^n}{(q;q^2)_n}, \tag{2.4}$$

where

$$D_6(q)=-\frac{1}{8}+\sum_{n=1}^{\infty}\frac{q^{4n}}{1-q^{4n}}+\frac{1}{2}\sum_{n=1}^{\infty}\frac{q^{2n-1}}{1-q^{2n-1}}$$
$$-\frac{1}{2}\sum_{n=1}^{\infty}\frac{q^{n(3n+1)/2}}{1-q^n},$$

and

$$D_2(q)=\sum_{n=1}^{\infty}\frac{(-1)^nq^{2n}}{1-q^{2n}}+\frac{1}{2}\sum_{n=1}^{\infty}\frac{q^{2n-1}}{1-q^{2n-1}}+\frac{1}{2}\sum_{n=1}^{\infty}\frac{q^{2n-1}}{1+q^{2n-1}}.$$

Proof. We consider (2.4) first. From Proposition 2.2 with $a = q$, $b = q$, $c = -q^{3/2}$, and then replacing q by q^2, we have

$$\sum_{n=0}^{\infty}\left(\frac{(q^2;q^4)_\infty^2}{(-q;q^2)_\infty}-\frac{(q^2;q^4)_n^2}{(-q;q^2)_{2n+1}}\right)$$
$$=\frac{(q^2;q^4)_\infty^2}{(-q;q^2)_\infty}\left(\sum_{n=1}^{\infty}\frac{(-1)^n q^{2n}}{1-q^{2n}}-\sum_{n=1}^{\infty}\frac{(-q^2;q^2)_{n-1}q^{2n}}{(q;q^2)_n(1-q^{2n})}\right). \tag{2.5}$$

On the other hand, from [2, Theorem 2] with $b = q$, $a = q^{1/2}$, and $c = -q$, we obtain

$$-\sum_{n=1}^{\infty}\frac{(-q^2;q^2)_{n-1}q^{2n}}{(q;q^2)_n(1-q^{2n})}$$
$$=\frac{1}{2}\frac{(-q^2;q^2)_\infty}{(q;q^2)_\infty}\sum_{n=0}^{\infty}\left(\frac{(q;q^2)_\infty}{(-q^2;q^2)_\infty}-\frac{(q;q^2)_n}{(-q^2;q^2)_n}\right)-\frac{1}{2}\sum_{n=1}^{\infty}\frac{(-1)^n q^n}{1-q^n}. \tag{2.6}$$

after replacing q by q^2. Also, from [2, Theorem 1] we find

$$\sum_{n=0}^{\infty}\left(\frac{(q;q^2)_\infty}{(-q^2;q^2)_\infty}-\frac{(q;q^2)_n}{(-q^2;q^2)_n}\right)$$
$$=\sum_{n=1}^{\infty}\frac{(-1;q^2)_n(-q)^n}{(q;q^2)_n}+\frac{(q;q^2)_\infty}{(-q^2;q^2)_\infty}\left(\sum_{n=1}^{\infty}\frac{q^{2n}}{1-q^{2n}}+\sum_{n=1}^{\infty}\frac{q^{2n-1}}{1+q^{2n-1}}\right), \tag{2.7}$$

from putting $t = q^{1/2}$, and $a = -q$, and then replacing q by q^2. Using the fact that

$$\frac{(q^2;q^4)_\infty^2}{(-q;q^2)_\infty}=\frac{(q;q^2)_\infty}{(-q^2;q^2)_\infty},$$

and combining (2.5), (2.6), and (2.7), we find that we obtain (2.4).

Now for the case (2.3) we take $a = 0$, $b = q$, $c = -q^{3/2}$, and then replace q by q^2 in (2.2) to get

$$\sum_{n=0}^{\infty}\left(\frac{(q^2;q^4)_\infty}{(-q;q^2)_\infty}-\frac{(q^2;q^4)_n}{(-q;q^2)_{2n+1}}\right)$$
$$=\frac{(q^2;q^4)_\infty}{(-q;q^2)_\infty}\left(\sum_{n=1}^{\infty}\frac{q^{4n}}{1-q^{4n}}-\sum_{n=1}^{\infty}\frac{(-q;q)_{2n-1}q^{2n}}{(1-q^{2n})}\right). \tag{2.8}$$

Note that

$$\sum_{n=1}^{\infty}\frac{(-q;q)_{2n-1}q^{2n}}{(1-q^{2n})}=\frac{1}{2}\left(\sum_{n=1}^{\infty}\frac{(-q;q)_{n-1}q^n}{(1-q^n)}+\sum_{n=1}^{\infty}\frac{(-q;q)_{n-1}(-1)^n q^n}{(1-q^n)}\right). \tag{2.9}$$

From [2, Theorem 2] with $a = 0$, $b = q$, and $c = -q$, we find

$$\sum_{n=0}^{\infty}\left(\frac{1}{(-q;q)_\infty}-\frac{1}{(-q;q)_n}\right)=\frac{1}{(-q;q)_\infty}\left(\sum_{n=1}^{\infty}\frac{q^n}{1-q^n}-2\sum_{n=1}^{\infty}\frac{(-q;q)_{n-1}q^n}{(1-q^n)}\right), \tag{2.10}$$

which by [2, Theorem 3, $i = 6$] is equal to

$$\frac{1}{(-q;q)_\infty}\left(-\frac{1}{2}+\sum_{n=1}^{\infty}\frac{q^n}{1-q^n}\right)-2\sum_{n=1}^{\infty}\frac{(-1)^n q^{n^2}}{(q;q^2)_n}. \tag{2.11}$$

Hence,

$$\begin{aligned}
&-(q;q^2)_\infty\sum_{n=1}^{\infty}\frac{(-q;q)_{n-1}q^n}{(1-q^n)}\\
&=-\frac{1}{2}(q;q^2)_\infty\sum_{n=1}^{\infty}\frac{q^n}{1-q^n}+\frac{1}{2}\sum_{n=0}^{\infty}\left(\frac{1}{(-q;q)_\infty}-\frac{1}{(-q;q)_n}\right)\\
&=-\frac{1}{2}(q;q^2)_\infty\sum_{n=1}^{\infty}\frac{q^n}{1-q^n}+\frac{1}{2(-q;q)_\infty}\left(-\frac{1}{2}+\sum_{n=1}^{\infty}\frac{q^n}{1-q^n}\right)-\sum_{n=1}^{\infty}\frac{(-1)^n q^{n^2}}{(q;q^2)_n}\\
&=-\frac{1}{4}(q;q^2)_\infty-\sum_{n=1}^{\infty}\frac{(-1)^n q^{n^2}}{(q;q^2)_n}.
\end{aligned} \tag{2.12}$$

Now we need Baileys' Lemma [9], which states that, if for every $n \geq 0$,

$$\beta_n=\sum_{r=0}^{n}\frac{\alpha_r}{(q)_{n-r}(aq)_{n+r}}, \tag{2.13}$$

then

$$\sum_{n=0}^{\infty}(x)_n(y)_n(aq/xy)^n\beta_n=\frac{(aq/x)_\infty(aq/y)_\infty}{(aq)_\infty(aq/xy)_\infty}\sum_{n=0}^{\infty}\frac{(x)_n(y)_n(aq/xy)^n\alpha_n}{(aq/x)_n(aq/y)_n}. \tag{2.14}$$

Also, we require the well-known Bailey pair (α_n, β_n) [9, B(1)] with respect to 1,

$$\beta_n=\frac{1}{(q)_n}, \tag{2.15}$$

$$\alpha_n=(-1)^n q^{n(3n-1)/2}(1+q^n). \tag{2.16}$$

Inserting (2.15)-(2.16) into (2.14) and differentiating with respect to x, setting $x = 1$, and putting $y = -1$, we find

$$\begin{aligned}
-2\sum_{n=1}^{\infty}\frac{(-q;q)_{n-1}(-1)^n q^n}{(1-q^n)}&=\sum_{n=1}^{\infty}\frac{q^n}{1-q^n}+\sum_{n=1}^{\infty}\frac{q^n}{1+q^n}-2\sum_{n=1}^{\infty}\frac{q^{n(3n+1)/2}}{1-q^n}\\
&=2\sum_{n=1}^{\infty}\frac{q^{2n-1}}{1-q^{2n-1}}-2\sum_{n=1}^{\infty}\frac{q^{n(3n+1)/2}}{1-q^n}.
\end{aligned} \tag{2.17}$$

Finally, we get

$$-(q;q^2)_\infty\sum_{n=1}^{\infty}\frac{(-q;q)_{2n-1}q^{2n}}{(1-q^{2n})}=-\frac{1}{8}(q;q^2)_\infty-\frac{1}{2}\sum_{n=1}^{\infty}\frac{(-1)^n q^{n^2}}{(q;q^2)_n}$$

$$+\frac{1}{2}(q;q^2)_\infty \sum_{n=1}^{\infty} \frac{q^{2n-1}}{1-q^{2n-1}} - \frac{1}{2}(q;q^2)_\infty \sum_{n=1}^{\infty} \frac{q^{n(3n+1)/2}}{1-q^n}. \tag{2.18}$$

After inserting (2.18) into (2.8) we obtain the desired identity. □

We prove Theorem 1.2 and since Theorem 1.1 directly falls out of the work of [2] and our sums of tails identity (2.3) we leave this to the reader.

***Proof of Theorem* 1.2.** For this theorem, we appeal to [4, Theorem 2.4]

$$\sum_{n=1}^{\infty} \frac{(-1;q^2)_n(-q)^n}{(q;q^2)_n} = \sum_{\substack{n=1 \\ -n\leq j\leq n-1}}^{\infty} (-1)^n q^{n(2n-1)-(j^2-j)}(1+q^{2n}).$$

Notice that

$$\frac{(q^2;q^4)_\infty^2}{(-q;q^2)_\infty}$$

vanishes to infinite order as $q \to 1$. Replace q by e^{-t} with $t \searrow 0$ in (2.4). Next, we define the coefficients $\sigma(n)$ and $\sigma'(n)$ by the following asymptotic expansions

$$-e^{t/8}\sum_{n=0}^{\infty} \frac{(1-e^{-2t})^2(1-e^{-6t})^2\cdots(1-e^{-(4n-2)t})^2}{(1+e^{-t})(1+e^{-3t})\cdots(1+e^{-(4n+1)t})} = \sum_{n=0}^{\infty}\sigma(n)t^n, \tag{2.19}$$

and

$$\sum_{\substack{a\subseteq O_K \\ N(a)\equiv -1 \pmod 8}} \chi(a)e^{-N(a)t/8} \sim \sum_{n=0}^{\infty}\sigma'(n)t^n, \tag{2.20}$$

where O_K is as in (1.4). We find that (2.4), (2.19)-(2.20), and (3.2) of [5], imply $\sigma(n) = \frac{1}{2}\sigma'(n)$ for all natural n. Taking the Mellin transform of the left hand side of (2.20) we find

$$\int_0^\infty \sum_{\substack{a\subseteq O_K \\ N(a)\equiv -1 \pmod 8}} \chi(a)e^{-N(a)t/8}t^{s-1}dt$$

$$= \sum_{\substack{a\subseteq O_K \\ N(a)\equiv -1 \pmod 8}} \chi(a)\int_0^\infty e^{-N(a)t/8}t^{s-1}dt$$

$$= 8^s\Gamma(s)\zeta_K(s). \tag{2.21}$$

On the other hand, we have

$$\int_0^\infty \sum_{\substack{a\subseteq O_K \\ N(a)\equiv -1 \pmod 8}} \chi(a)e^{-N(a)t/8}t^{s-1}dt$$

$$= \int_0^\infty \left(\sum_{n=0}^{N-1}\sigma'(n)t^n + O(t^N)\right)t^{s-1}dt = \sum_{n=0}^{N-1}\frac{\sigma'(n)}{(n+s)} + \lambda(s),$$

where $\lambda(s)$ is an analytic function for $\Re(s) > -N$. Lastly, we compute the residue at $s = -n$ of (2.21) to be

$$\sigma'(n) = \frac{8^{-n}(-1)^n}{n!}\zeta_K(-n). \qquad \square$$

***Proof of Theorem* 1.3.** For (1.6), we replace c by cb, and apply the operator ϵ to b on both sides of (2.1), and then take $a = -1$, $t = q^2$, $c = 0$. The resulting identity is

$$\sum_{n=0}^{\infty}\left(\frac{(-q;q)_\infty}{(-q^2;q^2)_\infty} - \frac{(-q;q)_n}{(-q^2;q^2)_n}\right)$$
$$= \frac{(-q;q)_\infty}{(-q^2;q^2)_\infty}\left(\sum_{n=1}^{\infty}\frac{q^n}{1-q^n} - \sum_{n=1}^{\infty}\frac{(-1;q^2)_n(q;q^2)_n q^{2n}}{1-q^{2n}}\right). \tag{2.22}$$

Notice the last series on the right side of (2.22) is twice the left side of (2.18) when q has been replaced by $-q$. Inserting (2.18) in (2.22) we find

$$\sum_{n=0}^{\infty}\left(\frac{(-q;q)_\infty}{(-q^2;q^2)_\infty} - \frac{(-q;q)_n}{(-q^2;q^2)_n}\right)$$
$$= \frac{(-q;q)_\infty}{(-q^2;q^2)_\infty}\sum_{n=1}^{\infty}\frac{q^n}{1-q^n} - \frac{1}{4}(-q;q^2)_\infty - \sum_{n=1}^{\infty}\frac{q^{n^2}}{(-q;q^2)_n}$$
$$- (-q;q^2)_\infty\sum_{n=1}^{\infty}\frac{q^{2n-1}}{1+q^{2n-1}} - (-q;q^2)_\infty\sum_{n=1}^{\infty}\frac{(-q)^{n(3n+1)/2}}{1-(-q)^n}. \tag{2.23}$$

The remaining details rely on the fact that $(-q;q)_\infty/(-q^2;q^2)_\infty$ vanishes to infinite order when $q \to -1$. $\square$

We give the proof of Theorem 1.4 and give brief details for Theorem 1.5.

***Proof of Theorem* 1.4.** In Proposition 2.2, take $a = q^2$, $c = q^{3/2}$, $b = q$, to get

$$\sum_{n=0}^{\infty}\left(\frac{(q^2;q^2)_\infty}{(q;q^2)_\infty} - \frac{(q^2;q^2)_{2n}}{(q;q^2)_{2n+1}}\right) = -\frac{(q^2;q^2)_\infty}{(q;q^2)_\infty}\sum_{n=1}^{\infty}\frac{q^{2n}(q;q^2)_n}{(q^2;q^2)_n(1-q^{4n})}.$$

Rewriting $1/(1-q^{4n})$ we get the right side is equal to

$$-\frac{1}{2}\frac{(q^2;q^2)_\infty}{(q;q^2)_\infty}\left(\sum_{n=1}^{\infty}\frac{q^{2n}(q;q^2)_n}{(q^2;q^2)_n(1-q^{2n})} + \sum_{n=1}^{\infty}\frac{q^{2n}(q;q^2)_n}{(q^2;q^2)_n(1+q^{2n})}\right). \tag{2.24}$$

The right most sum is easily identified to be the $z = q$, $a = -1$, $c = -q$, with $b = q^{1/2}$ case of [7, pg.13, eq.(1.4.3)]:

$$\sum_{n=1}^{\infty}\frac{(q;q^2)_n q^{2n}}{(q^2;q^2)_n(1+q^{2n})} = \frac{1}{2}\left(\frac{(q;q^2)_\infty}{(q^2;q^2)_\infty}\sum_{n=0}^{\infty}\frac{(-q;q^2)_n q^n}{(-q^2;q^2)_n} - 1\right). \tag{2.25}$$

The first sum on the right side of (2.24) is just equation (3.9) from [2]. That is,

$$\sum_{n=0}^{\infty}\left(\frac{(q^2;q^2)_\infty}{(q;q^2)_\infty}-\frac{(q^2;q^2)_n}{(q;q^2)_{n+1}}\right)=-\frac{(q^2;q^2)_\infty}{(q;q^2)_\infty}\sum_{n=1}^{\infty}\frac{(q;q^2)_n q^{2n}}{(q^2;q^2)_n(1-q^{2n})}. \tag{2.26}$$

Now from [3], we find that

$$\sum_{n=0}^{\infty}\frac{(q;q^2)_n(-q)^n}{(-q^2;q^2)_n}=\sum_{n=0}^{\infty}(-1)^n q^{n(2n+1)}(1-q^{2n+1})\sum_{j=-n}^{n}(-1)^j q^{-j^2}. \tag{2.27}$$

Hence, inserting (2.25), (2.26), and (2.27) in (2.24) we obtain the result. □

***Proof of Theorem* 1.5.** In order to prove this theorem, subtract [2, eq.(4.7)] from twice equation (1.7), and proceed as in the proof of other similar results, using the fact that $(q^2;q^2)_\infty/(q;q^2)_\infty$ vanishes to infinite order when $q\to -1$, and writing the right side of (2.27) in the form

$$\sum_{\substack{a\subseteq O_K\\ N(a)\equiv 1 \pmod 8}}(-1)^{\frac{N(a)-1}{8}}q^{\frac{N(a)-1}{8}}.$$ □

3. Concluding Remarks

We have restricted ourselves to the study of different identities from [2] and the work of [5]. One should also observe something interesting with the products that appear in each of the sums of tails identities. That is, for example, in (2.3) we have that

$$\frac{(q^2;q^4)_\infty}{(-q;q^2)_\infty}=(q;q^2)_\infty,$$

which led to a connection to [2, Theorem 6].

References

[1] G.E. Andrews, F.J. Dyson, and D. Hickerson, Partitions and indefinite quadratic forms, *Invent. Math.* 91 (1988), 391–407.

[2] G. E. Andrews, J. Jiménez-Urroz and K. Ono, q-series identities and values of certain L- functions, *Duke Math. J.* 108 (2001), 395–419.

[3] G.E. Andrews, Partitions with distinct evens, *Advances in Combinatorial Mathematics, Proceedings of the Waterloo Workshop in Computer Algebra 2008*, I. Kotsireas and E. Zima, eds., pp. 31–37. Springer, 2009.

[4] K. Bringmann and B. Kane, New identities involving sums of the tails related to real quadratic fields, *Special issue of the Ramanujan Journal in honor of G.E. Andrews's 70th birthday,* Ramanujan J., to appear.

[5] D. Corson, D. Favero, K. Liesinger, and S. Zubairy, Characters and q-series in $\mathbb{Q}(\sqrt{2})$, *J. Number Theory.* 107 (2004), 392–405.

[6] N. J. Fine, Basic hypergeometric series and applications, *Math. Surveys and Monographs,* Vol. 27, Amer. Math. Soc., Providence, 1988.

[7] G. Gasper and M. Rahman, *Basic hypergeometric series*, Cambridge Univ. Press, Cambridge, 1990.

[8] J. Lovejoy, K. Ono, Hypergeometric generating functions for values of Dirichlet and other L-functions *Proc. Nat. Acad. Sci. USA* 100(2003), 6904-6909.

[9] L. J. Slater, A new proof of Rogers transformations of infinite series, *Proc. London Math. Soc.* (2)53(1951), 460–475.

[10] D. Zagier, Vassiliev invariants and a strange identity related to the Dedekind eta-function, *Topology* 40(2001), 945–960.

In: Frontiers of Combinatorics and Number Theory, Volume 2 ISBN: 978-1-61942-918-5
Editor: Zhi-Wei Sun, pp. 177-185

Chapter 11

SMALL PRIMITIVE ROOTS AND MALLEABILITY OF RSA MODULI*

***Luis Dieulefait*[1,†] *and Jorge Jiménez Urroz*[2,‡]**
[1] Dept. Álgebra. Universitat de Barcelona
[2]Dept. Matemàtica Aplicada IV.
Universitat Politècnica de Catalunya
Barcelona, Spain

Abstract

In their paper [13], P. Paillier and J. Villar make a conjecture about the malleability of an RSA modulus. In this paper we show how subtle the notion of malleability is. Specifically, we demonstrate how to factor an RSA modulus n using some information on the factorization of a concrete n' coprime to n. However, we believe the conjecture might be true, when imposing some extra conditions on the auxiliary n' allowed to be used.

2010 Mathematics Subject Classification: Primary 11A07; Secondary 11A51, 94A60.

1. Introduction

When building a secure cryptosystem one needs a one-way function, i.e. a function which serves to easily encrypt clear messages as ciphertexts, but that also is very hard to invert, to recover the message from the ciphertext. The security of the encryption may be called one-wayness. One of the most popular ways to prove the security of the cryptosystem is through the concept of Semantic Security: a cryptosystem is semantically secure if an attacker cannot retrieve any information of the original message from the ciphertext without solving a difficult mathematical problem. And, in fact, sometimes getting some information and solving the problem are equivalent tasks. A higher level of security is when the attacker

*Partially supported by MEC of Spain, first author by grant MTM2006-04895, second by grants MTM2006-15038-C02-02, TSI2006-02731 and MTM2009-11068

†E-mail address: ldieulefait@ub.es

‡E-mail address: jjimenez@ma4.upc.edu

cannot get any information even when one provides him with several clear messages corresponding to ciphertexts conveniently chosen by him. This is called chosen ciphertext security, or indistinguishable under chosen ciphertext attack, in short, IND-CCA security. Unfortunately, it seems to exists a tradeoff between one-wayness and IND-CCA security, see for example [14, 16, 5]. In some sense, one cannot achieve one-way encryption with a level of security equivalent to a difficult problem, and also achieve IND-CCA security with respect to it.

Since this so called paradox was first observed, a number of authors have tried to formally prove it. However no one succeeded until very recently, when Paillier and Villar (cf. [13]) clarified the question for the case of factoring-based cryptosystems. More precisely, they reformulated the paradox in terms of key preserving black-box reductions, (with only one public key allowed), and proved that if factoring can be reduced in the Standard Model to breaking one-wayness of the cryptosystem then it is impossible to achieve chosen-cyphertext security. As an aside, already mentioned by the authors in their paper (cf. [13]), it is interesting to note that the combination of this result with the security proofs contained in [3, 4] gives a very interesting separation result between the Random Oracle Model and the Standard Model.

In order to extend this result the authors needed to assume an extra hypothesis, which they call "non-malleability" of the key generator. The precise definition of non-malleability is a little involved, (you can find it in [13, page 262]), but the idea is that factorization is malleable if there exist a polynomial time algorithm which can factor an integer n with higher probability when provided as input the factorization of some other integer n', conveniently chosen, but coprime with n. In fact, for the definition, Paillier and Villar restrict the set of integers n and n' only to RSA moduli. With the assumption that factoring is non-malleable they were able to extend the result from key preserving black box reductions to the case of arbitrary black box reductions, (where multiple public keys are allowed).

Hence, as the authors themselves stress in [13], it is very important to study non-malleability of key generators. In fact, they conjecture that most instance generators are non-malleable, and claim that they observed this property in many of them. Apart from that, no arguments are given to support this belief (see Sections 4 and 6 of [13]). The goal of this note is to shed some light on this open question.

Actually, the notion of non-malleability captures a very basic fact in arithmetic: the random behavior of prime numbers, observed many times in the literature, suggests that if the numbers n' are randomly selected their factorization is useless for the problem of factoring n (see [1]). However this might not be so relevant to malleability because we have the freedom to select cleverly the additional numbers n'.

Indeed, the result contained in this note goes against the non-malleability intuition, thus showing how subtle this notion is. Concretely, we are able to prove the existence of a polynomial time algorithm reducing factoring an RSA modulus n to a computational problem related to the factorization of certain explicit n', relatively prime to n. In other words, we show that factoring is, in this generality, a malleable problem. Let us stress that this might be compatible with the conjecture of [13] mentioned above because imposing extra conditions on the numbers n' may result in transforming the problem in a non-malleable one. In fact, it is our belief that malleability is a notion that depends strongly on these kind of extra conditions, and hence requires further research.

2. The Algorithm

Given an RSA modulus $n = pq$, we want to find n' such that factoring n', with the help of an oracle, will allow us in finding the factorization of n. In fact we will only need very partial information about the factorization of n' in order to get the complete factorization of n. From now on, and without loss of generality, we will make the assumption that $p < q$.

2.1. A Particular Case

In this section we assume that either $2^{p-1} \not\equiv 1 \pmod q$ or $2^{q-1} \not\equiv 1 \pmod p$. This is simpler than the general case, and illustrates the whole idea of the method. We consider $n' = 2^n + 1$. Observe that an efficient encoding of n' of size comparable to n is available since all these numbers in binary form have a 1 at the beginning and end, and the rest are precisely $n - 1$ zeros. In this way, the binary code of n is already a compressed way of writing n'. We will call the compressed form of n' the binary code of n. Let us assume the existence of an oracle $\mathcal{O}$ which, on input n', returns the residue classes modulo n of three prime factors $r|n'$. In fact, the only thing we need is the residue class of just one factor of n' modulo n different from 1 and 3 so, if convenient, one can admit an oracle answering any set $S \subset \{r \pmod n) \ : \ r \text{ prime }, r|n'\}$, $S \not\subset \{1, 3\}$ and polynomial size. We now present an algorithm which on the input and RSA modulus n in the conditions of this section, outputs a nontrivial factor of n.

Algorithm 1. 1. *Send* $n' = 2^n + 1$ *in compressed form to* $\mathcal{O}$.

2. $\mathcal{O}$ *will return a set* S *as defined above.*

3. *Take* $r \in S$, $r \neq 1, 3$, *and compute* $d = (r - 1, n)$.

Theorem 2.1. *Let* $n = pq$ *be and RSA modulus such that either* $2^{p-1} \not\equiv 1 \pmod q$ *or* $2^{q-1} \not\equiv 1 \pmod p$. *Then, the number* d *computed with the aim of Algorithm* 1 *is a prime divisor of* n. *Algorithm* 1 *runs in polynomial time in* $\log n$.

Proof. We will prove that at least one prime factor of n' is not 1 or 3 modulo n. Suppose r is a prime factor of n'. Then, $2^{2n} \equiv 1 \pmod r$ and so, either $r = 3$ which always divides n', or the order of 2 in $\mathbb{F}_r^*$ is $\text{ord}_r(2) = p, q, 2p, 2q, pq$ or $2pq$. In this case we just have to recall that the order of any element must divide the order of the group to conclude that either $p|(r-1)$, $q|(r-1)$ or $n|(r-1)$. Note, on the other hand that 9 never divides n' since $n \equiv \pm 1 \pmod 6$ and so $2^n \equiv 2$ or 5 modulo 9. Hence, If $n|(r-1)$ for any $r|n'/3$, then each factor of $n'/3$ is 1 modulo n and so $n'/3 \equiv 1 \pmod n$ which is the same as saying $2^{n-1} \equiv 1 \pmod n$. This is impossible since in particular $2^{n-1} \equiv 2^{p-1} \pmod q$ and $2^{n-1} \equiv 2^{q-1} \pmod p$. Hence there exists $r_0|n'$ such that $r_0 \not\equiv 1 \pmod n$. Observe also that any such factor verifies $r_0 \equiv 1 \pmod p$ or $r_0 \equiv 1 \pmod q$ and, in particular, $r_0 \not\equiv 3 \pmod n$. Finally, observe that appart from the call to the oracle, algorithm chosse and element from a polynomial size set, and then computes the Euclid algorithm which again is polynomic. The result follows. □

The previous algorithm would work, in particular, for any modulus $n = pq$ such that $(p-1, q-1) = D$ is small, for example $D \leq \log_2(n)$. Indeed, if $2^{p-1} \equiv 1 \pmod q$

and $2^{q-1} \equiv 1 \pmod p$, then $2^D \equiv 1 \pmod n$ which is impossible for $D \le \log_2(n)$. On the other hand, when D is large, pairs of such primes are rare. Indeed, both primes would be in the same residue class, 1, modulo D, and the number of such pairs of primes should decrease with D, since primes are well distributed among arithmetic progressions. In particular by the Prime Number Theorem for arithmetic progressions the number of pairs of primes up to z with such property is roughly $\frac{z^2}{\varphi(D)(\log z)^2}$. While this is valid for any D fixed, we will prove a more useful result to make D vary. To this purpose we will use the Barban-Davenport-Halberstam Theorem for primes in arithmetic progressions, (cf. page 421, [8]), which we now include for convenience. We need some notation first. For any positive x, y, and integer d, let

$$\Psi(d; x, y) = \sum_{\substack{m \equiv 1 \pmod d \\ x \le m < y}} \Lambda(m),$$

where, $\Lambda(\cdot)$ is the usual Von Mangoldt function. Given $z > 0$, we will write $E(d; z) = \Psi(d; z, 2z) - \frac{1}{\varphi(d)} z$, to measure the error in the approximation of the number of primes in the congruence 1 modulo d, by the total number of primes divided by the number of congruences. Note that, trivially,

$$\Psi(d; z, 2z) = \sum_{\substack{p \equiv 1 \pmod d \\ z \le p < 2z}} \log p + O(\sqrt{z} \log z).$$

Theorem 2.2 (Barban-Davenport-Halberstam). *We have*

$$\sum_{d \le z^{1-\varepsilon}} (E(d; z))^2 \ll z^2/(\log z)^A,$$

for any $A > 0$, and $\varepsilon > 0$, where the implied constant only depends on A and ε.

Remark 2.1. This is not exactly the theorem as it appears in page 421 of [8]. However, Theorem 2.2 easily follows from that one by noting that

$$\begin{aligned} E(d; z)^2 &= \left(\left(\psi(2z; d, 1) - \frac{2z}{\varphi(d)}\right) - \left(\psi(z; d, 1) - \frac{z}{\varphi(d)}\right)\right)^2 \\ &\le \left(\psi(2z; d, 1) - \frac{2z}{\varphi(d)}\right)^2 + \left(\psi(z; d, 1) - \frac{z}{\varphi(d)}\right)^2. \end{aligned}$$

where, as in page 121 of [8],

$$\psi(x; q, a) = \sum_{\substack{m \equiv a \pmod q \\ m < x}} \Lambda(m),$$

for any real x and coprime integers a, q.

The previous theorem leads to the interesting observation that even the probability that $D > \log_2(n)$ tends to zero with n. This is the content of the following proposition

Theorem 2.3. *We have*

$$\sum_{\substack{z\le p,q<2z\\(p-1,q-1)>\log z}} 1 = O\left(\frac{z^2}{(\log z)^3}\right),$$

where the sum runs over the prime numbers in the interval.

Remark 2.2. Before proving the theorem, let us observe that we just have to use the Prime Number Theorem to obtain $\sum_{z\le p,q<2z} 1 \sim (z/\log z)^2$ and, hence, the probability of finding a pair of primes in the interval $[z, 2z]$ which do not satisfy the conditions in Theorem 2.1 tends to zero faster than $1/\log z$. Also note that even if $(p-1, q-1)$ would be big, we still would need 2 to have order D modulo p and modulo q which one expects to be false for many pairs of primes by Artin's conjecture, (cf. [11]).

Proof of Theorem 2.3. Given a positive z, with the notation as above we have that the number of pairs of primes $z \le p, q < 2z$ such that $(p-1, q-1) = d > \log z$ is bounded above by

$$\begin{aligned}\sum_{\log z<d<z}\ \sum_{\substack{p,q\equiv 1 \pmod d\\ z\le p<q<2z}} 1 &< \frac{1}{(\log z)^2}\sum_{\log z<d<z}\ \sum_{\substack{m,k\equiv 1 \pmod d\\ z\le m<k<2z}} \Lambda(m)\Lambda(k)\\ &< \frac{1}{(\log z)^2}\sum_{\log z<d<z^\alpha}\Psi(d;z,2z)^2 + \frac{1}{(\log z)^2}\sum_{z^\alpha<d<z}\Psi(d;z,2z)^2\\ &= S_1 + S_2,\end{aligned}$$

for any $0 < \alpha < 1$. For the second term we get trivially the bound $S_2 < 4z^{3-2\alpha}$. To estimate S_1 we see that

$$\begin{aligned}S_1 &= \frac{1}{(\log z)^2}\sum_{\log z<d<z^\alpha}\left(\frac{z}{\varphi(d)} + E(d,z)\right)^2\\ &= \left(\frac{1}{\log z}\right)^2\left(\sum_{\log z<d<z^\alpha}\frac{z^2}{\varphi(d)^2} + \sum_{\log z<d<z^\alpha}(E(d,z))^2 + 2\sum_{\log z<d<z^\alpha}\frac{z}{\varphi(d)}E(d,z)\right).\end{aligned}$$

We can use now Cauchy-Schwartz inequality to get, for the last sum above

$$\sum_{\log z<d<z^\alpha}\frac{z}{\varphi(d)}E(d,z) \le \left(\sum_{\log z<d<z^\alpha}\left(\frac{z}{\varphi(d)}\right)^2\right)^{1/2}\left(\sum_{\log z<d<z^\alpha}(E(d,z))^2\right)^{1/2}. \tag{2.1}$$

We are in the correct position to use Theorem 2.2. Indeed, substituting the above inequality in S_1, and putting $A = 2$ and $\varepsilon = 1/4$ in Theorem 2.2 we get for some constant C

$$S_1 \le \left(\frac{z}{\log z}\right)^2\sum_{\log z<d<z}\frac{1}{\varphi(d)^2} + \frac{Cz^2}{(\log z)^4} + \frac{Cz^2}{(\log z)^3}\left(\sum_{\log z<d<z}\frac{1}{\varphi(d)^2}\right)^{1/2}. \tag{2.2}$$

In [9], Mirsky proved that

$$S(x) = \sum_{d \le x} \frac{d^2}{\varphi(d)^2} = Bx + O(\log x), \tag{2.3}$$

where

$$B = \prod_p \left(1 + \frac{2}{p(p-1)}\right)\left(1 + \frac{1}{(p-1)(p^2-p+2)}\right).$$

Hence, partial summation gives

$$\sum_{\alpha<d<\beta} \frac{1}{\varphi(d)^2} = \sum_{\alpha<d<\beta} \frac{1}{d^2}\frac{d^2}{\varphi(d)^2} = \frac{S(\beta)}{\beta^2} - \frac{S(\alpha)}{\alpha^2} + 2\int_\alpha^\beta \frac{S(t)}{t^3}dt.$$

Letting $\beta \to \infty$, and using (2.3) we get

$$\sum_{\alpha<d} \frac{1}{\varphi(d)^2} = \frac{B}{\alpha} + O\left(\frac{\log \alpha}{\alpha^2}\right).$$

We just have to choose $\alpha = \log z$, and plug it in (2.2) to get the result.

Remark 2.3. From 2.2, one could use Mertens Theorem (cf. p.34, [8]) and the fact that

$$\varphi(d) = d\prod_{p|d}(1-1/p) \ge d\prod_{p\le d}(1-1/p) > Cd/\log d, \tag{2.4}$$

to get

$$\sum_{\log z<d<z} \frac{1}{\varphi(d)^2} \le C\sum_{\log z<d}\left(\frac{\log d}{d}\right)^2 \le C_1\frac{(\log\log z)^2}{\log z},$$

for some constants C, C_1, which already would give probability tending to zero for the number of pairs of primes with $D > \log z$.

2.2. The General Case

For a few pairs of primes, it could happen that the order of 2 in $\mathbb{F}_q^*$ and $\mathbb{F}_p^*$ is a divisor of D and, in that case, 2^n is indeed 2 modulo n which could make Algorithm 1 fail. To avoid this problem, instead of 2, we will choose a primitive root of $\mathbb{F}_q^*$, m, to build our test number $n' = m^n + 1$. It is very easy to see that the number of primitive roots of $\mathbb{F}_q^*$ is $\varphi(q-1)$, hence, the probability for an integer m to be a primitive root is

$$\frac{\varphi(q-1)}{q-1} = \prod_{p|(q-1)}\left(1-\frac{1}{p}\right) \ge \prod_{p<\omega(q-1)}\left(1-\frac{1}{p}\right) > \prod_{p<\log_2 q}\left(1-\frac{1}{p}\right) \sim \frac{e^{-\gamma}}{\log\log q},$$

again by Mertens theorem, and the trivial fact that for any integer n, the number of distinct prime factors of n is at most $\log_2 n$. In other words, a set of size $C\log\log q$ of integers contains a primitive root modulo q with probability as close to one as we want, making the constant C big enough. To see this, note that the probability for a randomly chosen set of this size to contain no primitive roots would be $(1-1/(e^{\gamma}\log\log q))^{C\log\log q}\sim e^{-C/e^{\gamma}}$. For a set of consecutive integers Bach, in [2], made a much more accurate heuristic argument to claim that the least primitive root modulo p, should be less than $e^{\gamma}\log p(\log\log p)^2(1+\varepsilon)$ for almost all p. Although this fact is not yet proved, there are conditional results consistent with the statement. In particular we will mention the following result of V. Shoup in [15] proved under the Grand Riemann Hypothesis, GRH from now on.

Theorem 2.4 (Shoup). *Let p be a prime number. Then, if GRH is true, the least primitive root modulo p is $O((\log p)^6)$.*

Observe that, although far from the expected result, $O((\log p)^6)$ is still of polynomial size and, hence, good enough for our purposes. Let us recall that Artin's Conjecture was proved by Hooley in [7], under GRH and also, unconditionally, Heath-Brown was able to prove in [6] that among $2,3,5$ there is a primitive root for infinitely many primes p. Let us now describe the algorithm.

For convenience we will call $c\in\{0,1\}^{n+2}$ the binary encoding of 2^n+1. We will take advantage of the fact that the m-ary representation of the numbers m^n+1 is always c, independent of m. Let $n'_m=m^n+1$ and consider the function $\omega(n)$ counting the number of distinct prime factors of n. Assume the existence of an oracle $\mathcal{O}$ which, on input (c,m), returns a set of residue classes S of size $|S|=\omega(m)+2$ when such a set S exists, and otherwise returns $\emptyset$. Again, the only thing we need is the residue class of just one factor of n'_m modulo n different from 1 and the classes of the prime divisors of $m+1$. Hence, if convenient, we can consider the set S to be of polynomial size such that $S\subset\{r\,(\mathrm{mod}\,n)\,:\,r \text{ prime}\,, r|n'_m\}$, $S\not\subseteq S_m\cup\{1\}$ where $S_m=\{r\,(\mathrm{mod}\,n)\,:\,r\text{ prime}\,, r|(m+1)\}$. The following algorithm on the input of an RSA modulus n outputs a nontrivial factor of n.

Algorithm 2.

1. $m=2$
2. *Send (c,m) to $\mathcal{O}$.*
3. *$\mathcal{O}$ will return a set S as defined above.*
4. *If $S=\emptyset\Rightarrow m=m+1$ and go to (2). Else,*
5. *Take $r\in S$, $r\neq S_m\cup\{1\}$, and compute $d=(r-1,n)$.*

Theorem 2.5. *Let $n=pq$ be an RSA modulus. If GRH is true then the Algorithm 2 runs in polynomial time and the number d given by it is a prime divisor of n.*

Proof. By Theorem 2.4 we can assume that m is a primitive root modulo q, at a polynomial time cost. Then $m^{p-1}\not\equiv 1\,(\mathrm{mod}\,q)$, since $p<q$. Hence, in a similar way as in the proof of Theorem 2.1 we have to prove that a certain prime factor r of n'_m belongs to a residue class modulo n not in $S_m\cup\{1\}$. We will use the following straightforward lemma.

Lemma 2.1. *Let n be an RSA modulus. For any integer m, such that $(m+1,n)=1$ we have $((m^n+1)/(m+1),m+1)=1$.*

Proof. Observe that if $r|(m+1)$, then

$$(m^n+1)/(m+1)=\sum_{j=0}^{n-1}(-m)^j\equiv\sum_{j=0}^{n-1}1\,(\mathrm{mod}\,r)\equiv n\,(\mathrm{mod}\,r)\,. \qquad\square$$

Now, analogously to what we did in the proof of Theorem 2.1, if $r|n'_m$ then $m^{2n} \equiv 1 \pmod r$, and so $\text{ord}_r(m) = 2, p, q, 2p, 2q, pq$ or $2pq$ and clearly $\text{ord}_r(m) \neq 2$ for any r a prime factor of $(m^n+1)/(m+1)$. To see this use Lemma 2.1 and observe that if $r|(m-1)$ then $m^n + 1 \equiv 2 \pmod r$. Hence, as in the previous section, for any $r|(m^n+1)/(m+1)$ then either $p|(r-1)$, $q|(r-1)$ or $n|(r-1)$. If $r \equiv 1 \pmod n$ for any $r|(m^n+1)/(m+1)$ then $m^{n-1} \equiv 1 \pmod n$, which is impossible for m a primitive root modulo q since $m^{n-1} \equiv m^{p-1} \pmod q$. The proof of the theorem concludes as in Theorem 2.1. □

Remark 2.4. In order to prove Theorem 2.5 we do not need m to be a primitive root modulo q, but only we need an integer m such that $m^{q-1} \not\equiv 1 \pmod p$ or $m^{p-1} \not\equiv 1 \pmod q$. The problem of finding such an m could have been handled by means of the Ankeny-Montgomery theorem, (see [10]), which in fact would give a better bound for the running time of the algorithm, (see [11]). However, we prefer to show, yet in another context, the explicit relation of RSA with the easy to understand notion of primitive roots.

Acknowledgment

We would like to express our sincere gratitude to the anonymous referee for his/her very many valuable comments to improve notably the paper. Among them, she/he provided us with the proof to remove the $(\log\log z)^2$ factor from Theorem 2.3, and pointed out the unnecessary use of primitive roots to prove Theorem 2.5.

References

[1] E. Bach, How to generate factored random numbers, *SIAM J. Comp.,* 17(1988), 179–193.

[2] E. Bach, Comments on search procedures for primitive roots, *Math. Comp.* 66(1997), 1719–1727.

[3] M. Bellare and P. Rogaway, Random oracles are practical: A paradigm for designing efficient protocols, in: *Proc. of 1st ACM Computer and Communications Security (CCS'93),* pp. 62–73, 1993.

[4] D. Boneh, Simplified OAEP for the RSA and Rabin functions, in: Proceedings of Crypto, 2001, *Lect. Notes in Comput. Sci.*, Vol. 2139, Springer, pp. 275–291, 2001.

[5] S. Goldwasser, S. Micali and R. L. Rivest, A digital signature scheme secure against adaptive chosen-message attacks, *SIAM J. Comp.* 17(1988), 281–308.

[6] R. Heath-Brown, Artin's conjecture for primitive roots, *Quart. J. Math. Oxford Ser.* 37(1986), 27–38.

[7] C. Hooley, Artin's conjecture for primitive roots, *J. Reine Angew. Math.* 225(1967), 209–220.

[8] H. Iwaniec and E. Kowalski, Analytic Number Theory, *Colloquium Publications,* Vol 53, Amer. Math. Soc., Providence, 2004.

[9] L. Mirsky, Summation formulae involving arithmetic functions, *Duke Math. J.* 16(1949), 261–272.

[10] H. L. Montgomery, Topics in Multiplicative Number Theory, *Lect. Notes in Math.*, Vol. 227, Springer, 1971.

[11] M. R. Murty, Artin's conjecture for primitive roots, *Math. Intelligencer,* 10(1988), no.4, 59–67.

[12] P. Pailler, *IPAM Workshop: Number Theory and Cryptography–Open problems*, IPAM, UCLA, October 9-13, 2006.

[13] P. Paillier and J. L. Villar, Trading one-wayness against chosen-ciphertext security in factoring-based encryption, in : Advances in Cryptology (Asiacrypt 2006), *Lect. Notes in Comput. Sci.,* Vol 4284, Springer, pp. 252–266, 2006.

[14] M. O. Rabin, *Digital signatures and public key functions as intractable as factorization*, Technical Report, MIT, Laboratory for Computer Science, 1979.

[15] V. Shoup, Searching for primitive roots in finite fields, *Math. Comp.* 58(1992), 369–380.

[16] H. C. Williams, A modification of the RSA public-key encryption procedure, *IEEE Trans. Inform. Theory* 26(1980), 726–729, 1980.

In: Frontiers of Combinatorics and Number Theory, Volume 2 ISBN: 978-1-61942-918-5
Editor: Zhi-Wei Sun, pp. 187-195

Chapter 12

COMBINATORIAL IDENTITIES INVOLVING MERTENS FUNCTION THROUGH RELATIVELY PRIME SUBSETS

Mohamed El Bachraoui*
Department of Mathematical Sciences,
United Arab Emirates University,
Al-Ain, United Arab Emirates

Abstract

In this note we give some identities which involve the Mertens function $M(n)$. Our proofs are combinatorial with relatively prime subsets as a main tool.

2010 Mathematics Subject Classification: Primary 11A25; Secondary 11B05, 11B75.

1. Introduction

The Mertens function given by

$$M(n) = \sum_{d=1}^{n} \mu(d),$$

where μ denotes the Möbius mu function, is an important function in (analytic) number theory. Most of mathematical identities where $M(n)$ appears are either recursive formulas for $M(n)$ or formulas with an analytic flavor relating $M(n)$ to other functions. For a survey on identities involving $M(n)$ we refer to [1, 4] and for a survey on combinatorial identities involving other arithmetical functions we refer to [7, 10] and their references. In this work we will give some other identities involving the function $M(n)$. Our proofs are combinatorial and based on relatively prime subsets of sets of positive integers. We list two of the identities which we intend to prove. Let $\lfloor x \rfloor$ denote the floor of x.

*E-mail address: melbachraoui@uaeu.ac.ae

1) If $n > 3$, then

$$\sum_{d=1}^{n} \mu(d) 2^{\lfloor \frac{n}{d} \rfloor - \lfloor \frac{n-3}{d} \rfloor} = \begin{cases} 3 + M(n), & \text{if } n \text{ is even,} \\ 4 + M(n), & \text{if } n \text{ is odd.} \end{cases}$$

2) When $1 < m < n$, we have

$$\sum_{d=1}^{n+1} \mu(d) 2^{\lfloor \frac{n+1}{d} \rfloor - \lfloor \frac{n-1}{d} \rfloor + \lfloor \frac{m}{d} \rfloor - \lfloor \frac{m-1}{d} \rfloor}$$
$$= \begin{cases} 2 + M(n+1), & \text{if } (m,n) > 1 \text{ and } (m,n+1) > 1, \\ 3 + M(n+1), & \text{if } (m,n) = 1 \text{ and } (m,n+1) > 1 \\ & \quad \text{or } (m,n) > 1 \text{ and } (m,n+1) = 1, \\ 4 + M(n+1), & \text{if } (m,n) = (m,n+1) = 1. \end{cases}$$

2. Relatively Prime Subsets of $[l_1, m_1] \cup [l_2, m_2]$

Throughout this section let k, l, m, l_1, l_2, m_1, and m_2 be positive integers such that $l \leq m$, $l_1 \leq m_1 \leq l_2 \leq m_2$, let $[l, m] = \{l, l+1, \ldots, m\}$, and let A be a nonempty finite set of positive integers. The set A is called *relatively prime* if $\gcd(A) = 1$.

Definition 1. *Let*

$$f(A) = \#\{X \subseteq A : \ X \neq \emptyset \text{ and } \gcd(X) = 1\}$$

and

$$f_k(A) = \#\{X \subseteq A : \ \#X = k \text{ and } \gcd(X) = 1\}.$$

Nathanson in [8] introduced among other functions $f(n)$ and $f_k(n)$ (in our terminology $f([1,n])$ and $f_k([1,n])$ respectively) and found

$$f([1,n]) = \sum_{d=1}^{n} \mu(d)(2^{\lfloor \frac{n}{d} \rfloor} - 1) \text{ and } f_k([1,n]) = \sum_{d=1}^{n} \mu(d) \binom{\lfloor \frac{n}{d} \rfloor}{k}. \tag{2.1}$$

Formulas for $f([m,n])$ and $f_k([m,n])$ are found in [5, 9]. Recently Ayad and Kihel in [3] considered relatively prime subsets of sets which are in arithmetic progression and obtained formulas for $f([l,m])$ and $f_k([l,m])$ as consequences since the integer interval $[l,m]$ is in arithmetic progression. However the authors' argument seems not to extend to unions of integer intervals. In this section we will give formulas for $f([l_1,m_1] \cup [l_2,m_2])$ and for $f_k([l_1,m_1] \cup [l_2,m_2])$. For the sake of completeness we include the following result which is a natural extension of [5, Theorem 2 (a)] on Möbius inversion for arithmetical functions of several variables. For simplicity of notation we let

$$(\overline{m}_a, \overline{n}_b) = (m_1, m_2, \ldots, m_a, n_1, n_2, \ldots, n_b)$$

and

$$\left(\frac{\overline{m}_a}{d}, \left\lfloor \frac{\overline{n}_b}{d} \right\rfloor\right) = \left(\frac{m_1}{d}, \frac{m_2}{d}, \ldots, \frac{m_a}{d}, \left\lfloor \frac{n_1}{d} \right\rfloor, \left\lfloor \frac{n_2}{d} \right\rfloor, \ldots, \left\lfloor \frac{n_b}{d} \right\rfloor\right).$$

Theorem 2.1. *If F and G are arithmetical of $a+b$ variables, then*

$$G(\overline{m}_a, \overline{n}_b) = \sum_{d|(m_1, m_2, \ldots, m_a)} F\left(\frac{\overline{m}_a}{d}, \left\lfloor \frac{\overline{n}_b}{d} \right\rfloor\right)$$

if and only if

$$F(\overline{m}_a, \overline{n}_b) = \sum_{d|(m_1, m_2, \ldots, m_a)} \mu(d) G\left(\frac{\overline{m}_a}{d}, \left\lfloor \frac{\overline{n}_b}{d} \right\rfloor\right).$$

We need the following three lemmas the proofs of which can be obtained using the same sort of idea and therefore we prove only the first one.

Lemma 2.1. *Let*

$$g(m_1, l_2, m_2) = \#\{X \subseteq [1, m_1] \cup [l_2, m_2] : \ l_2 \in X \text{ and } \gcd(X) = 1\},$$

$$g_k(m_1, l_2, m_2) = \#\{X \subseteq [1, m_1] \cup [l_2, m_2] : \ l_2 \in X, \ |X| = k, \text{ and } \gcd(X) = 1\}.$$

Then

$$(a) \quad g(m_1, l_2, m_2) = \sum_{d|l_2} \mu(d) 2^{\lfloor m_1/d \rfloor + \lfloor m_2/d \rfloor - l_2/d},$$

$$(b) \quad g_k(m_1, l_2, m_2) = \sum_{d|l_2} \mu(d) \binom{\lfloor m_1/d \rfloor + \lfloor m_2/d \rfloor - l_2/d}{k-1}.$$

Proof. (a) Let $\mathcal{P}(m_1, l_2, m_2)$ denote the set of subsets of $[1, m_1] \cup [l_2, m_2]$ containing l_2 and let $\mathcal{P}(m_1, l_2, m_2, d)$ be the set of subsets X of $[1, m_1] \cup [l_2, m_2]$ such that $l_2 \in X$ and $\gcd(X) = d$. It is clear that the set $\mathcal{P}(m_1, l_2, m_2)$ of cardinality $2^{m_1+m_2-l_2}$ can be partitioned using the equivalence relation of having the same gcd (dividing l_2). Moreover, the mapping $A \mapsto \frac{1}{d}X$ is a one-to-one correspondence between $\mathcal{P}(m_1, l_2, m_2, d)$ and the set of subsets Y of $[1, \lfloor m_1/d \rfloor] \cup [l_2/d, \lfloor m_2/d \rfloor]$ such that $l_2/d \in Y$ and $\gcd(Y) = 1$. Then

$$2^{m_1+m_2-l_2} = \sum_{d|l_2} \#\mathcal{P}(m_1, l_2, m_2, d) = \sum_{d|l_2} g(\lfloor m_1/d \rfloor, l_2/d, \lfloor m_2/d \rfloor),$$

which by Theorem 2.1 is equivalent to

$$g(m_1, l_2, m_2) = \sum_{d|l_2} \mu(d) 2^{\lfloor m_1/d \rfloor + \lfloor m_2/d \rfloor - l_2/d}.$$

(b) Similarly

$$\binom{m_1+m_2-l_2}{k-1} = \sum_{d|l_2} g_k(\lfloor m_1/d\rfloor, l_2/d, \lfloor m_2/d\rfloor),$$

which by Theorem 2.1 is equivalent to

$$g_k(m_1,l_2,m_2) = \sum_{d|l_2} \mu(d)\binom{\lfloor m_1/d\rfloor + \lfloor m_2/d\rfloor - l_2/d}{k-1}. \qquad \square$$

Lemma 2.2. *Let*

$$h^{(1)}(l_1,m_1) = \#\{X \subseteq [l_1,m_1] : \ l_1 \in X \text{ and } \gcd(X)=1\},$$

$$h_k^{(1)}(l_1,m_1) = \#\{X \subseteq [l_1,m_1] : \ l_1 \in X, \ \#X = k, \text{ and } \gcd(X)=1\}.$$

Then

$$(a) \quad h^{(1)}(l_1,m_1) = \sum_{d|l_1} \mu(d) 2^{\lfloor m_1/d\rfloor - l_1/d},$$

$$(b) \quad h_k^{(1)}(l_1,m_1) = \sum_{d|l_1} \mu(d)\binom{\lfloor m_1/d\rfloor - l_1/d}{k}.$$

Lemma 2.3. *Let*

$$h^{(2)}(l_1,m_1,l_2,m_2) = \#\{X \subseteq [l_1,m_1]\cup[l_2,m_2] : \ l_1,l_2 \in X \text{ and } \gcd(X)=1\},$$

$$h_k^{(2)}(l_1,m_1,l_2,m_2) = \#\{X \subseteq [l_1,m_1]\cup[l_2,m_2] : \ l_1,l_2 \in X, \ |X| = k, \text{ and } \gcd(X)=1\}.$$

Then

$$(a) \quad h^{(2)}(l_1,m_1,l_2,m_2) = \sum_{d|(l_1,l_2)} \mu(d) 2^{\lfloor \frac{m_1}{d}\rfloor + \lfloor \frac{m_2}{d}\rfloor - \frac{l_1+l_2}{d}},$$

$$(b) \quad h_k^{(2)}(m_1,l_2,m_2) = \sum_{d|(l_1,l_2)} \mu(d)\binom{\lfloor \frac{m_1}{d}\rfloor + \lfloor \frac{m_2}{d}\rfloor - \frac{l_1+l_2}{d}}{k-2}.$$

We further need the following special case to prove the main theorem of this section.

Theorem 2.2. *We have*

$$(a) \quad f([1,m_1]\cup[l_2,m_2]) = \sum_{d=1}^{m_2} \mu(d)\left(2^{\lfloor \frac{m_1}{d}\rfloor + \lfloor \frac{m_2}{d}\rfloor - \lfloor \frac{l_2-1}{d}\rfloor} - 1\right),$$

$$(b) \quad f_k([1,m_1]\cup[l_2,m_2]) = \sum_{d=1}^{m_2} \mu(d)\binom{\lfloor \frac{m_1}{d}\rfloor + \lfloor \frac{m_2}{d}\rfloor - \lfloor \frac{l_2-1}{d}\rfloor}{k}.$$

Proof. (a) Clearly

$$
\begin{aligned}
f([1,m_1]\cup[l_2,m_2]) &= f([1,m_2]) - \sum_{i=m_1+1}^{l_2-1} g(m_1,i,m_2)\\
&= \sum_{d=1}^{m_2}\mu(d)(2^{\lfloor m_2/d\rfloor}-1) - \sum_{i=m_1+1}^{l_2-1}\sum_{d|i}\mu(d)2^{\lfloor \frac{m_1}{d}\rfloor+\lfloor \frac{m_2}{d}\rfloor-\frac{i}{d}}\\
&= \sum_{d=1}^{m_2}\mu(d)(2^{\lfloor m_2/d\rfloor}-1) - \sum_{d=1}^{l_2-1}\mu(d)2^{\lfloor \frac{m_1}{d}\rfloor+\lfloor \frac{m_2}{d}\rfloor}\sum_{j=\lfloor \frac{m_1}{d}\rfloor+1}^{\lfloor \frac{l_2-1}{d}\rfloor} 2^{-j}\\
&= \sum_{d=1}^{m_2}\mu(d)(2^{\lfloor m_2/d\rfloor}-1)\\
&\quad - \sum_{d=1}^{l_2-1}\mu(d)2^{\lfloor \frac{m_1}{d}\rfloor+\lfloor \frac{m_2}{d}\rfloor}2^{-\lfloor \frac{m_1}{d}\rfloor}\left(1-2^{-\lfloor \frac{l_2-1}{d}\rfloor+\lfloor \frac{m_1}{d}\rfloor}\right)\\
&= \sum_{d=1}^{m_2}\mu(d)(2^{\lfloor m_2/d\rfloor}-1) - \sum_{d=1}^{m_2}\mu(d)2^{\lfloor \frac{m_2}{d}\rfloor}\left(1-2^{-\lfloor \frac{l_2-1}{d}\rfloor+\lfloor \frac{m_1}{d}\rfloor}\right)\\
&= \sum_{d=1}^{m_2}\mu(d)(2^{\lfloor \frac{m_1}{d}\rfloor+\lfloor \frac{m_2}{d}\rfloor-\lfloor \frac{l_2-1}{d}\rfloor}-1),
\end{aligned}
$$

where the second identity follows by (2.1) and Lemma 2.1.
(b) We have

$$
\begin{aligned}
f_k([1,m_1]\cup[l_2,m_2]) &= f_k([1,m_2]) - \sum_{i=m_1+1}^{l_2-1} g_k(m_1,i,m_2)\\
&= \sum_{d=1}^{m_2}\mu(d)\binom{\lfloor m_2/d\rfloor}{k} - \sum_{i=m_1+1}^{l_2-1}\sum_{d|i}\mu(d)\binom{\lfloor \frac{m_1}{d}\rfloor+\lfloor \frac{m_2}{d}\rfloor-\frac{i}{d}}{k-1}\\
&= \sum_{d=1}^{m_2}\mu(d)\binom{\lfloor m_2/d\rfloor}{k} - \sum_{d=1}^{m_2}\mu(d)\sum_{j=\lfloor \frac{m_1}{d}\rfloor+1}^{\lfloor \frac{l_2-1}{d}\rfloor}\binom{\lfloor \frac{m_1}{d}\rfloor+\lfloor \frac{m_2}{d}\rfloor-\frac{i}{d}}{k-1}\\
&= \sum_{d=1}^{m_2}\mu(d)\binom{\lfloor m_2/d\rfloor}{k} - \sum_{d=1}^{m_2}\mu(d)\sum_{i=\lfloor \frac{m_1}{d}\rfloor+\lfloor \frac{m_2}{d}\rfloor-\frac{l_2-1}{d}}^{\lfloor \frac{m_2}{d}\rfloor-1}\binom{i}{k-1}\\
&= \sum_{d=1}^{m_2}\mu(d)\binom{\lfloor m_2/d\rfloor}{k}\\
&\quad - \sum_{d=1}^{m_2}\mu(d)\left(\binom{\lfloor m_2/d\rfloor}{k} - \binom{\lfloor \frac{m_1}{d}\rfloor+\lfloor \frac{m_2}{d}\rfloor-\frac{l_2-1}{d}}{k}\right)\\
&= \sum_{d=1}^{m_2}\mu(d)\binom{\lfloor \frac{m_1}{d}\rfloor+\lfloor \frac{m_2}{d}\rfloor-\frac{l_2-1}{d}}{k}.
\end{aligned}
$$

This completes the proof. □

We are now ready to prove the main theorem of this section.

Theorem 2.3. *We have*

$$(a)\quad f([l_1,m_1]\cup[l_2,m_2]) = \sum_{d=1}^{m_2}\mu(d)(2^{\lfloor\frac{m_1}{d}\rfloor+\lfloor\frac{m_2}{d}\rfloor-\lfloor\frac{l_1-1}{d}\rfloor-\lfloor\frac{l_2-1}{d}\rfloor}-1),$$

$$(b)\quad f_k([l_1,m_1]\cup[l_2,m_2]) = \sum_{d=1}^{m_2}\mu(d)\binom{\lfloor\frac{m_1}{d}\rfloor+\lfloor\frac{m_2}{d}\rfloor-\lfloor\frac{l_1-1}{d}\rfloor-\lfloor\frac{l_2-1}{d}\rfloor}{k}.$$

Proof. (a) Clearly

$$\begin{aligned}
&f([l_1,m_1]\cup[l_2,m_2])\\
=&f([1,m_1]\cup[l_2,m_2]) - \sum_{i=1}^{l_1-1}\sum_{j=l_2}^{m_2} h^{(2)}(i,m_1,j,m_2) - \sum_{i=1}^{l_1-1} h^{(1)}(i,m_1)\\
=&\sum_{d=1}^{m_2}\mu(d)(2^{\lfloor\frac{m_1}{d}\rfloor+\lfloor\frac{m_2}{d}\rfloor-\lfloor\frac{l_2-1}{d}\rfloor}-1)\\
&-\sum_{i=1}^{l_1-1}\sum_{j=l_2}^{m_2}\sum_{d|(i,j)}\mu(d)2^{\lfloor\frac{m_1}{d}\rfloor+\lfloor\frac{m_2}{d}\rfloor-\frac{i+j}{d}} - \sum_{i=1}^{l_1-1}\sum_{d|i}\mu(d)2^{\lfloor\frac{m_1}{d}\rfloor-\frac{i}{d}},
\end{aligned}\tag{2.2}$$

where the second identity follows by Theorem 2.2, Lemma 2.2, and Lemma 2.3. Rearranging the triple summation in identity (2.2), we get

$$\begin{aligned}
&\sum_{i=1}^{l_1-1}\sum_{j=l_2}^{m_2}\sum_{d|(i,j)}\mu(d)2^{\lfloor\frac{m_1}{d}\rfloor+\lfloor\frac{m_2}{d}\rfloor-\frac{i+j}{d}}\\
=&\sum_{d=1}^{m_2}\mu(d)2^{\lfloor\frac{m_1}{d}\rfloor+\lfloor\frac{m_2}{d}\rfloor}\sum_{i=1}^{\lfloor\frac{l_1-1}{d}\rfloor}2^{-i}\sum_{j=\lfloor\frac{l_2-1}{d}\rfloor+1}^{\lfloor\frac{m_2}{d}\rfloor}2^{-j}\\
=&\sum_{d=1}^{m_2}\mu(d)2^{\lfloor\frac{m_1}{d}\rfloor+\lfloor\frac{m_2}{d}\rfloor-\lfloor\frac{l_2-1}{d}\rfloor}(1-2^{-\lfloor\frac{m_2}{d}\rfloor+\lfloor\frac{l_2-1}{d}\rfloor})(1-2^{-\lfloor\frac{l_1-1}{d}\rfloor}).
\end{aligned}\tag{2.3}$$

Similarly the last double summation in identity (2.2) gives

$$\begin{aligned}
&\sum_{i=1}^{l_1-1}\mu(d)2^{\lfloor\frac{m_1}{d}\rfloor}\sum_{i=1}^{\lfloor\frac{l_1-1}{d}\rfloor}2^{-\frac{i}{d}}\\
=&\sum_{d=1}^{l_1-1}\mu(d)2^{\lfloor\frac{m_1}{d}\rfloor}(1-2^{-\lfloor\frac{l_1-1}{d}\rfloor})\\
=&\sum_{d=1}^{m_2}\mu(d)2^{\lfloor\frac{m_1}{d}\rfloor}(1-2^{-\lfloor\frac{l_1-1}{d}\rfloor}).
\end{aligned}\tag{2.4}$$

Appealing to identities (2.2)-(2.4) we find

$$f([l_1,m_1]\cup[l_2,m_2])=\sum_{d|n}\mu(d)2^{\lfloor\frac{m_1}{d}\rfloor+\lfloor\frac{m_2}{d}\rfloor-\lfloor\frac{l_1-1}{d}\rfloor-\lfloor\frac{l_2-1}{d}\rfloor}.$$

This completes the proof of part (a). Part (b) follows by similar arguments. □

Corollary 2.1. (Ayad-Kihel [3]) *We have*

$$(a)\quad f([l,m])=\sum_{d=1}^{m}\mu(d)(2^{\lfloor\frac{m}{d}\rfloor-\lfloor\frac{l-1}{d}\rfloor}-1),$$

$$(b)\quad f_k([l,m])=\sum_{d=1}^{m}\mu(d)\binom{\lfloor\frac{m}{d}\rfloor-\lfloor\frac{l-1}{d}\rfloor}{k}.$$

Proof. Use Theorem 2.3 with $l_1=l$, $m_1=m-1$, and $l_2=m_2=m$. □

3. Combinatorial Identities

Theorem 3.1. *If $n>1$, then*

$$\sum_{d=1}^{n+1}\mu(d)2^{\lfloor\frac{n+1}{d}\rfloor-\lfloor\frac{n-1}{d}\rfloor}=1+M(n+1).$$

Proof. Apply Corollary 2.1 to the interval $[n,n+1]$ and use the obvious fact that

$$f([n,n+1])=1.$$ □

Theorem 3.2. *For $n>3$ we have*

$$\sum_{d=1}^{n}\mu(d)2^{\lfloor\frac{n}{d}\rfloor-\lfloor\frac{n-3}{d}\rfloor}=\begin{cases}3+M(n), & \textit{if } n \textit{ is even},\\ 4+M(n), & \textit{if } n \textit{ is odd}.\end{cases}$$

Proof. Combine Corollary 2.1 applied to the interval $[n-2,n]$ and the fact that

$$f([n-2,n])=\begin{cases}3, & \text{if } n \text{ is even},\\ 4, & \text{if } n \text{ is odd}.\end{cases}$$ □

Theorem 3.3. *(a) If $1<m<n$, then*

$$\sum_{d=1}^{n+1}\mu(d)2^{\lfloor\frac{n+1}{d}\rfloor-\lfloor\frac{n-1}{d}\rfloor+\lfloor\frac{m}{d}\rfloor-\lfloor\frac{m-1}{d}\rfloor}$$

$$=\begin{cases}2+M(n+1), & \textit{if } (m,n)>1 \textit{ and } (m,n+1)>1,\\ 3+M(n+1), & \textit{if } (m,n)=1 \textit{ and } (m,n+1)>1\\ & \quad \textit{or } (m,n)>1 \textit{ and } (m,n+1)=1,\\ 4+M(n+1), & \textit{if } (m,n)=(m,n+1)=1.\end{cases}$$

(b) If $1 < n < m-1$*, then*

$$\sum_{d=1}^{m} \mu(d) 2^{\lfloor \frac{n+1}{d} \rfloor - \lfloor \frac{n-1}{d} \rfloor + \lfloor \frac{m}{d} \rfloor - \lfloor \frac{m-1}{d} \rfloor}$$

$$= \begin{cases} 2 + M(m), & \textit{if } (m,n) > 1 \textit{ and } (m,n+1) > 1, \\ 3 + M(m), & \textit{if } (m,n) = 1 \textit{ and } (m,n+1) > 1 \\ & \quad \textit{or } (m,n) > 1 \textit{ and } (m,n+1) = 1, \\ 4 + M(m), & \textit{if } (m,n) = (m,n+1) = 1. \end{cases}$$

Proof. (a) Clearly

$$f([m,m] \cup [n,n+1]) = \begin{cases} 2, & \text{if } (m,n) > 1 \text{ and } (m,n+1) > 1, \\ 3, & \text{if } (m,n) = 1 \text{ and } (m,n+1) > 1 \\ & \text{or } (m,n) > 1 \text{ and } (m,n+1) = 1, \\ 4, & \text{if } (m,n) = (m,n+1) = 1. \end{cases}$$

Combine this identity with Theorem 2.3(a) applied to $[m,m] \cup [n,n+1]$.

(b) Similar to part (a) with an application of Theorem 2.3(a) to $[n,n+1] \cup [m,m]$. □

Acknowledgment

The author is grateful to the referee for valuable comments and interesting suggestions.

References

[1] T. M. Apostol, Introduction to Analytic Number Theory, *Undergraduate Texts in Mathematics,* Springer, 1st Edition, 1976.

[2] Mohamed Ayad and Omar Kihel, On the Number of Subsets Relatively Prime to an Integer, *Integer Seq.* 11 (2008), Article 08.5.5.

[3] Mohamed Ayad and Omar Kihel, On Relatively Prime Sets, *Integers* 9 (2009), 343-352.

[4] F. Dress, Fonction sommatiore de la fonction de Möbius. I. Majorations expérimentales, *Experiment. Math.* 2 (1993), 89-98.

[5] Mohamed El Bachraoui, The number of relatively prime subsets and phi functions for sets $\{m, m+1, \ldots, n\}$, *Integers* 7 (2007), A43, 8pp.

[6] Mohamed El Bachraoui, On the Number of Subsets of $[1, m]$ Relatively Prime to n and Asymptotic Estimates, *Integers* 8 (2008), A41, 5 pp.

[7] M. Hall, *Combinatorial Theory*, Second edition, Wiley, New York, 1986.

[8] M. B. Nathanson, Affine invariants, relatively prime sets, and a phi function for subsets of $\{1, 2, \ldots, n\}$, *Integers* 7 (2007), A1, 7pp.

[9] M. B. Nathanson and Brooke Orosz, Asymptotic estimates for phi functions for subsets of $\{m+1, m+2, \ldots, n\}$, *Integers* 7 (2007), A54, 5pp.

[10] J. Riordan, *Combinatorial Identities*, Wiley, New York, 1968.

In: Frontiers of Combinatorics and Number Theory, Volume 2 ISBN: 978-1-61942-918-5
Editor: Zhi-Wei Sun, pp. 197-207

Chapter 13

GENERALIZATIONS OF EHRHART-MACDONALD RECIPROCITY

Robin Chapman*
Mathematical Research Institute, University of Exeter,
Exeter, UK

Abstract

We extend the combinatorial method of Sam's proof of Ehrhart's reciprocity theorem in two ways: we avoid having to triangulate the polytope in question, and the results hold for valuations invariant under translations by lattice vectors. As a consequence we obtain a simple proof of Stanley's reciprocity theorem.

2010 Mathematics Subject Classification: Primary 05A15; Secondary 52B20

1. Introduction

Throughout this paper the term polytope will always mean a convex polytope.

A *polytope* A in a finite-dimensional real vector space V is the convex hull of a finite set of points. Its *dimension* $\dim A$ is the dimension of its affine span, and its *interior* A° is its relative interior, that is the interior of A considered as a topological subspace of its affine span. If A is the convex hull of $v_1, \ldots, v_k \in V$ then

$$A = \left\{ \sum_{j=1}^{k} t_j v_j : t_j \geq 0, \sum_{j=1}^{k} t_j = 1 \right\}$$

and

$$A^\circ = \left\{ \sum_{j=1}^{k} t_j v_j : t_j > 0, \sum_{j=1}^{k} t_j = 1 \right\}.$$

Received: December 2, 2009; Accepted: September 29, 2010
*E-mail address: R.J.Chapman@ex.ac.uk

Fix a field F of characteristic zero. For a subset A of V, let $[A] : V \to F$ denote its indicator function:

$$[A](x) = \begin{cases} 1 & \text{if } x \in A, \\ 0 & \text{if } x \notin A. \end{cases}$$

Define $\mathcal{P}_V$, the space of polytopes in V, as the F-linear span of the $[A]$ where A ranges over all polytopes in V. Note that if A is a polytope then $[A^\circ] \in \mathcal{P}_V$ for $[A] - [A^\circ]$ is the sum of $[B^\circ]$ over all proper faces B of A and these have smaller dimension than A.

A *valuation* on $\mathcal{P}_V$ is an F-linear map from $\mathcal{P}_V$ to F. Define for $a \in V$ the translation map $\tau_a : \mathcal{P}_V \to \mathcal{P}_V$ by $\tau_a(f) : x \mapsto f(x-a)$. Then $\tau_a([A]) = [A+a]$.

If μ is a valuation we write $\mu(A)$ for $\mu([A])$. Let Γ be an additive subgroup of V. A valuation μ is Γ-*invariant* if $\mu(\tau_a(f)) = \mu(f)$ for all $f \in \mathcal{P}_V$ and $a \in \Gamma$.

For $t \in \mathbb{R}$ and $A \subseteq V$ define $tA = \{ta : a \in A\}$. For a polytope A and $t \in \mathbb{R}$ define

$$[A]_t = \begin{cases} [tA] & \text{if } t \geq 0, \\ (-1)^{\dim A}[tA^\circ] & \text{if } t < 0. \end{cases}$$

We can now state our main theorem.

Theorem 1.1. *Let Γ be an additive subgroup of the real vector space V, let A be a polytope in V with vertices in Γ, and let μ be a Γ-invariant valuation on $\mathcal{P}_V$. For each $c \in \mathbb{R}$ the map $p : \mathbb{Z} \to \mathbb{R}$ defined by*

$$p(t) = \mu([A]_{c+t})$$

equals a polynomial.

When we say that the map $p : \mathbb{Z} \to \mathbb{R}$ equals a polynomial, we mean that there is a polynomial P such that $p(t) = P(t)$ for all $t \in \mathbb{Z}$.

The classical example of a Γ-invariant valuation on $\mathcal{P}_V$ has $V = \mathbb{R}^n$, $\Gamma = \mathbb{Z}^n$ and $\mu(A) = |A \cap \mathbb{Z}^n|$. Then Theorem 1.1 with $c = 0$ gives the result usually known as Ehrhart's reciprocity law, namely that when A is a polytope with vertices in $\mathbb{Z}^n$ then $p : \mathbb{Z} \to \mathbb{Z}$ defined by

$$p(t) = \begin{cases} |tA \cap \mathbb{Z}^n| & \text{if } t \geq 0, \\ (-1)^{\dim A}|tA^\circ \cap \mathbb{Z}^n| & \text{if } t < 0 \end{cases}$$

is a polynomial function. (It is straightforward then that its degree is the dimension of A).

Recall there is a unique linear map $\chi : \mathcal{P}_V \to \mathbb{R}$, the *Euler characteristic*, satisfying $\chi([A]) = 1$ whenever A is a polytope [4, Theorem 5.2.1]. Also $\chi([A^\circ]) = (-1)^{\dim A}$ [1, Lemma VI.3.1]. For a linear map $T : V \to W$ of finite-dimensional real vector spaces there is an induced linear map $\mathcal{T} : \mathcal{P}_V \to \mathcal{P}_W$ with the property that $\mathcal{T}([A]) = [T(A)]$ for every polytope A in V [1, Theorem I.8.1]. When $f \in \mathcal{P}_V$ and $y \in W$, $(\mathcal{T}f)(y) = \chi(f_y)$ where f_y is the valuation on $\mathcal{P}_V$ defined by $f_y(x) = f(x)[T^{-1}(y)](x)$, that is

$$f_y(x) = \begin{cases} f(x) & \text{if } T(x) = y, \\ 0 & \text{if } T(x) \neq y. \end{cases}$$

There is an $\mathbb{R}$-bilinear map called *convolution* on $\mathcal{P}_V$ which makes $\mathcal{P}_V$ into a commutative $\mathbb{R}$-algebra [1, §I.8] defined by

$$(f * g)(x) = \int f(y)g(x-y)d\chi(y).$$

The integral notation is meant to draw a parallel with the classical notion of convolution of real-valued functions on $\mathbb{R}^n$. The value $(f * g)(x)$ is χ applied to the function $y \mapsto f(y)g(x-y)$. Then for polytopes A and B

$$[A] * [B] = [A + B]$$

where $A + B = \{x + y : x \in A, y \in B\}$ is the Minkowski sum of A and B. (This formula does not apply to general sets with $[A]$, $[B] \in \mathcal{P}_V$.) As a consequence

$$\tau_a(f) = f * [a]$$

for all $f \in \mathcal{P}_V$.

2. An Identity in $\mathcal{P}_V$

The main ingredient in the proof is the following identity.

Theorem 2.1. *Let A be a polytope in V with vertices $v_1, \dots, v_s$. For $I \subseteq [s] = \{1, \dots, s\}$ set $v_I = \sum_{i \in I} v_i$. Then for each $c \in \mathbb{R}$*

$$\sum_{I \subseteq [s]} (-1)^{|I|} \tau_{-v_I}([A]_{c+|I|}) = 0.$$

Proof. We first prove the theorem in the case where A is a standard simplex. We adapt an argument of Sam [6, Theorems 4 and 5].

Let Δ be the $(s-1)$-simplex in $\mathbb{R}^s$ with vertices the s coordinate vectors $e_1, \dots, e_s$. Define $\phi : \mathbb{R}^s \to F$ by

$$\phi(\xi_1, \dots, \xi_s) = \begin{cases} 1 & \text{if all } \xi_i \geq 0, \\ (-1)^{s-1} & \text{if all } \xi_i < 0, \\ 0 & \text{otherwise} \end{cases}$$

and $\sigma : \mathbb{R}^s \to F$ by $\sigma(\xi_1, \dots, \xi_s) = \sum_{i=1}^{s} \xi_i$. Then

$$[\Delta]_c(x) = \begin{cases} \phi(x) & \text{if } \sigma(x) = c, \\ 0 & \text{otherwise.} \end{cases}$$

Thus for $I \subseteq [s]$,

$$\tau_{-e_I}([\Delta]_{c+|I|})(x) = [\Delta]_{c+|I|}(x + e_I) = \begin{cases} \phi(x + e_I) & \text{if } \sigma(x) = c \\ 0 & \text{otherwise} \end{cases}$$

where $e_I = \sum_{i \in I} e_i$. Consequently

$$\sum_{I \subseteq [s]} (-1)^{|I|} \tau_{-e_I}([\Delta]_{c+|I|})(x) = \begin{cases} \sum_{I \subseteq [s]} (-1)^{|I|} \phi(x + e_I) & \text{if } \sigma(x) = c, \\ 0 & \text{otherwise.} \end{cases}$$

To complete the proof for $A = \Delta$ it suffices to prove that

$$\sum_{I \subseteq [s]} (-1)^{|I|} \phi(x + e_I) = 0 \tag{2.1}$$

for all $x \in \mathbb{R}^s$.

Observe that $\phi(x) = \phi_+(x) - \phi_-(x)$ where

$$\phi_+(\xi_1, \ldots, \xi_s) = \prod_{i=1}^{s} \psi(\xi_i)$$

and

$$\phi_-(\xi_1, \ldots, \xi_s) = \prod_{i=1}^{s} (\psi(\xi_i) - 1)$$

and where ψ is the indicator function of the interval $[0, \infty)$. Then for $x = (\xi_1, \ldots, \xi_s)$

$$\sum_{I \subseteq [s]} (-1)^{|I|} \phi_+(x + e_I) = \prod_{i=1}^{s} (\psi(\xi_i) - \psi(\xi_i + 1))$$

and

$$\begin{aligned} \sum_{I \subseteq [s]} (-1)^{|I|} \phi_-(x + e_I) &= \prod_{i=1}^{s} ((\psi(\xi_i) - 1) - (\psi(\xi_i + 1) - 1)) \\ &= \prod_{i=1}^{s} (\psi(\xi_i) - \psi(\xi_i + 1)) \end{aligned}$$

from which (2.1) follows.

We deduce the general case from the case of the simplex. There is a unique linear map $T : \mathbb{R}^s \to V$ with $T(e_i) = v_i$. Then $A = T(\Delta)$ and $A^\circ = T(\Delta^\circ)$. Consider the induced map $\mathcal{T} : \mathcal{P}_{\mathbb{R}^s} \to \mathcal{P}_V$. We claim that $\mathcal{T}([\Delta]_c) = [A]_c$ for all $c \in \mathbb{R}$. This is apparent for $c > 0$. When $c = 0$, $[\Delta]_0 = [0]$ and $[A]_0 = [0]$ so the result is clear there. For $c < 0$ we reduce to the case $c = -1$ where we need to prove that $\mathcal{T}([(-1)^{s-1}\Delta^\circ]) = (-1)^d [A^\circ]$ where $d = \dim A$. For $y \in V$, $\mathcal{T}[\Delta^\circ](y)$ equals $\chi([T^{-1}(y) \cap \Delta^\circ])$. The set $T^{-1}(y) \cap \Delta^\circ$ is nonempty if and only if $y \in A^\circ$. Let $H = \{x \in \mathbb{R}^s : \sigma(x) = 1\}$. Then $\Delta \subseteq H$ and so $T^{-1}(y) \cap \Delta^\circ = T^{-1}(y) \cap H \cap \Delta^\circ$. The set $T^{-1}(y) \cap H$ is an affine subspace of H with dimension $s - 1 - d$. If $y \in A^\circ$ then $T^{-1}(y) \cap H \cap \Delta^\circ$ contains a point in the relative interior of Δ°. As $T^{-1}(y) \cap H$ is an affine space then $T^{-1}(y) \cap H \cap \Delta^\circ$ is the relative interior of a polytope of dimension $s - 1 - d$. Thus $\chi([T^{-1}(y) \cap \Delta^\circ]) = (-1)^{s-1-d}$ and the assertion follows.

Applying $\mathcal{T}$ to (2.1) gives

$$\sum_{I \subseteq [s]} (-1)^{|I|} \tau_{-v_I}([A]_{c+|I|}) = 0$$

as required. □

Sam's Theorem 5 in [6] is essentially the special case of this result when A is a simplex. Sam deduces Ehrhart reciprocity from this special case by triangulating the polytope and dealing with the overlaps between the various simplices by using the Möbius function of the face lattice of the triangulation. Proving the properties needed of this Möbius functions uses homology theory [8, Proposition 3.8.9]. We avoid these problems by using the formalism of the space of polytopes, in particular the map $\mathcal{T} : \mathcal{P}_{\mathbb{R}^s} \to \mathcal{P}_V$.

Before proving Theorem 1.1 we draw some corollaries.

Corollary 2.1. *Let A be a polytope. Then for all t, u* $\in \mathbb{R}$, $[A]_{t+u} = [A]_t * [A]_u$.

Proof. When $t, u \geq 0$,

$$[A]_t * [A]_u = [tA] * [uA] = [tA + uA] = [(t+u)A]$$

by the convexity of A. Also $[A]_0 = [0]$ which is the identity with respect to convolution. To complete the proof, it suffices to prove that when $t > 0$, $[A]_{-t} = (-1)^{\dim A}[-tA^\circ]$ is the inverse of $[A]_t = [tA]$ with respect to convolution. This reduces immediately to the case $t = 1$.

Let A have vertices $v_1, \ldots, v_s$. Apply Theorem 2.1 both with $c = 0$ and $c = -1$. Separating out the term with $[I] = \emptyset$ gives

$$[0] = -\sum_{I \subseteq [s], I \neq \emptyset} (-1)^{|I|} \tau_{-v_I}([A]_{|I|})$$

and

$$[A]_{-1} = -\sum_{I \subseteq [s], I \neq \emptyset} (-1)^{|I|} \tau_{-v_I}([A]_{|I|-1}).$$

Convolving the second equation with $[A]$ gives

$$[A]_{-1} * [A] = -\sum_{I \subseteq [s], I \neq \emptyset} (-1)^{|I|} \tau_{-v_I}([A]_{|I|}) = [0].$$

This completes the proof. □

We can now restate Theorem 2.1 as a convolutional identity.

Corollary 2.2. *Let A be a polytope in V with vertices* $v_1, \ldots, v_s$. *Then*

$$([A] - [v_1]) * \cdots * ([A] - [v_s]) = 0.$$

Proof. In terms of convolution, Theorem 2.1 states that

$$\begin{aligned} 0 &= \sum_{I \subseteq S} (-1)^{|I|} [A]_c * [A]_{|I|} * [-v_I] \\ &= [A]_c * ([0] - [-v_1] * [A]) * \cdots * ([0] - [-v_s] * [A]). \end{aligned}$$

Setting $c = 0$ gives

$$([0] - [-v_1] * [A]) * \cdots * ([0] - [-v_s] * [A]) = 0$$

which on convolving by $(-1)^s [v_1] * \cdots * [v_s]$ gives

$$([A] - [v_1]) * \cdots * ([A] - [v_s]) = 0$$

as claimed. □

3. Proof of Theorem 1.1

We now prove the main theorem.
Proof. Apply μ to the identity in Theorem 2.1. This gives, using the Γ-invariance of μ,

$$\sum_{j=0}^{s}(-1)^j\binom{s}{j}\mu([A]_{c+j}) = 0$$

for all $c \in \mathbb{R}$. Replacing c by $c + t$ gives

$$\sum_{j=0}^{s}(-1)^j\binom{s}{j}p(t+j) = 0.$$

By the calculus of finite differences this implies that p is a polynomial with degree at most $s - 1$. □

A polytope can be triangulated without introducing new vertices [2, Appendix]. It follows that if A is a d-dimensional polytope then the polynomial p has degree at most d.

The classical application is when $\Gamma = \mathbb{Z}^n$ and $\mu(A) = |A \cap \mathbb{Z}^n|$. We recover the theorem of Ehrhart [3] that when A is a polytope with integer coefficients, then there are polynomials p and q such that $p(t) = \mu(tA)$ and $q(t) = \mu((-t)A^\circ)$ for integers $t \geq 0$, and the reciprocity theorem conjectured by Ehrhart and proved by Macdonald [5] that $q(t) = (-1)^{\dim A}p(-t)$.

Another example of a valuation is that taking $[A]$ to its area; of course the conclusion of Theorem 1.1 is trivial in this case. A more interesting example is the solid angle enumerator discussed by Sam [6]. This takes A to the sum of the solid angles subtended by A at each point of $A \cap \mathbb{Z}^n$.

A more general version of Ehrhart's theorem applies to polytopes with rational vertices. If A has vertices in $\mathbb{Q}^n$ then for some positive integer m, $B = mA$ has vertices in $\mathbb{Z}^n$. Applying Theorem 1.1 to B we find that for $j \in \{0, \dots, m-1\}$, $p_j(t) = \mu([B]_{t+j/m}) = \mu([A]_{tm+j})$ is a polynomial in t for $t \in \mathbb{Z}$. Thus for such j there is a polynomial P_j such that $P_j(t) = \mu([A]_t)$ for $t \in j + m\mathbb{Z}$. These polynomials can be assembled into a *quasipolynomial* $P(t) = \sum_{i=0}^{k} a_i(t)t^i$ where the coefficients $a_i(t)$ are not constant, but periodic with period m. Thus $t \mapsto \mu([A]_t)$ is a quasipolynomial on $\mathbb{Z}$.

We can prove a reciprocity law in several variables.

Theorem 3.1. *Let $A_1, \dots, A_m$ be polytopes all of whose vertices lie in a subgroup Γ of the vector space V. Let μ be a Γ-invariant valuation. Define*

$$p(t_1, \dots, t_n) = \mu([A_1]_{t_1} * \cdots * [A_m]_{t_m}).$$

Then there is a polynomial P in m variables such that

$$p(t_1, \dots, t_m) = P(t_1, \dots, t_m)$$

for all $(t_1, \dots, t_m) \in \mathbb{Z}^m$.

Before proving this we remark that when all $t_i \geq 0$,

$$[A_1]_{t_1} * \cdots * [A_m]_{t_m} = [t_1 A_1 + \cdots + t_m A_m]$$

and when all $t_i < 0$

$$[A_1]_{t_1} * \cdots * [A_m]_{t_m} = (-1)^d [(t_1 A_1 + \cdots + t_m A_m)^\circ]$$

where d is the dimension of $t_1 A_1 + \cdots + t_m A_m$.
Proof. We show that for fixed $t_2, \ldots, t_m$, $p(t, \ldots, t_m)$ is a polynomial function of t for $t \in \mathbb{Z}$. Let $h = [A_2]_{t_2} * \cdots * [A_m]_{t_m}$ and set

$$\phi(t) = p(t, t_2, \ldots, t_m) = \mu([A_1]_t * h).$$

If $v_1, \ldots, v_s$ are the vertices of A_1 then

$$[A_1]_t * \sum_{I \subseteq [s]} (-1)^{|I|} [A_1]_{|I|} * [-v_I] = 0.$$

Convolving with h and noting that μ is invariant under convolution by each $[-v_I]$, we get on applying μ,

$$\sum_{j=0}^{s} (-1)^j \binom{s}{j} \phi(t+j) = 0.$$

This recurrence implies that the restriction of ϕ to $\mathbb{Z}$ is a polynomial of degree at most s.

Repeating the above argument with an arbitrary t_i playing the role of t_1 we see that on $\mathbb{Z}^m$, $p(t_1, \ldots, t_m)$ is a polynomial of degree less than S in t_i whenever all the other arguments are fixed, where S is the maximum number of vertices of any of the polytopes A_j. We claim that this implies that p is a polynomial on $\mathbb{Z}^n$. The values of p on $\mathbb{Z}^m$ are determined by its values on the $(t_1, \ldots, t_m) \in \mathbb{Z}^m$ with $0 \leq t_i < S$. But by Lagrange interpolation there is a unique polynomial P of degree less than S in all variables, with $P(t_1, \ldots, t_m) = p(t_1, \ldots, t_m)$ whenever each $0 \leq t_i < S$. It follows that $P(t_1, \ldots, t_m) = p(t_1, \ldots, t_m)$ whenever $t_i \in \mathbb{Z}$. □

4. Stanley's Reciprocity Theorem

Stanley's reciprocity theorem concerns certain generating functions for integer points in rational cones. Before stating Stanley's theorem we recast Theorem 1.1 in terms of generating functions.

Let Γ be an additive subgroup of the vector space V and let $\psi : \Gamma \to F^*$ be a character, that is a group homomorphism. We say that the valuation $\mu : \mathcal{P}_V \to F$ has *character* ψ if $\mu(\tau_a(f)) = \psi(a)\mu(f)$ for all $a \in \Gamma$ and $f \in \mathcal{P}_V$.

Theorem 4.1. *Let A be a polytope in a real vector space V with vertices in a subgroup Γ of V, μ be a valuation on $\mathcal{P}_V$ with character ψ and $c \in \mathbb{R}$. Then*

$$G(T) = \sum_{m=0}^{\infty} \mu([A]_{c+m}) T^m$$

and

$$H(T) = \sum_{m=1}^{\infty} \mu([A]_{c-m})T^{-m} = \sum_{m=-\infty}^{-1} \mu([A]_{c+m})T^{m}$$

are Laurent series of rational functions. Furthermore $G(T) = -H(T)$ *as rational functions.*

Proof. Let $p(t) = \mu([A]_t)$ and denote the vertices of A by $v_1, \ldots, v_s$. Applying μ to the identity of Theorem 2.1 gives

$$0 = \sum_{I\subseteq[s]} (-1)^{|I|}\mu(\tau_{-v_I}([A]_{c+|I|})) = \sum_{I\subseteq[s]} (-1)^{|I|}\psi(-v_I)p(c+|I|).$$

Replacing I by $[s] \setminus I$, c by $c - s$ and multiplying by $(-1)^s\psi(v_{[s]})$ gives

$$0 = \sum_{I\subseteq[s]} (-1)^{|I|}\psi(v_I)p(c-|I|). \tag{4.1}$$

Define

$$D(T) = \sum_{I\subseteq[s]} (-1)^{|I|}\psi(v_I)T^{|I|}.$$

From (4.1) it is apparent that $D(T)G(T)$ is a polynomial of degree at most $s - 1$. Also $D(T)H(T)$ is a polynomial of degree at most $s - 1$. Indeed (4.1) states that multiplying the formal Laurent series $G(T) + H(T)$ by $D(T)$ gives zero. Hence $D(T)G(T) = -D(T)H(T)$ and so $G(T) = -H(T)$ as rational functions.

We remark that

$$D(T) = \prod_{i=1}^{s}(1 - \psi(v_i)T).$$ □

To state Stanley's reciprocity theorem we need some notation. We follow [2, §4.2]. Let $X_1, \ldots, X_n$ be indeterminates and for $a = (a_1, \ldots, a_n) \in \mathbb{Q}^n$ set $X^a = X_1^{a_1} \cdots X_n^{a_n}$. For a subset $A \subseteq \mathbb{R}^n$ define

$$\sigma_A(X_1, \ldots, X_n) = \sum_{a\in A\cap\mathbb{Z}^n} X^a.$$

At present this is just a formal Laurent series in n variables.

A *rational cone* K in $\mathbb{R}^n$ is the set of nonnegative real combinations of a finite set of points $v_1, \ldots, v_k \in \mathbb{Z}^n$ (called generators), that is

$$K = \left\{ \sum_{j=1}^{k} t_j v_j : t_j \geq 0 \right\}.$$

Its relative interior is

$$K^\circ = \left\{ \sum_{j=1}^{k} t_j v_j : t_j > 0 \right\}.$$

When K is a rational cone which is *pointed*, that is there is a hyperplane meeting the cone in just the origin, then there is a polynomial $v(X_1,\ldots,X_n)$ such that the formal product

$$u(X_1,\ldots,X_n) = v(X_1,\ldots,X_n)\sigma_K(X_1,\ldots,X_n)$$

is a Laurent polynomial [2, Corollary 3.7]. Thus we can identify σ_K with the rational function $u(X_1,\ldots,X_n)/v(X_1,\ldots,X_n)$. Likewise we can identify σ_{K° with a rational function. Also when K is a pointed rational cone, one can assign nonzero values in $\mathbb{C}$ to $X_1,\ldots,X_n$ in such a way that $X^{v_1},\ldots,X^{v_k}$ take values of modulus strictly less than 1. Considered as a numerical series, $\sigma_K(X_1,\ldots,X_n)$ then converges, to its value as considered as a rational function. Similar considerations apply to σ_{K°.

Theorem 4.2 (Stanley's reciprocity theorem). *[7, Proposition 7.1] Let K be a pointed rational cone in $\mathbb{R}^n$. Then*

$$\sigma_K(1/X_1,\ldots,1/X_n) = (-1)^{\dim K}\sigma_{K^\circ}(X_1,\ldots,X_n)$$

as rational functions.

Proof. We can rewrite the claimed identity as

$$\sigma_K(X_1,\ldots,X_n) = (-1)^{\dim K}\sigma_{-K^\circ}(X_1,\ldots,X_n). \tag{4.2}$$

It will be more convenient to work in $\mathbb{R}^{n+1}$ with the typical point denoted by $a = (a_0,\ldots,a_n)$. Let $w_1,\ldots,w_s$ be generators of K. We first prove that we can reduce to the case where each w_i has the form $w_j = (1, v_j)$ where $v_j \in \mathbb{Q}^n$.

As K is a pointed cone then there is a hyperplane $\mathcal{H}$ meeting K only at the origin. Perturbing $\mathcal{H}$ slightly, we may assume that the equation of $\mathcal{H}$ has rational coefficients, and then by homogeneity, that it has integer coefficients with greatest common divisor 1. Thus we may assume that $\mathcal{H}$ has equation $\sum_{j=0}^n a_j x_j = 0$ where $(a_0,\ldots,a_n) \in \mathbb{Z}^{n+1}$ satisfies $\gcd(a_0,\ldots,a_n) = 1$. Moreover, by negating all a_i if necessary, we may assume that

$$K \subseteq \left\{ (x_0,\ldots,x_n) \in \mathbb{R}^{n+1} : \sum_{j=0}^{n} a_j x_j \geq 0 \right\} = \mathcal{H}^+.$$

There is a basis $b_0,\ldots,b_n$ of $\mathbb{Z}^{n+1}$ such that $b_1,\ldots,b_n$ generates $\mathbb{Z}^{n+1}\cap\mathcal{H}$ and that $b_0 \in \mathcal{H}^+$. Let $B : \mathbb{R}^{n+1} \to \mathbb{R}^{n+1}$ be the unique linear map taking b_j to e_j. Then B is a bijection, $B(\mathbb{Z}^{n+1}) = \mathbb{Z}^{n+1}$ and $B(K) = L$ where L is a pointed rational cone contained in $\{x_0 \geq 0\}$ and meeting the hyperplane $\{x_0 = 0\}$ only at the origin.

Then

$$\sigma_K(X_0,\ldots,X_n) = \sigma_L(X^{b_0},\ldots,X^{b_n})$$

and

$$\sigma_{K^\circ}(X_0,\ldots,X_n) = \sigma_{L^\circ}(X^{b_0},\ldots,X^{b_n}).$$

It follows that K satisfies (4.2) if and only if L does. In the sequel we may assume that K is contained in $\{x_0 \geq 0\}$ and meets $\{x_0 = 0\}$ only at the origin. Then K has generators of the form $(1, v_1),\ldots,(1, v_s)$ with the $v_j \in \mathbb{Q}^n$.

Let A be a polytope with vertices $v_1, \ldots, v_s \in \mathbb{Q}^n$. Let Γ be the additive group generated by $\mathbb{Z}^n$ together with the v_j. Then Γ is a lattice in $\mathbb{R}^n$ and $H = \Gamma/\mathbb{Z}^n$ is a finite group. There is a positive integer r with the property that $rv_j \in \mathbb{Z}^n$ for each j so that $\Gamma \subseteq (\frac{1}{r}\mathbb{Z})^n$. Each character $\psi : H \to \mathbb{C}^*$ lifts to a character of Γ which we shall also denote by ψ. Define a new character of H by

$$\psi^*(a) = \psi(a)X^a.$$

This takes values in the field $F = \mathbb{C}(X_1^{1/r}, \ldots, X_n^{1/r})$. The map $\mu_\psi : \mathcal{P}_{\mathbb{R}^n} \to F$ defined by

$$\mu_\psi(f) = \sum_{a\in\Gamma} f(a)\psi^*(a) = \sum_{a\in\Gamma} f(a)\psi(a)X^a$$

is a valuation. Also μ_ψ has character ψ^* with respect to Γ.

Applying Theorem 4.1 in the special case $c = 0$ to the polytope A and valuation μ_ψ gives $G_\psi(T) = -H_\psi(T)$ as rational functions over F where

$$G_\psi(T) = \sum_{m=0}^{\infty} \mu_\psi([A]_m)T^m$$

and

$$H_\psi(T) = \sum_{m=-\infty}^{-1} \mu_\psi([A]_m)T^m.$$

Then

$$G_\psi(T) = \sum_{m=0}^{\infty} T^m \sum_{a\in mA\cap\Gamma} \psi(a)X^a$$

and

$$H_\psi(T) = (-1)^{\dim A} \sum_{m=-\infty}^{-1} T^m \sum_{a\in mA^\circ\cap\Gamma} \psi(a)X^a$$

Summing over the character group $\widehat{H}$ of H gives

$$\begin{aligned}\sum_{\psi\in\widehat{H}} G_\psi(T) &= \sum_{m=0}^{\infty} T^m \sum_{a\in mA\cap\Gamma} X^a \sum_{\psi\in\widehat{H}} \psi(a) \\ &= |H| \sum_{m=0}^{\infty} T^m \sum_{a\in mA\cap\mathbb{Z}^n} X^a\end{aligned}$$

as $\sum_{\psi\in\widehat{H}} \psi(u) = 0$ whenever u is a nonzero element of the finite abelian group H. As $K \cap \mathbb{Z}^{n+1}$ consists of the elements $b = (m, a)$ where $m \geq 0$ is a nonnegative integer and $a \in mA \cap \mathbb{Z}^n$ then

$$\sum_{\psi\in\widehat{H}} G_\psi(X_0) = |H| \sum_{b\in K\cap\mathbb{Z}^{n+1}} X^b = |H|\sigma_K(X_0, \ldots, X_n).$$

Similarly

$$\sum_{\psi \in \widehat{H}} H_\psi(X_0) = (-1)^{\dim A} |H| \sigma_{-K^\circ}(X_0, \ldots, X_n).$$

By Theorem 4.1, $G_\psi(X_0) = -H_\psi(X_0)$ as rational functions for each ψ. It follows that

$$\sigma_K(X_0, \ldots, X_n) = -(-1)^{\dim A} \sigma_{-K^\circ}(X_0, \ldots, X_n)$$

as rational functions. But then (4.2) follows from the observation that $\dim K = 1 + \dim A$. □

Acknowledgement

The author is grateful to the anonymous referee for detailed comments which helped to considerably improve the standard of the paper.

References

[1] Alexander Barvinok, *A Course in Convexity*, American Mathematical Society, Providence, 2002.

[2] Matthias Beck & Sinai Robins, *Computing the Continuous Discretely*, Springer-Verlag, New York, 2007.

[3] Eugène Ehrhart, *Sur les polyèdres rationnels homothétiques à n dimensions'*, C. R. Acad. Sci. Paris 254 (1962), 616–618.

[4] Daniel A. Klain & Gian-Carlo Rota, *Introduction to Geometric Probability*, Cambridge University Press, Cambridge, 1997.

[5] Ian G. Macdonald *Polynomials associated with finite cell-complexes*, J. London Math. Soc. (2) 4 (1971), 181–192.

[6] Steven V. Sam, *A bijective proof for a theorem of Ehrhart*, Amer. Math. Monthly 116 (2009), 688–701.

[7] Richard P. Stanley, *Combinatoral reciprocity theorems*, Advances in Math. 14 (1974), 194–253.

[8] Richard P. Stanley, *Enumerative Combinatorics I*, Cambridge University Press, Cambridge, 1997.

In: Frontiers of Combinatorics and Number Theory, Volume 2 ISBN: 978-1-61942-918-5
Editor: Zhi-Wei Sun, pp. 209-217

Chapter 14

A NEW CLASS OF THETA FUNCTION IDENTITIES IN TWO VARIABLES

***Robin Chapman*[1,*], *William B. Hart*[2,†] *and Pee Choon Toh*[3,‡]**
[1] Mathematics Research Institute, University of Exeter
Exeter, UK
[2] Mathematics Institute, University of Warwick
Coventry, UK
[3] Department of Mathematics, National University of Singapore
S17, 10 Lower Kent Ridge Road, Singapore

Abstract

We describe a new class of identities, which hold for certain general theta series, in two completely independent variables. We provide explicit examples of these identities involving the Dedekind eta function, Jacobi theta functions, and various theta functions of Ramanujan.

2010 Mathematics Subject Classification: Primary 11F27; Secondary 11E39, 33E05.

1. Introduction

Let $z \in \mathcal{H} = \{x + iy \mid x, y \in \mathbb{R}, y > 0\}$ and set $q = e(z)$ where $e(x) = \exp(2\pi i x)$. The Dedekind eta-function is defined by

$$\eta(z) = q^{1/24} \prod_{n=1}^{\infty} (1 - q^n).$$

Up to normalisation it is a 24-th root of the discriminant of an elliptic curve $\mathbb{C}/\Lambda$ where Λ is the lattice $[1, z]$. The Dedekind eta function also appears in the Chowla-Selberg formula, which gives a closed form expression for the value at 1 of the L-series associated to

Received April 20, 2010; Accepted September 17, 2010.
*E-mail address: R.J.Chapman@exeter.ac.uk
†E-mail address: W.B.Hart@warwick.ac.uk; supported by EPSRC grants EP/G004870/1, EP/D079543/1
‡E-mail address: mattpc@nus.edu.sg

unramified Abelian extensions of imaginary quadratic number fields. As such, the Dedekind eta function is intimately tied up with the theory of complex multiplication and explicit class field theory.

Relations in one variable for the Dedekind eta function (known as modular equations) have played an important part in construction of class invariants (generators for class field extensions of imaginary quadratic number fields). See [7] and [15] for details.

In [3] the following striking identity in two variables was proved using Jacobi's triple product identity.

$$\begin{aligned} 27\eta^3(3z)\eta^3(3w) &= \eta^3\left(\frac{z}{3}\right)\eta^3\left(\frac{w}{3}\right) + i\eta^3\left(\frac{z+1}{3}\right)\eta^3\left(\frac{w+1}{3}\right) \\ &\quad -\eta^3\left(\frac{z+2}{3}\right)\eta^3\left(\frac{w+2}{3}\right), \end{aligned} \tag{1}$$

for all $z, w \in \mathcal{H}$.

Köhler [11] subsequently also used the triple product identity to give another proof of the above.

Two variable relations such as the above can be viewed as giving rise to *permanent identities*, i.e. infinite families of one variable identities. For example, one may make the substitution $w = \tau$, $z = n\tau$. The identity then becomes a relation between $\eta(\tau)$ and $\eta(n\tau)$ which does not vary as n varies. In fact the above identity was discovered in the second author's thesis [9] by noting such a permanent family of identities for the eta function.

In this paper we generalize identity (1) for a very general class of theta functions. In particular we define a general theta function in the following way:

$$\Theta_{\nu,a,b,m,\psi}(z) = \sum_{n=-\infty}^{\infty} \psi(n)(an+b)^{\nu} q^{\frac{(an+b)^2}{m}}, \tag{2}$$

where $z \in \mathcal{H}$, $\nu, a, b, m \in \mathbb{Z}$, $m > 0$, $(a, b) = 1$, and ψ is an additive character modulo a. Many well known functions in number theory, including $\eta^3(z)$, fall into this class of theta function. We will describe some examples in Section 3

The general theta function bears some resemblance to the theta function defined by Shimura [13]. However we note that unlike Shimura's theta function, our definition includes functions which are not modular forms (see Example 5 below).

2. Generalized Identities

Fix two sets ν, a, b, m, ψ and ν', a', b', m', ψ', satisfying the above conditions and, for simplicity, denote

$$f(z) = \Theta_{\nu,a,b,m,\psi}(z), \quad g(z) = \Theta_{\nu',a',b',m',\psi'}(z).$$

The main result that we prove in this paper is the following identity.

Theorem 1. *Let p be an odd prime such that $a \mid (p-1)$ and $a' \mid (p-1)$, then*

$$\sum_{j=0}^{p-1} f\left(\frac{z+mj}{p}\right) g\left(\frac{z'-m'jk}{p}\right) = \psi(c)\psi'(c')\, p^{\nu+\nu'+1} f(pz)g(pz'),$$

for all $z, z' \in \mathcal{H}$, where k is any quadratic nonresidue modulo p, $c = b(p-1)/a$ and $c' = b'(p-1)/a'$.

In particular when $p \equiv 3 \pmod 4$ then

$$\sum_{j=0}^{p-1} f\left(\frac{z+mj}{p}\right) g\left(\frac{z'+m'j}{p}\right) = \psi(c)\psi'(c')\, p^{\nu+\nu'+1}\, f(pz)g(pz'),$$

since -1 is a quadratic nonresidue.

When the theta function involved is a half integer weight modular form, the identity can be interpreted in terms of the half integer weight modular form being annihilated by the Hecke operator T_p. See [10] for details. However, as noted, our definition includes theta functions which are not modular forms, so no general proof along these lines is known to the authors.

Proof of Theorem 1. Let

$$f_j(z) = f\left(\frac{z+mj}{p}\right),$$

and note that

$$f_j(z) = \sum_{n=-\infty}^{\infty} \psi(n)(an+b)^{\nu} e((an+b)^2 j/p) q^{\frac{(an+b)^2}{mp}}.$$

Thus f_j depends only on the congruence class of j modulo p. The vector

$$\alpha(z) = \big(f_0(z), f_1(z), \ldots, f_{p-1}(z)\big) \in \mathbb{C}^p,$$

lies in the linear subspace V_p of $\mathbb{C}^p$ spanned by the vectors

$$v_d = \big(1, e(d^2/p), e(2d^2/p), \ldots, e((p-1)d^2/p)\big),$$

where d runs through the integers.

There are $(p+1)/2$ distinct squares modulo p, so there are $(p+1)/2$ distinct v_d; these are linearly independent. Hence V_p has dimension $(p+1)/2$.

Now let

$$f_\infty(z) = \psi(c)p^{\nu+1/2} f(pz) = \psi(c)p^{\nu+1/2} \sum_{n=-\infty}^{\infty} \psi(n)(an+b)^{\nu} q^{\frac{(an+b)^2 p}{m}},$$

where $c = b(p-1)/a$.

We express f_∞ in terms of f_j, i.e. we derive an identity involving only one variable z, for these functions. Consider

$$\begin{aligned}\sum_{j=0}^{p-1} f_j(z) &= \sum_{n=-\infty}^{\infty} \psi(n)(an+b)^{\nu} q^{\frac{(an+b)^2}{mp}} \sum_{j=0}^{p-1} e((an+b)^2 j/p) \\ &= p \sum_{\substack{n=-\infty \\ p \mid (an+b)}}^{\infty} \psi(n)(an+b)^{\nu} q^{\frac{(an+b)^2}{mp}}.\end{aligned}$$

But $p \mid (an+b)$ if and only if $n = ps + c$ where $s \in \mathbb{Z}$. Thus

$$\begin{aligned}\sum_{j=0}^{p-1} f_j(z) &= p \sum_{s=-\infty}^{\infty} \psi(ps+c)(aps+bp)^\nu q^{\frac{(aps+bp)^2}{mp}} \\ &= p^{\nu+1}\psi(c) \sum_{s=-\infty}^{\infty} \psi(s)(as+b)^\nu q^{\frac{(as+b)^2 p}{m}} \\ &= \sqrt{p}\, f_\infty(z).\end{aligned}$$

Now we see that

$$F(z) = \big(f_0(z), f_1(z), \ldots, f_{p-1}(z), f_\infty(z)\big)$$

lies in the vector subspace W_p of $\mathbb{C}^{p+1}$ spanned by the vectors

$$w_0 = \big(1, 1, 1, \ldots, 1, \sqrt{p}\big)$$

and

$$w_d = \big(1, e(d^2/p), e(2d^2/p), \ldots, e((p-1)d^2/p), 0\big),$$

where d runs through the integers prime to p. Again W_p has dimension $(p+1)/2$.

Now, if we let

$$g_j(z') = g\left(\frac{z' - m'jk}{p}\right) \qquad \text{and} \qquad g_\infty(z') = \psi'(c')p^{\nu'+1/2} g(pz'),$$

a similar argument proves that

$$G(z') = \big(g_0(z'), g_1(z'), \ldots, g_{p-1}(z'), g_\infty(z')\big) \in W'_p$$

where W'_p is the subspace of $\mathbb{C}^{p+1}$ spanned by

$$w'_0 = w_0 = \big(1, 1, 1, \ldots, 1, \sqrt{p}\big)$$

and

$$w'_d = \big(1, e(-kd^2/p), e(-2kd^2/p), \ldots, e(-(p-1)kd^2/p), 0\big).$$

Next, for $v, u \in \mathbb{C}^{p+1}$, we define

$$B(v, u) = \sum_{j=0}^{p-1} v_j u_j - v_\infty u_\infty,$$

a bilinear form on $\mathbb{C}^{p+1}$.

Clearly for d_1, d_2 not both zero modulo p,

$$B(w_{d_1}, w'_{d_2}) = \sum_{j=0}^{p-1} e\left(\frac{j(d_1^2 - kd_2^2)}{p}\right) = 0,$$

since k is a quadratic nonresidue of p.

Also

$$B(w_{d_1}, w_0') = \sum_{j=0}^{p-1} e(d_1^2 j/p) = 0$$

for d_1 nonzero modulo p. Similarly $B(w_0, w_{d_2}') = 0$ for d_2 nonzero modulo p.
Finally,

$$B(w_0, w_0') = \sum_{j=0}^{p-1} 1 - (\sqrt{p})^2 = 0.$$

Thus $B(v, u) = 0$ for all $v \in W_p, u \in W_p'$. In particular, $B(F(z), G(z')) = 0$, i.e.

$$\sum_{j=0}^{p-1} f_j(z) g_j(z') = \psi(c)\psi'(c')\, p^{\nu+\nu'+1}\, f(pz)g(pz'),$$

as was to be shown. □

Theorem 1 required that $a \mid (p-1)$ and $a' \mid (p-1)$. We can make a small modification to deal with the case where $a \mid (p+1)$. Indeed we simply note that for the original theta function we defined,

$$\Theta_{\nu,a,b,m,\psi}(z) = (-1)^\nu \sum_{n=-\infty}^{\infty} \psi(-n)(an-b)^\nu q^{\frac{(an-b)^2}{m}}.$$

Now our argument goes through much the same as before, except that we now require $\psi(n) = \psi(-n)$, etc., i.e. ψ and ψ' must now be real-valued characters taking only the values ± 1. We thus have the following.

Theorem 1.1. *Let p be an odd prime such that $a \mid (p+1)$ and $a' \mid (p+1)$, then*

$$\sum_{j=0}^{p-1} f\left(\frac{z+mj}{p}\right) g\left(\frac{z'-m'jk}{p}\right) = -\psi(c)\psi(c')\,(-p)^{\nu+\nu'+1}\, f(pz)g(pz'),$$

for all $z, z' \in \mathcal{H}$, where k is any quadratic nonresidue modulo p, $c = b(p+1)/a$ and $c' = b'(p+1)/a'$.

Theorem 1.2. *Let p be an odd prime such that $a \mid (p-1)$ and $a' \mid (p+1)$, then*

$$\sum_{j=0}^{p-1} f\left(\frac{z+mj}{p}\right) g\left(\frac{z'-m'jk}{p}\right) = (-1)^{\nu'}\psi(c)\psi(c')\, p^{\nu+\nu'+1}\, f(pz)g(pz'),$$

for all $z, z' \in \mathcal{H}$, where k is any quadratic nonresidue modulo p, $c = b(p-1)/a$ and $c' = b'(p+1)/a'$.

Remark. We note that the space V_p is essentially the same as $\mathcal{Q}$ in [5], which played a rôle as complex analogue of the quadratic residue code.

3. Examples

The theta function that we introduced in (2) is clearly a generalization of the null values of the classical Jacobi theta functions [16, p. 463]:

$$\begin{aligned}
\theta_1'(q) &= \sum_{n=-\infty}^{\infty} i^{(2n-1)}(2n+1)q^{(2n+1)^2/4}, \\
\theta_2(q) &= \sum_{n=-\infty}^{\infty} q^{(2n+1)^2/4}, \\
\theta_3(q) &= \sum_{n=-\infty}^{\infty} q^{n^2}, \\
\theta_4(q) &= \sum_{n=-\infty}^{\infty} (-1)^n q^{n^2}.
\end{aligned}$$

The Jacobi theta functions have several important applications to number theory. For example θ_3^3 can be used to compute class numbers of imaginary quadratic number fields, see [14] for details.

We shall describe below, several important functions satisfying the definition in (2).

Example 1: Jacobi's formula gives,

$$\eta^3(z) = \sum_{n=-\infty}^{\infty} (4n+1)q^{(4n+1)^2/8},$$

corresponding to $\Theta_{\nu,a,b,m,\psi}(z)$ with $\nu = 1$, $a = 4$, $b = 1$, $m = 8$ and ψ the trivial character.

Theorem 1 and 1.1, with $f(z) = \eta^3(z)$ and $g(z') = \eta^3(z')$, then yields:

$$p^3\eta^3(pz)\eta^3(pz') = \sum_{j=0}^{p-1} \eta^3\left(\frac{z+8j}{p}\right)\eta^3\left(\frac{z'-8jk}{p}\right), \tag{3}$$

for any odd prime p.

The case $p = 3$ and $k = -1$ reduce to identity (1) on using the fact that $\eta^3(z+1) = e(1/8)\eta^3(z)$.

Example 2: Let $f(z) = \eta(z)$ and $g(z') = \eta(z')$. Then by Euler's pentagonal number formula,

$$\eta(z) = q^{1/24}\sum_{n=-\infty}^{\infty} (-1)^n q^{n(3n+1)/2} = \sum_{n=-\infty}^{\infty} (-1)^n q^{(6n+1)^2/24},$$

corresponding to $\Theta_{\nu,a,b,m,\psi}(z)$ with $\nu = 0$, $a = 6$, $b = 1$, $m = 24$ and $\psi(n) = (-1)^n$.

It follows that for $p > 3$,

$$p\eta(pz)\eta(pz') = \sum_{j=0}^{p-1} \eta\left(\frac{z+24j}{p}\right)\eta\left(\frac{z'-24jk}{p}\right), \tag{4}$$

where k is a quadratic nonresidue modulo p.

Example 3: If we let $f(z) = \eta(z)$ and $g(z') = \eta^3(z')$. Since -3 is a quadratic nonresidue for primes $p \equiv 11 \pmod{12}$, we have

$$p^2\eta(pz)\eta^3(pz') = \sum_{j=0}^{p-1} \eta\left(\frac{z+24j}{p}\right)\eta^3\left(\frac{z'+24j}{p}\right). \tag{5}$$

Example 4: In [2, p. 114] Entry 8 (ix) and (x), Ramanujan studied two theta functions which are consequences of the quintuple product identity [2, p. 83], [6]. Using his notation for $\phi(q), \psi(q)$ and $f(-q)$, these are

$$\begin{aligned} q^{1/24}\phi^2(-q)f(-q) &= \frac{\eta^5(z)}{\eta^2(2z)} = \sum_{n=-\infty}^{\infty} (6n+1)q^{\frac{(6n+1)^2}{24}}, \\ q^{1/3}\psi(q^2)f^2(-q) &= \frac{\eta^2(4z)\eta^2(z)}{\eta(2z)} = \sum_{n=-\infty}^{\infty} (3n+1)q^{\frac{(3n+1)^2}{3}}. \end{aligned}$$

A simple transform of the latter function gives the following,

$$\frac{\eta^5(2z)}{\eta^2(z)} = \sum_{n=-\infty}^{\infty} (-1)^n(3n+1)q^{\frac{(3n+1)^2}{3}}.$$

All three clearly satisfy the definition given in (2).

Example 5: The preceding examples all involved the case where $\nu = 0$ or 1. Ramanujan [12, p. 369], [1, p. 355–362] also studied theta functions involving higher values for ν. Let

$$\begin{aligned} U_{2\alpha+1}(q) &= \sum_{n=-\infty}^{\infty} (4n+1)^{2\alpha+1}q^{\frac{(4n+1)^2}{8}}, \\ V_{2\alpha}(q) &= \sum_{n=-\infty}^{\infty} (-1)^n(6n+1)^{2\alpha}q^{\frac{(6n+1)^2}{24}}. \end{aligned}$$

Ramanujan showed that

$$\begin{aligned} U_{2\alpha+1}(q) &= \eta^3(z) \sum_{i+2j+3k=\alpha} a_{ijk}E_2^iE_4^jE_6^k, \\ V_{2\alpha}(q) &= \eta(z) \sum_{i+2j+3k=\alpha} b_{ijk}E_2^iE_4^jE_6^k, \end{aligned}$$

where a_{ijk} and b_{ijk} are rational numbers and E_n are the normalized Eisenstein series of weight n. See [4] and references therein for more information about these two classes of theta functions.

Note that this example includes theta functions under our definition which are not modular forms, due to the presence of E_2.

Summary and Concluding Remarks

We have used elementary methods to prove our main identity for a general class of theta functions and given several well known examples of such theta functions.

Our general theorem can be used to construct infinite families of identities for such theta functions. It is unknown whether there is a link between such permanent identities and the identities of Nathan Fine [8].

Acknowledgements

Many thanks to Bruce Berndt for suggesting that we check the applicability of the method of this paper to the theta functions of Example 4, as defined by Ramanujan.

References

[1] G. E. Andrews and B. C. Berndt, *Ramanujan's lost notebooks. Part II.*, Springer-Verlag, New York, 2009.

[2] B. C. Berndt, *Ramanujan's notebooks. Part III.*, Springer-Verlag, New York, 1991.

[3] B. C. Berndt and W. B. Hart, An identity for the Dedekind eta-function involving two independent complex variables, *Bull. London Math. Soc.* 39 (2007), 345–347.

[4] H. H. Chan, S. Cooper and P. C. Toh, Ramanujan's Eisenstein series and powers of Dedekind's eta-function, *J. Lond. Math. Soc.* 75 (2007), 225–242.

[5] R. Chapman, Determinants of Legendre symbol matrices, *Acta Arith.* 115 (2004), 231–244.

[6] S. Cooper, The quintuple product identity, *Int. J. Number Theory* 2 (2006), 115–161.

[7] D. A. Cox, *Primes of the form* $x^2 + ny^2$, Wiley-Interscience, 1989.

[8] N. J. Fine, Basic hypergeometric series and applications, *Mathematical Surveys and Monographs,* vol. 27, Amer. Math. Soc., Providence, 1988.

[9] W. B. Hart, *Explicit evaluation of the Dedekind eta function* PhD. Thesis, 2004, Macquarie University, Sydney.

[10] N. Koblitz, Introduction to Elliptic Curves and Modular Forms, *Graduate Texts in Math.*, vol. 97, Springer, New York, 1984.

[11] G. Köhler, Note on an eta identity presented by B. C. Berndt and W. B. Hart, *Bull. Lond. Math. Soc.* 40 (2008), 172–173.

[12] S. Ramanujan, The lost notebook and other unpublished papers, *Narosa*, New Delhi, 1988.

[13] G. Shimura, On modular forms of half integral weight, *Ann. of Math.* 97 (1973), 440-481.

[14] G. N. Watson, Generating functions of class-numbers, *Compos. Math.* 1 (1935), 39–68.

[15] H. Weber, *Lehrbuch der Algebra*, Dritter Band, 3rd ed., Chelsea, New York, 1961.

[16] E. T. Whittaker and G.N. Watson, *A course of modern analysis*, 4th ed., Cambridge University Press, Cambridge, 1966.

In: Frontiers of Combinatorics and Number Theory, Volume 2 ISBN: 978-1-61942-918-5
Editor: Zhi-Wei Sun, pp. 219-225

Chapter 15

ON UNIFORM BOUNDS FOR LATTICE POINTS IN INTERSECTIONS OF HYPERBOLIC PLANE REGIONS

Patrick X. Rault*
Department of Mathematics, State University of New York
Geneseo, NY, US

Abstract

We show that for any indefinite binary quadratic forms f and g the number of primitive integral lattice points in the (nonconvex and nonsmooth) plane region bounded by the four hyperbolas $|f(X,Y)| = B$ and $|g(X,Y)| = C$ is at most $32\frac{(BC)^{1/2}}{|R|^{1/4}} + 8$. Here R is the resultant of f and g. When f and g each have coprime integral coefficients, then R is also an integer and we arrive at the uniform bound $32(BC)^{1/2} + 8$, which is independent of the choice of forms f and g.

2000 Mathematics Subject Classification: Primary 11P21; Secondary 11H16.

1. Introduction

Number theorists have uncovered many relations between the area of plane regions and the number of lattice points contained therein. Jarnik showed that, for a closed rectifiable Jordan curve of length L, the area A enclosed and the number N of integral lattice points within the curve are related by $|N - A| < L$ (see [5]). This appears to show that the area and number of lattice points are relatively close; however it gives little information when the length of the curve is large. Take for example the ellipse $aX^2 + Y^2/a = 1$ which for large a has length roughly $2\sqrt{a}$, area π, and contains only two primitive lattice points $(0, \pm 1)$; Jarnik's theorem only provides us with the trivial inequality $|2 - \pi| < 2\sqrt{a}$.

Other methods are needed for regions with large perimeter, and in fact the problem has been most studied for such convex domains. For example, lemma 2 in [2], states that the number of primitive integral lattice points inside an ellipse is at most $4(\text{Area} + 1)$. A similar result for arbitrary smooth convex contours is given in [3].

Received: May 20, 2010; Accepted: August 21, 2010
*E-mail address: rault@geneseo.edu; Website: `http://geneseo.edu/~rault`

For nonconvex regions one cannot make use of this machinery. Indeed, the region bounded by the curves $X = 0$, $X = 1$, $Y = 0$, and $Y = \sqrt{\tan(\pi x/2)}$ has area $\sqrt{2}$ and yet has infinitely many primitive integral lattice points on the boundary.

The aim of this paper is to prove the following upper bound for the number of integral lattice points in the star-shaped region bounded by two symmetric pairs of hyperbolas; this bound is an essential ingredient in [4] where the author proves a uniform bound for the number of rational points of bounded height on rational curves.

Theorem 1.1. *Let f and g be indefinite binary quadratic forms such that fg is squarefree. Let $R(f, g)$ denote the resultant of f and g and let B and C be positive real numbers.*
Then:

1. *The area of the region bounded by the curves $|f(X, Y)| = B$ and $|g(X, Y)| = C$ is at most $16\frac{(BC)^{1/2}}{|R(f,g)|^{1/4}}$.*
2. *The number of primitive integral lattice points (x, y) satisfying $|f(x, y)| < B$ and $|g(x, y)| < C$ is at most $32\frac{(BC)^{1/2}}{|R(f,g)|^{1/4}} + 8$.*

The content of our theorem is twofold. First, because $R(f, g)$ is a nonzero integer we give a uniform upper bound independent of the particular forms f and g in question; indeed, for fixed hyperbolas, one could establish a sharp asymptotic formula. Second, we give an inverse dependence on the coefficients defining the hyperbolas by way of an explicit formula involving a resultant.

Note that the theorem does not provide a formula for the area of the region; indeed we do not claim a lattice point bound of the form $2(\text{Area}+4)$. For a computation of these areas by a student, see [1].

2. Preliminaries

In this section we outline the proof of Theorem 1.1 and introduce the essential lemmas. We also formally define the concepts of the preceding introduction.

Let f and g be rational indefinite binary quadratic forms such that the product fg is squarefree. Let B and C be positive real numbers. Consider the region

$$\mathcal{R}_{(f,g)}(B, C) = \{(x, y) \in \mathbb{R}^2 : |f(x, y)| < B, |g(x, y)| < C\}$$

and let $A_{(f,g)}(B, C)$ denote the area of $\mathcal{R}_{(f,g)}(B, C)$. Recall that a nonzero lattice point $(x, y) \in \mathbb{Z}^2$ is primitive if $\gcd(x, y) = 1$. Let $N_{(f,g)}(B, C)$ count the number of primitive integral lattice points (x, y) in $\mathcal{R}_{(f,g)}(B, C)$, that is

$$N_{(f,g)}(B, C) := \#\{(x, y) \in \mathbb{Z}^2 : \gcd(x, y) = 1, |f(x, y)| < B, |g(x, y)| < C\}$$

In the following graph, the lightly-shaded region denotes the region within either $|f(X, Y)| = B$ or $|g(X, Y)| = C$, whereas the darkly-shaded region denotes $\mathcal{R}_{(f,g)}(B, C)$.

Observe that, by varying f and g so that their asymptotes become closer, points in $\mathcal{R}_{(f,g)}(B, C)$ may be arbitrarily distant from the origin; for example, the solutions to Pell's equation yield infinitely many primitive points (x, y) satisfying $-2 < x^2 - 2y^2 < 2$.

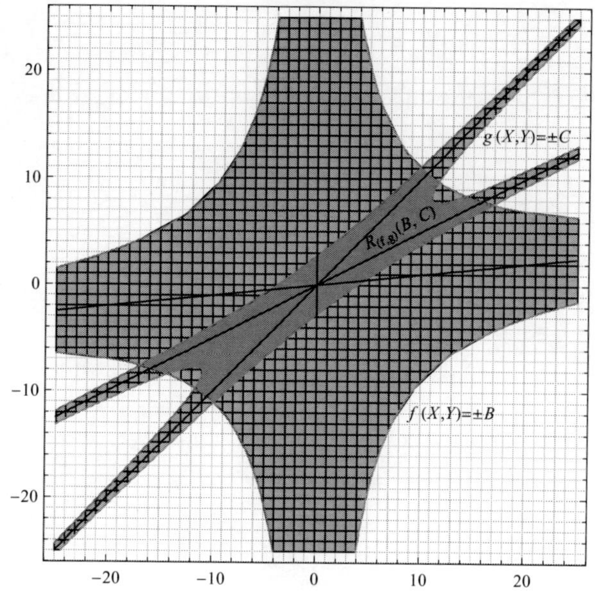

While studying $N_{(f,g)}(B,C)$ the following invariants of binary forms will be instrumental in our calculations.

Definition 1.

I. Let $h(X,Y) = \alpha_1 X + \alpha_2 Y$ and $k(X,Y) = \beta_1 X + \beta_2 Y$ be binary linear forms. We define their resultant $R(h,k)$ to be the determinant of their coefficient matrix:

$$R(h,k) = \alpha_1\beta_2 - \alpha_2\beta_1$$

II. Let f and g (resp.) be binary forms of degrees d_1 and d_2 (resp.) that factor linearly over $\overline{\mathbb{Q}}$ as $f = \prod_{i=1}^{d_1} h_i$ and $g = \prod_{j=1}^{d_2} k_j$. We define the resultant $R(f,g)$ to be the following symmetric polynomial of resultants of the linear forms:

$$R(f,g) = \prod_{i=1}^{d_1}\prod_{j=1}^{d_2} R(h_i,k_j)$$

Note that if f and g have rational coefficients then the factors of $R(f,g)$ are permuted by the Galois group; hence $R(f,g)$ is rational. In fact, if f and g have integral coefficients then $R(f,g)$ will also be an integer. Note further that $R(\lambda f, \lambda g) = \lambda^{d_1 d_2} R(f,g)$.

The invariants of interest to us behave well under linear changes of coordinates.

Lemma 2.1. *Let f and g be binary quadratic forms and let (B,C) be a pair of positive real numbers. Then for any matrices $S \in SL_2(\mathbb{Z})$ and $T \in GL_2(\mathbb{Q})$ we have*

1. $N_{(f\circ S, f\circ S)}(B,C) = N_{(f,g)}(B,C)$

2. $R(f\circ T, g\circ T) = R(f,g)\,|\det T|^4$

Proof. It is clear that S is one-to-one and sends coprime integral lattice points to coprime integral lattice points by

$$\begin{bmatrix} x \\ y \end{bmatrix} \mapsto \begin{bmatrix} u \\ v \end{bmatrix} = S\begin{bmatrix} x \\ y \end{bmatrix}.$$

Therefore, the number of primitive lattice points counted in the image $N_{(f,g)\circ S}(B,C) = \#\{(x,y)\in\mathbb{Z}^2 : \gcd(x,y)=1, |(f\circ S)(x,y)|<B, |(g\circ S)(x,y)|<C\}$ is exactly equal to the number of points counted in the pre-image $N_{(f,g)}(B,C) = \#\{(u,v)\in\mathbb{Z}^2 : \gcd(u,v)=1, |f(u,v)|<B, |g(u,v)|<C\}$. This proves the first conclusion of the lemma.

We now tackle the conclusion concerning resultants. We write the linear factors of f and g as $f(X,Y) = (\alpha_1 X+\alpha_2 Y)(\alpha_3 X+\alpha_4 Y)$ and $g(X,Y) = (\beta_1 X+\beta_2 y)(\beta_3 X+\beta_4 Y)$. Let $p(X,Y) = XY$, so that $f = p\circ\begin{bmatrix} \alpha_1 & \alpha_2 \\ \alpha_3 & \alpha_4 \end{bmatrix}$ and $g = p\circ\begin{bmatrix} \beta_1 & \beta_2 \\ \beta_3 & \beta_4 \end{bmatrix}$.

We compute that

$$R(f,g) = \det\begin{bmatrix} \alpha_1 & \alpha_2 \\ \beta_1 & \beta_2 \end{bmatrix}\cdot\begin{bmatrix} \alpha_1 & \alpha_2 \\ \beta_3 & \beta_4 \end{bmatrix}\cdot\begin{bmatrix} \alpha_3 & \alpha_4 \\ \beta_1 & \beta_2 \end{bmatrix}\cdot\begin{bmatrix} \alpha_3 & \alpha_4 \\ \beta_3 & \beta_4 \end{bmatrix}.$$

Let $T\in \mathrm{GL}_2(\mathbb{Q})$. Then we have, for example, $f\circ T = p\circ\left(\begin{bmatrix} \alpha_1 & \alpha_2 \\ \alpha_3 & \alpha_4 \end{bmatrix}\cdot T\right)$. Hence

$$R(f\circ T, g\circ T) = \det\left(\begin{bmatrix} \alpha_1 & \alpha_2 \\ \beta_1 & \beta_2 \end{bmatrix}\cdot T\right)\cdots\left(\begin{bmatrix} \alpha_3 & \alpha_4 \\ \beta_3 & \beta_4 \end{bmatrix}\cdot T\right) = R(f,g)\cdot\det T^4. \quad\square$$

Our technique for calculating the area $A_{(f,g)}(B,C)$ and number of lattice points $N_{(f,g)}(B,C)$ involves covering the region $\mathcal{R}_{(f,g)}(B,C)$ with polygons whose area and number of lattice points are easy to calculate. The following lemma is our principal tool for using the area to count the lattice points.

Lemma 2.2. *Let P be a closed convex plane region symmetric about the origin with area A and containing L primitive integral lattice points. Then $L\le 2A+2$.*

Proof. Let H denote the convex hull of all integral lattice points contained in P. Clearly both P and H contain the origin.

Let A_H denote the area of H; clearly $A_H\le A$. Let L denote the number of primitive integral lattice points in P, which is also the number of primitive integral lattice points in H.

If all lattice points in P are collinear, then H is a line segment containing the origin. Hence there are at most two primitive lattice points and the formula $L\le 2A+2$ is trivially true.

If H is a polygon, it follows from Pick's Theorem that $A_H\ge \frac{|H\cap\mathbb{Z}^2|}{2}-1$, or equivalently $|H\cap\mathbb{Z}^2|\le 2A_H+2$. Since $A_H\le A$, we have $L\le 2A+2$. $\square$

We first apply this tool to bound the area and number of lattice points in the intersection of the region $\mathcal{R}_{(f,g)}(B,C)$ with the vertical strip $-b<x<b$ (for any positive number b).

Lemma 2.3. *Let g be a real indefinite binary quadratic form, and let C and b be positive real numbers. The region bounded by the curves $X = -b$, $X = b$, and $|g(X,Y)| = C$ is covered by two parallelograms, each containing the origin, the sum of whose areas is at most $8b\sqrt{\frac{C}{|R(g,X)|}}$.*

Proof. Let g factor linearly over $\mathbb{R}$ as $g = hk$, with $h(X,Y) = \alpha X + \beta Y$ and $k(X,Y) = \gamma X + \delta Y$. If $|g(x,y)| < C$, then either $|h(x,y)| < \sqrt{\left|\frac{\beta C}{\delta}\right|}$ or $|k(x,y)| < \sqrt{\left|\frac{\delta C}{\beta}\right|}$. Let P_h denote the parallelogram bounded by the four lines $X = \pm b$, $h(X,Y) = \pm\sqrt{\left|\frac{\beta C}{\delta}\right|}$ and similarly let P_k denote the parallelogram bounded by the lines $X = \pm b$, $k(X,Y) = \pm\sqrt{\left|\frac{\delta C}{\beta}\right|}$. Then these two parallelograms cover the region of points (x,y) satisfying $|g(x,y)| < C$ and $-b \le x \le b$. An example is shown in the following graph.

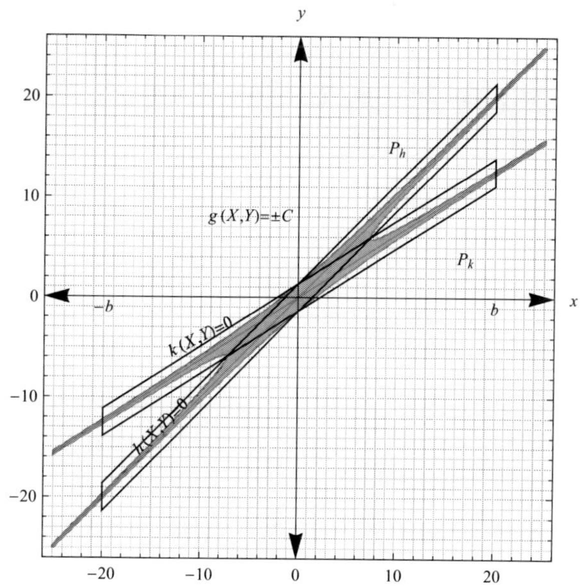

The y-intercepts of the parallelograms occur at $y = \pm\sqrt{\frac{C}{|\beta\delta|}} = \pm\sqrt{\frac{C}{|R(g,X)|}}$. Hence each parallelogram has height $2\sqrt{\frac{C}{|R(g,X)|}}$ and width $2b$; it follows that their areas are both equal to $4b\sqrt{\frac{C}{|R(g,X)|}}$.

We conclude the proof by noting that we have shown that the region is covered by two parallelograms, each containing the origin, whose areas sum to $8b\sqrt{\frac{C}{|R(g,X)|}}$. □

We are now ready to prove Theorem 1.1. As we shall see in the next section, lemma 2.3 is the main component in the proof.

3. Proof of Theorem 1.1

3.1. Special Case: $f(X,Y) = XY$

We first work under the assumption that $f(X,Y) = XY$. By lemma 2.3, for any $b > 0$ the region obtained by intersecting $\mathcal{R}_{(f,g)}(B,C)$ with the vertical strip $-b < x < b$ is covered by two parallelograms whose areas sum to at most $8b\sqrt{\frac{C}{|R(g,X)|}}$. Similarly, the intersection of $\mathcal{R}_{(f,g)}(B,C)$ with the horizontal strip $-c < y < c$ is covered by two parallelograms whose areas sum to at most $8c\sqrt{\frac{C}{|R(g,Y)|}}$. Observe that these two strips cover $\mathcal{R}_{(f,g)}(B,C)$ if and only if $bc \geq B$. Hence if (b,c) is a point on the curve $XY = B$, then the region $\mathcal{R}_{(f,g)}(B,C)$ is covered by four parallelograms whose areas sum to at most $8b\sqrt{\frac{C}{|R(g,X)|}} + 8c\sqrt{\frac{C}{|R(g,Y)|}}$.

A trivial calculus computation reveals that this area bound is minimized for points (b,c) on the curve $XY = B$ when $b = B^{1/2}\left|\frac{R(g,X)}{R(g,Y)}\right|^{1/4}$ and $c = B^{1/2}\left|\frac{R(g,Y)}{R(g,X)}\right|^{1/4}$. Hence $\mathcal{R}_{(f,g)}(B,C)$ is covered by four parallelograms whose sum of areas is bounded as $A_{(f,g)}(B,C) < 16\frac{(BC)^{1/2}}{|R(f,g)|^{1/4}}$. By lemma 2.2, the number of primitive integral lattice points is at most $32\frac{(BC)^{1/2}}{|R(f,g)|^{1/4}} + 8$.

3.2. General Case

Let $f(X,Y) = (aX + bY)(cX + dY)$, for some real numbers a, b, c, d, and write $T = \begin{bmatrix} a & b \\ c & d \end{bmatrix}$. Let $f'(X,Y) = XY$, so that $f = f' \circ T$. By the assumption that fg is squarefree, $\det T$ is nonzero. Note that the transformation T sends the set $\{(x,y) : |(ax + by)(cx + dy)| < B\}$ to the set $\{(x,y) : |xy| < B\}$, scales areas by a factor of $|\det T|$, and sends parallelograms to parallelograms. Hence the inverse transformation is well defined and we can write $g' = g \circ T^{-1}$.

The previous case shows that the region bounded by $|f'(X,Y)| < B$ and $|g'(X,Y)| < C$ is covered by four parallelograms, each containing the origin, whose areas sum to at most $16\frac{(BC)^{1/2}}{|R(f',g')|^{1/4}}$. Hence the region bounded by $|f(X,Y)| < B$ and $|g(X,Y)| < C$ is covered by at most four parallelograms, each containing the origin, whose areas sum to at most $16\frac{(BC)^{1/2}}{|R(f',g')|^{1/4} \cdot |\det T|}$. By lemma 2.1, this quantity is equal to $16\frac{(BC)^{1/2}}{|R(f,g)|^{1/4}}$.

We conclude by using lemma 2.2 to bound the number of lattice points contained in these four parallelograms:

$$N_{(f,g)}(B,C) < 32\frac{(BC)^{1/2}}{|R(f,g)|^{1/4}} + 8.$$

This completes the proof of Theorem 1.1.

References

[1] M. Bennett and P. X. Rault, *On areas of intersections of hyperbolic and elliptic plane regions*, In preparation.

[2] D. R. Heath-Brown, Diophantine approximation with square-free numbers, *Math. Z.* 187 (1984), 335–344.

[3] D. Hensley, The number of lattice points within a contour and visible from the origin, *Pacific J. Math.* 166 (1994), 295–304.

[4] P. X. Rault, *On uniform bounds for rational points on rational curves and thin sets*, In preparation.

[5] H. Steinhaus, Sur un theorem de M. V. Jarnik, *Colloq. Math.* 1 (1947), 1–5.

In: Frontiers of Combinatorics and Number Theory, Volume 2 ISBN: 978-1-61942-918-5
Editor: Zhi-Wei Sun, pp. 227-239

Chapter 16

CYCLES AND FIXED POINTS OF HAPPY FUNCTIONS

Kathryn Hargreaves* **and Samir Siksek**[†]
Mathematics Institute, University of Warwick,
Coventry, United Kingdom

Abstract

Let $\mathbb{N} = \{1, 2, 3, \cdots\}$ denote the natural numbers. Given integers $e \geq 1$ and $b \geq 2$, let $x = \sum_{i=0}^{n} a_i b^i$ with $0 \leq a_i \leq b - 1$ (thus a_i are the digits of x in base b). We define the *happy function* $S_{e,b} : \mathbb{N} \longrightarrow \mathbb{N}$ by

$$S_{e,b}(x) = \sum_{i=0}^{n} a_i^e.$$

A positive integer x is then said to be (e, b)-happy if $S_{e,b}^r(x) = 1$ for some $r \geq 0$, otherwise we say it is (e, b)-unhappy. In this paper we investigate the cycles and fixed points of the happy functions $S_{e,b}$. We give an upper bound for the size of elements belonging to the cycles of $S_{e,b}$. We also prove that the number of fixed points of $S_{2,b}$ is

$$\begin{cases} 2\,\tau\left(\frac{b^2+1}{2}\right) - 1 & \text{if } b \text{ is odd,} \\ \tau(b^2+1) - 1 & \text{if } b \text{ is even,} \end{cases}$$

where $\tau(m)$ denotes the number of positive divisors of m. We use our results to determine the densities of happy numbers in a small handful of cases.

2000 Mathematics Subject Classification: Primary 11A63; Secondary 11B05.

1. Introduction

Guy [5] introduces **Problem E34 Happy numbers** as follows:

Received October 8, 2009; Accepted September 22, 2010
*E-mail address: k.hargreaves@warwick.ac.uk
†E-mail address: s.siksek@warwick.ac.uk

Reg. Allenby's daughter came home from school in Britain with the concept of **happy numbers**. The problem may have originated in Russia. If you iterate the process of summing the squares of decimal digits of a number, then it's easy to see that you either reach the cycle

$$4 \to 16 \to 37 \to 58 \to 89 \to 145 \to 42 \to 20 \to 4$$

or you arrive at 1. In the latter case you started from a happy number.

Guy asks several questions about happy numbers which can be paraphrased as follows:

(a) It seems that about $1/7$ of all numbers are happy, but what bounds on the density can be proved?

(b) How many consecutive happy numbers can you have? Can there be arbitrarily many?

(c) We define the *height* of a happy number to be the least number of iterations needed to reach 1. How big is the least happy number of height h?

(d) What if we replace squares by cubes, fourth powers, fifth powers etc., or we work in different bases?

Most of the work done so far on happy numbers concerns question (b). For example, El-Sedy and Siksek [2] show that there are arbitrarily many consecutive happy numbers. This was generalised by others, notably Grundman and Teeple [3] and [4], Pan [7] and Zhou and Cai [9], to cover other bases and exponents. In all these proofs the cycles and fixed points of the iteration process play a crucial rôle. This paper is concerned with the cycles and fixed points of the iteration process.

Let $\mathbb{N} = \{1, 2, 3, \cdots\}$ denote the natural numbers. Given integers $e \geq 1$ and $b \geq 2$, let $x = \sum_{i=0}^{n} a_i b^i$ with $0 \leq a_i \leq b - 1$ (thus a_i are the digits of x in base b). We define the *happy function* $S_{e,b} : \mathbb{N} \longrightarrow \mathbb{N}$ by

$$S_{e,b}(x) = \sum_{i=0}^{n} a_i^e.$$

A positive integer x is then said to be (e, b)-happy if $S_{e,b}^r(x) = 1$ for some $r \geq 0$ (sometimes, we simply say x is happy), otherwise we say it is (e, b)-unhappy. It is usual to exclude 0 from the domain of $S_{e,b}$.

By a *cycle* of $S_{e,b}$ of length n we mean a sequence of positive integers $x_1, x_2, \ldots, x_n$ such that

$$S_{e,b}(x_1) = x_2, \quad S_{e,b}(x_2) = x_3, \ldots, \quad S_{e,b}(x_{n-1}) = x_n, \quad S_{e,b}(x_n) = x_1.$$

Where the pair (e, b) is clear from the context, we shall use the following clearer notation to denote the above cycle:

$$x_1 \to x_2 \to x_3 \to \cdots \to x_n \to x_1.$$

By a *fixed point* of $S_{e,b}$ we mean a positive integer x such that $S_{e,b}(x) = x$. Of course a fixed point is nothing more than a cycle of length 1.

Write $D_{e,b}$ for the union of all cycles of $S_{e,b}$. Thus, for example,

$$D_{2,10} = \{1\} \cup \{4, 16, 37, 58, 89, 145, 42, 20\}.$$

This paper is concerned in part with giving an upper bound for the elements of $D_{e,b}$.

Theorem 1.1. *Let $e \geq 1$ and $b \geq 2$. If $b > e$, let $M = eb^e$. If $b \leq e$, let r be the smallest positive integer satisfying $e + r \leq b^r$, and let $M = b^{e+r}$. For any $x \in \mathbb{N}$, there is some k_0 such that $S^k_{e,b}(x) < M$ for all $k \geq k_0$.*

In the paper [4], it is shown that for every $x \in \mathbb{N}$ there is some k such that $S^k_{e,b}(x) < b^{e+1}$. Note that our result is stronger in the sense that we prove $S^k_{e,b}(x) < M$ for all sufficiently large k.

The proof of Theorem 1.1 is given in Section 2. The following corollary is immediate.

Corollary 1.1. $D_{e,b} \subseteq \{1, 2, \ldots, M - 1\}$.

Again our conclusion is stronger than that in [4]. There, the conclusion is that every cycle contains some $x < b^{e+1}$. Our conclusion is that every x belonging to some cycle satisfies $x < M$.

In Section 3 we explain a simple and practical algorithm for computing $D_{e,b}$ based on this corollary. Guy's book [5, pages 358–359] gives partial lists of cycles for $S_{4,10}$ and $S_{5,10}$. We complete these lists using a program based on our algorithm.

In Section 4 we shall give some modest applications to densities of happy numbers. For example, as we will see, every positive integer is $(2, 4)$-happy. We will also show that the $(5, 4)$-happy integers are precisely those $\equiv 1 \pmod 3$, and for $b \geq 2$, that the $(1, b)$-happy numbers are precisely those $\equiv 1 \pmod{(b - 1)}$. So far as we know, these are the only cases where the densities of happy numbers have been determined. In Section 5 we show that the cycles of happy functions can be arbitrarily large.

We also pay special attention to the fixed points of $S_{e,b}$. In Section 6 we show that any fixed point x of $S_{e,b}$ satisfies $x < \min\{e, b\} \cdot b^e$. In the particular case $e = 2$ we have an exact formula for the number of fixed points.

Theorem 1.2. *For a positive integer m, denote the number of positive divisors of m by $\tau(m)$.*

1. *If b is odd then $S_{2,b}$ has precisely $2\,\tau\left(\frac{b^2+1}{2}\right) - 1$ fixed points.*

2. *If b is even then $S_{2,b}$ has precisely $\tau(b^2 + 1) - 1$ fixed points.*

In either case, exactly one of the fixed points has a single digit in base b; this is 1. The remaining fixed points have precisely two digits in base b.

Theorem 1.2 is proved in Section 6. In Section 7 we deduce that the number of fixed points of $S_{2,b}$ is arbitrarily large.

It is not immediately obvious, for a given pair (e, b), whether there should exist any unhappy numbers. If $e = 2$ then the above theorem shows that there must exist unhappy numbers whenever $b^2 + 1$ is composite. This is not necessarily true if $b^2 + 1$ is prime; as we noted earlier, all integers are $(2, 4)$-happy.

The paper concludes with some further questions.

2. Proof of Theorem 1.1

Before we launch into the proof of Theorem 1.1, we point out another, arguably more obvious approach, to determining $D_{e,b}$. Observe that every cycle must contain some element x satisfying $x \leq S_{e,b}(x)$. It is easy to see that

$$x < (b-1)^e(\log_b(x)+1).$$

There are standard (but slightly complicated) bounds for solutions to such inequalities; see for example [8, Appendix B] which gives an explicit bound $x < M'$. We have found that the bounds M' obtained via this approach are worse than our bound M given in Theorem 1.1. Moreover, all we know via this approach is that the cycle contains some element $x < M'$; we do not know that the entire cycle satisfies the bound. Thus our approach below, although longer, gives stronger results.

Theorem 1.1 is trivially true if $e = 1$ or $b = 2$, so for the remainder of this section we suppose $e \geq 2$ and $b \geq 3$. The theorem follows straightaway from the following two lemmas.

Lemma 2.1. *Let M be as in Theorem 1.1. If $y < M$ then $S_{e,b}(y) < M$.*

Lemma 2.2. *Let M be as in Theorem 1.1. For any $x \in \mathbb{N}$ there exists some ℓ such that $S_{e,b}^{\ell}(x) < M$.*

For the proofs of Lemmas 2.1 and 2.2 we shall need the following lemma.

Lemma 2.3. *Let s, t be positive integers with $s < b$. Suppose $y < sb^t$. Then*

$$S_{e,b}(y) \leq t(b-1)^e + (s-1)^e.$$

Proof. Consider the representation of y in base b. Clearly this has at most $t+1$ digits. Moreover, if y has precisely $t+1$ digits, then the leading digit is at most $s-1$. The lemma follows from the definition of $S_{e,b}$. □

Proof of Lemma 2.1. Suppose $y < M$. We consider the cases $b > e$ and $b \leq e$ separately.

Suppose first that $b > e$. In this case $M = eb^e$. Then $y < eb^e$. By Lemma 2.3,

$$S_{e,b}(x) \leq e(b-1)^e + (e-1)^e$$

Now observe, using the Binomial Theorem, that

$$eb^e = e((b-1)+1)^e > e(b-1)^e + e^2(b-1)^{e-1} > e(b-1)^e + (e-1)^e,$$

where we have made use of $b > e$. Therefore $S_{e,b}(x) < eb^e = M$ as required.

We now turn to the case $b \leq e$. Here $M = b^{e+r}$ where r satisfies $e + r \leq b^r$. As $y < M$, by Lemma 2.3

$$S_{e,b}(y) \leq (e+r)(b-1)^e \leq b^r(b-1)^e < b^{e+r} = M,$$

as required. □

The proof of Lemma 2.2 is more complicated, and we shall need the following preparatory lemma.

Lemma 2.4. *Let* $B = b^{e+1}$. *If* $y < B^m$ *then* $S_{e,b}(y) < m(B-1)$.

Proof. Note $y < B^m = b^{m(e+1)}$; by Lemma 2.3, $S_{e,b}(y) \le m(e+1)(b-1)^e$. Note that

$$\begin{aligned} B - 1 &= b^{e+1} - 1 = ((b-1)+1)^{e+1} - 1 \\ &= \left((b-1)^{e+1} + (e+1)(b-1)^e + \cdots + 1\right) - 1 \qquad \text{by the Binomial Theorem} \\ &> (e+1)(b-1)^e. \end{aligned}$$

The lemma follows. □

Proof of Lemma 2.2. Let $B = b^{e+1}$. Define

$$B_1 = B, \qquad \text{and} \qquad B_{i+1} = B^{B_i} \text{ for } i \ge 2.$$

Let $x \in \mathbb{N}$ be given. We would like to show that $S^{\ell}_{e,b}(x) < M$ for some ℓ where M is as in the statement of Theorem 1.1. Clearly

$$x < B^{1+B_k},$$

for some $k \ge 2$. Then

$$\begin{aligned} S_{e,b}(x) &\le (1+B_k)(B-1) \quad \text{by Lemma 2.4} \\ &= (1+B^{B_{k-1}})(B-1) \\ &= B^{1+B_{k-1}} + B - B^{B_{k-1}} - 1 \\ &< B^{1+B_{k-1}}. \end{aligned}$$

Thus

$$S^{k-1}_{e,b}(x) < B^{1+B}.$$

Applying Lemma 2.4 again gives

$$S^{k}_{e,b}(x) < (1+B)(B-1) < B^2,$$

and again

$$S^{k+1}_{e,b}(x) < 2B = 2b^{e+1}.$$

Hence, by Lemma 2.3,

$$S^{k+2}_{e,b}(x) \le 1 + (e+1)(b-1)^e. \tag{2.1}$$

We now consider the cases $b \le e$, $b = e+1$ and $b \ge e+2$ separately. Suppose first that $b \le e$. Recall that we have defined r to be the least positive integer such that $e + r \le b^r$, and in this case $M = b^{e+r}$. Then

$$S^{k+2}_{e,b}(x) \le 1 + (e+1)(b-1)^e \le 1 + (e+r)(b-1)^e < b^{e+r} = M.$$

This completes the proof of the lemma for $b \le e$.

Suppose now $b = e+1$. In this case $M = eb^e$. By (2.1),

$$S^{k+2}_{e,b}(x) \le 1 + (e+1)(b-1)^e = 1 + b(b-1)^e < (b-1)b^e = M,$$

as required.

Finally suppose that $b \geq e+2$. Again, $M = eb^e$. By (2.1),

$$S_{e,b}^{k+2}(x) \leq 1 + (e+1)(b-1)^e < (e+1)b^e.$$

Applying Lemma 2.3 we obtain

$$S_{e,b}^{k+3}(x) \leq e(b-1)^e + e^e$$

By the Binomial Theorem applied to $b^e = ((b-1)+1)^e$ we have

$$eb^e > e(b-1)^e + e^2(b-1)^{e-1} > S_{e,b}^{k+3}(x),$$

since $b \geq e+2$. This completes the proof. □

3. Computing $D_{e,b}$

Corollary 1.1 gives an efficient method of computing $D_{e,b}$. Let

$$U_0 = \{1, 2, \ldots, M-1\}$$

where M is as in Theorem 1.1. Let $U_{i+1} = S_{e,b}(U_i)$. Let k be the smallest non-negative integer such that $U_{k+1} = U_k$. Then $D_{e,b} = U_k$.

3.1. Examples

We wrote a short program in MAGMA to compute $D_{e,b}$ using the above method. Here are two examples.

Example 1. Take $e = 2$ and $b = 10$. Here $M = 200$. We start with $U_0 = \{1, \ldots, 199\}$ and after 11 iterations find

$$U_{10} = U_{11} = \{1, 4, 16, 20, 37, 42, 58, 89, 145\}.$$

This is $D_{2,10}$ in agreement with Guy's statement in the introduction.

Example 2. Take $e = 5$ and $d = 10$. Guy [5, page 359] lists 12 non-trivial cycles found by Somjit Datta. Thus together with the trivial cycle $\{1\}$ he has found 13 cycles; no claim is made that this is the complete list of cycles. Grundman and Teeple [4, page 1141] list nine cycles and mention that there are an additional seven longer cycles (that are not listed). Here $M = 5 \times 10^5$. We ran our program on a 2.8 GHz Dual-Core AMD Opteron. It took our program 58 iterations and just 9 seconds to compute $D_{5,10}$. The cycles of $S_{5,10}$ are

$$C_1 : 1 \to 1, \qquad C_2 : 4150 \to 4150, \qquad C_3 : 4151 \to 4151, \qquad C_4 : 54748 \to 54748,$$
$$C_5 : 92727 \to 92727, \qquad C_6 : 93084 \to 93084, \qquad C_7 : 194979 \to 194979,$$
$$C_8 : 89883 \to 157596 \to 89883, \qquad C_9 : 58618 \to 76438 \to 58618,$$
$$C_{10} : 10933 \to 59536 \to 73318 \to 50062 \to 10933,$$
$$C_{11} : 8299 \to 150898 \to 127711 \to 33649 \to 68335 \to 44155 \to 8299,$$

$$C_{12}:\begin{cases} 8294 \to 92873 \to 108899 \to 183635 \to 44156 \to \\ 12950 \to 62207 \to 24647 \to 26663 \to 23603 \to 8294, \end{cases}$$

$$C_{13}:\begin{cases} 9044 \to 61097 \to 83633 \to 41273 \to 18107 \to 49577 \\ \to 96812 \to 99626 \to 133682 \to 41063 \to 9044, \end{cases}$$

$$C_{14}:\begin{cases} 24584 \to 37973 \to 93149 \to 119366 \to 74846 \to 59399 \to 180515 \\ \to 39020 \to 59324 \to 63473 \to 26093 \to 67100 \to 24584, \end{cases}$$

$$C_{15}:\begin{cases} 9045 \to 63198 \to 99837 \to 167916 \to 91410 \to 60075 \to 27708 \to 66414 \\ \to 17601 \to 24585 \to 40074 \to 18855 \to 71787 \to 83190 \to 92061 \to 66858 \\ \to 84213 \to 34068 \to 41811 \to 33795 \to 79467 \to 101463 \to 9045, \end{cases}$$

$$C_{16}:\begin{cases} 244 \to 2080 \to 32800 \to 33043 \to 1753 \to 20176 \to 24616 \to 16609 \\ \to 74602 \to 25639 \to 70225 \to 19996 \to 184924 \to 93898 \to 183877 \\ \to 99394 \to 178414 \to 51625 \to 14059 \to 63199 \to 126118 \to 40579 \\ \to 80005 \to 35893 \to 95428 \to 95998 \to 213040 \to 1300 \to 244. \end{cases}$$

Datta overlooked the cycles C_5, C_7 and C_8.

Example 3. Let $e = 4$ and $b - 10$. The cycles are

$$C_1 : 1 \to 1, \qquad C_2 : 1634 \to 1634, \qquad C_3 : 8208 \to 8208,$$
$$C_4 : 9474 \to 9474, \qquad C_5 : 2178 \to 6514 \to 2178,$$
$$C_6 : 1138 \to 4179 \to 9219 \to 13139 \to 6725 \to 4338 \to 4514 \to 1138.$$

Guy's book [5, page 358] mentions C_1, C_3, C_5, C_6 but not C_2 and C_4.

Example 4. We used our program to compute $D_{2,b}$ for all $b \leq 1000$ and its decomposition into cycles. The entire computation took just under four hours on our 2.8 GHz Dual-Core AMD Opteron. The largest $D_{2,b}$ found was $D_{2,816}$ which has 1676 elements. This decomposes into 11 cycles of lengths 1, 2, 2, 3, 5, 7, 33, 47, 210, 630, 736.

The largest cycle for $S_{2,b}$ with $b \leq 1000$ is for $b = 974$. This cycle has length 1005 and has least element 1858 and largest element 1792145. It is interesting to note that precisely 495 elements of this cycle satisfy the inequality $x < S_{2,974}(x)$ and the remainder satisfy $x > S_{2,974}(x)$.

We also took the opportunity test to how good our bound for the elements of $D_{e,b}$ is. By Theorem 1.1 and Corollary 1.1 we know that $\max D_{2,b} \leq 2b^2 - 1$. For each $3 \leq b \leq 1000$ we computed the value

$$m(b) = \frac{\max D_{2,b}}{2b^2 - 1}.$$

The largest value for $m(b)$ we found was for $b = 981$. Here $2b^2 - 1 = 1924721$ and $\max D_{2,981} = 1916884$, giving

$$m(981) = \frac{1916884}{1924721} = 0.9959\cdots.$$

This shows the our bound M is indeed sharp.

4. A Remark on the Densities of Happy Numbers

As quoted in the introduction, Guy asks for the density of happy numbers in the case $e = 2$, $b = 10$. This appears to us to be a difficult problem. However, there are a few pairs (e, b) where have been able to compute the precise density of happy numbers.

Let $e \geq 1$ and $b \geq 2$. Let

$$\mathcal{A} = \{p \text{ prime } : \ p \mid (b-1) \text{ and } (p-1) \mid (e-1)\}, \qquad P = \prod_{p \in \mathcal{A}} p.$$

By convention, if $\mathcal{A} = \emptyset$ then $P = 1$. As observed in [9], for all positive integers n, we have $S_{e,b}(n) \equiv n \pmod{P}$; this congruence is easy to prove using Fermat's Little Theorem. Let

$$D'_{e,b} = \{x \in D_{e,b} : x \equiv 1 \pmod{P}\}.$$

If $D'_{e,b} = \{1\}$, then it is easy to see that the (e, b)-happy integers are precisely those that are $\equiv 1 \pmod{P}$. In this case the density of happy numbers is $1/P$. We can only find a handful of examples where this observation is useful.

Example 5. Let $b = 2$ (and $e \geq 1$ arbitrary). Then $D_{2,e} = \{1\}$ and so every positive integer is $(2, e)$-happy.

Example 6. Let $e = 1$ (and $b \geq 2$ arbitrary). It is easy to see that $S_{1,b}(n) \equiv n \pmod{b-1}$ for all n. Moreover,

$$D_{1,b} = \{1, 2, \ldots, b-1\}.$$

Thus the happy numbers are precisely those $\equiv 1 \pmod{b-1}$, and the density of happy numbers is $1/(b-1)$.

Example 7. Let $(e, b) = (2, 4)$. We found that $D_{2,4} = \{1\}$. Hence every number is $(2, 4)$-happy.

Example 8. Let $(e, b) = (5, 4)$. Then

$$D_{5,4} = \{1, 3, 32, 33, 243, 245, 246, 276, 308, 309, 488, 519, 729\}.$$

However $P = 3$ in this case and so $D'_{5,4} = \{1\}$. By the above observation we know that the $(5, 4)$-happy numbers are precisely the ones $\equiv 1 \pmod{3}$. Hence exactly $1/3$ of numbers are $(5, 4)$-happy.

We ran a search with $1 \leq e \leq 6$ and $2 \leq b \leq 10$ for (e, b) satisfying the condition $D'_{e,b} = \{1\}$. The above examples were the only ones found.

5. Large Cycles

As we saw, $S_{2,10}$ has two cycles, one of length 1 and the other of length 8. An obvious question is how large can the cycles of happy functions be. In fact they can be arbitrarily large as we shall see now.

Let $r, e \geq 1$. Let $b = 2^{e^{r-1}} - 1$. Consider the cycle starting with 2:

$$2 \to 2^e \to 2^{e^2} \to \cdots \to 2^{e^{r-1}}.$$

But $2^{e^{r-1}} = b + 1$, so in base b this has the digits 11, hence $S_{e,b}(2^{e^{r-1}}) = 2$. This shows that we obtain a cycle of length r starting with 2, where r is as large as we please.

6. Fixed Points of Happy Functions

As we saw before, fixed points are cycles of length 1. We will see that fixed points are much easier to analyse than arbitrary cycles and so we can say much more about them. For $e = 2$ we show that there can be arbitrarily many fixed points.

It is easy to see that if $e = 1$ then the only fixed points are the single digit numbers $1, 2, \ldots, b-1$. Also, if $b = 2$ then the only fixed point is 1. We shall restrict to $e \geq 2$ and $b \geq 3$. Of course, the only single-digit fixed point is 1.

6.1. Fixed points for exponent $e = 2$

We now restrict to $e = 2$ which turns out to be much easier to analyse than the general case.

Lemma 6.1. *Let $b \geq 3$. Then the happy function $S_{2,b}$ has no fixed points with at least three digits.*

Proof. Suppose that x is a non-zero fixed point of $S_{2,b}$. By Theorem 1.1, $x < 2b^2$. We may write $x = ub^2 + vb + w$ where $0 \leq u, v, w \leq (b-1)$ and $u = 0$ or 1. It is sufficient to show that $u = 0$.

Suppose $u = 1$. As x is a fixed point,

$$b^2 + vb + w = 1 + v^2 + w^2.$$

Thus

$$(b-1)^2 \geq w^2 = b^2 + v(b-v) + w - 1 \geq b^2 - 1 = (b-1)(b+1),$$

giving a contradiction. □

6.2. Proof of Theorem 1.2

Here $e = 2$. The theorem is obvious for $b = 2$ so we suppose $b \geq 3$. By Lemma 6.1, $S_{2,b}$ has no fixed points with three or more digits in base b, and so every fixed point has at most two digits.

A non-negative number with at most two digits in base b has the form $ub + v$ with $0 \leq u,\ v < b$. This is a fixed point for $S_{2,b}$ if and only if $u^2 + v^2 = ub + v$. Let

$$U = \{(u, v) \in \mathbb{Z}^2 \quad : \quad 0 < u,\ v < b \quad \text{and} \quad u^2 + v^2 = ub + v\}.$$

It is easy to see that $1 + \#U$ is the number of fixed points of $S_{2,b}$. Thus to prove Theorem 1.2 we need to show that

$$\#U = \begin{cases} 2\,\tau\left(\frac{b^2+1}{2}\right) - 2 & \text{if } b \text{ is odd,} \\ \tau(b^2+1) - 2 & \text{if } b \text{ is even.} \end{cases}$$

Starting with $u^2 + v^2 = ub + v$ and completing the squares we obtain

$$(2u - b)^2 + (2v - 1)^2 = b^2 + 1.$$

Let $n = b^2 + 1$ and let

$$X = \{(x, y) \in \mathbb{Z}^2 \quad : \quad y \geq 1 \text{ and odd}, \quad x \neq \pm b \quad \text{and} \quad x^2 + y^2 = n\}.$$

As we will see, the number of elements of X can be determined by relating it to the number of representations of n as the sum of two squares. For now we would like to show that $\#U = \#X$. Define $\phi : U \to X$ and $\psi : X \to U$ by

$$\phi(u, v) = (2u - b, 2v - 1), \qquad \psi(x, y) = \left(\frac{x + b}{2}, \frac{y + 1}{2} \right).$$

It is clear that ϕ and ψ are mutual inverses (and thus bijections) provided they are well-defined. Thus to show that $\#U = \#X$ all we have to do is to show that ϕ and ψ are well-defined.

There is no difficulty in seeing that ϕ is well-defined. Let us show that ψ is well-defined. Suppose $(x, y) \in X$. As y is odd and $x^2 + y^2 = b^2 + 1$, we see that x and b have the same parity. Thus $\psi(x, y) \in \mathbb{Z}^2$. Let $(u, v) = \psi(x, y)$. From $x^2 + y^2 = b^2 + 1$ we deduce that $u^2 + v^2 = ub + v$. Moreover, $-b \leq x \leq b$ and as $x \neq \pm b$, we have $-b < x < b$ so that $0 < u < b$. Also, $1 \leq y \leq b$, so that $1 \leq v \leq (b + 1)/2 < b$ as $b \geq 3$. Hence $(u, v) \in U$. This shows that ψ is well-defined. Thus $\#U = \#X$.

Now let

$$Y = \{(x, y) \in \mathbb{Z}^2 \quad : \quad y \geq 1 \text{ and odd}, \quad \text{and} \quad x^2 + y^2 = n\}.$$

Clearly $\#Y = 2 + \#X = 2 + \#U$. Hence to complete the proof of Theorem 1.2 it is sufficient to show that

$$\#Y = \begin{cases} 2\,\tau\left(\frac{n}{2}\right) & \text{if } b \text{ is odd,} \\ \tau(n) & \text{if } b \text{ is even.} \end{cases}$$

To do this we introduce the following set which contains Y,

$$Z = \{(x, y) \in \mathbb{Z}^2 : x^2 + y^2 = n\}.$$

The number of representations of an a positive integer n as the sum of two squares is well-known:

$$\#Z = 4 \sum_{d \mid n} \left(\frac{-4}{d} \right);$$

see for example [1, Theorem 5.4.15]. However, all odd divisors d of $n = b^2 + 1$ satisfy $d \equiv 1 \pmod 4$. Thus, in this case, $\#Z$ is 4 times the number of odd divisors of n. If b is even then n is odd. If b is odd then $n/2$ is odd. Hence

$$\#Z = \begin{cases} 4\,\tau(n/2) & \text{if } b \text{ is odd,} \\ 4\,\tau(n) & \text{if } b \text{ is even.} \end{cases}$$

To complete the proof of Theorem 1.2 we must relate $\#Y$ to $\#Z$.

Suppose first that b is odd, and so $n \equiv 2 \pmod{4}$. Every solution to $x^2 + y^2 = n$ must have x and y odd. Thus Y is the subset of pairs (x, y) in Z with $y \geq 1$. Since there are no solutions with $y = 0$, we see that

$$\#Y = \frac{\#Z}{2} = 2\,\tau(n/2),$$

as required. This completes the proof of Theorem 1.2 for b odd.

Suppose now that b is even. Then $n \equiv 1 \pmod{4}$. If (x, y) is any solution to $x^2 + y^2 = n$ then precisely one of x, y is odd and the other is even. Considering the fixed-point free involutions

$$(x, y) \mapsto (y, x), \qquad (x, y) \mapsto (x, -y)$$

on Z, we see that precisely $1/4$ of the elements of Z belong to Y. Hence

$$\#Y = \frac{\#Z}{4} = \tau(n),$$

completing the proof of Theorem 1.2.

7. Arbitrarily Many Fixed Points

In this section we use Theorem 1.2 to deduce that the number of fixed points for $S_{2,b}$ can be arbitrarily large.

Corollary 7.1. *$S_{2,b}$ can have arbitrarily many fixed points.*

Proof. Let $r \geq 1$ be given. We will show that there exists a value of b with $b^2 + 1$ having at least r odd prime factors. Then by Theorem 1.2, $S_{2,b}$ has at least $2^r - 1$ fixed points. Letting r be arbitrarily large proves the theorem.

Let $p_1, \ldots, p_r$ be distinct primes $\equiv 1 \pmod{4}$. There exist integers m_i with $m_i^2 \equiv -1 \pmod{p_i}$ for $i = 1, \ldots, r$. By the Chinese Remainder Theorem, there is some positive integer b such that $b \equiv m_i \pmod{p_i}$ for $i = 1, \ldots, r$. Then $b^2 + 1$ is divisible by $p_1, p_2, \ldots, p_r$ as required. □

8. Specific Fixed Points

By Theorem 1.2 we know that if b is odd then $S_{2,b}$ has at least three fixed points. If b is even then we can conclude that there is at least one fixed point, but of course 1 is always a fixed point, so the conclusion in this case does not add anything new. It is natural to wonder whether for odd b there is a formula for the other two fixed points that are guaranteed to exist. This is indeed the case.

Corollary 8.1. *Let $b \geq 3$ be odd and write $b = 2u + 1$. Then 1, $2u^2 + 2u + 1$ and $2(u+1)^2$ are fixed points of $S_{2,b}$, which respectively have the digits 1, $(u, u+1)$ and $(u+1, u+1)$ in base b. If $b^2 + 1$ is twice a prime then these are the only fixed points.*

Proof. Note that $2u^2 + 2u + 1$ and $2(u+1)^2$ respectively have digits $(u, u+1)$ and $(u+1, u+1)$ in base b. So

$$S_{2,b}(2u^2 + 2u + 1) = u^2 + (u+1)^2 = 2u^2 + 2u + 1.$$
$$S_{2,b}(2(u+1)^2) = 2(u+1)^2.$$

If $b^2 + 1$ is twice a prime then by Theorem 1.2 we know there are exactly three fixed points, and this completes the proof. □

If $e = 2$ and b is odd then, in the notation of Section 4, $P = 2$ and we know that every happy number must be odd. Note by the above that $2u^2 + 2u + 1$ is an unhappy number which is odd. The following corollary is interesting in view of Section 4.

Corollary 8.2. *If $b \geq 3$ is odd then the $(2, b)$-happy numbers are a proper subset of the odd numbers.*

9. Further Questions

The following two questions are very interesting and deserve further attention:

(I) Can one obtain sharp estimates for the number of fixed points of $S_{e,b}$ (for $e \geq 3$) in terms of e and b?

(II) Can one obtain sharp estimates for the number and sizes of the cycles of $S_{e,b}$ in terms of e and b?

Of course by Corollary 1.1 we know that $\#D_{e,b} < M$. But the experiments described in Section 3 indicate that this bound is very far from being sharp. Can one obtain better bounds? In fact for $e = 2$ we can do somewhat better. Recall in this case that $M = 2b^2$ and that every element of $x \in D_{2,b}$ satisfies $x < 2b^2$. However, if $x \in D_{2,b}$ then $x = S_{2,b}(y)$ for some $y \in D_{2,b}$. As $y < 2b^2$, it has at most three digits in base b, and if it has exactly three digits then the leading digit is 1. Hence $x = S_{2,b}(y)$ has the form $u^2 + v^2$ or $1 + u^2 + v^2$. Denote by $N_2(X)$ the number of positive integers $n \leq X$ that can be written as the sum of two squares. A well-known theorem of Landau (see [1, Proposition 5.4.10] and [6]) states that

$$N_2(X) = (C + o(1)) \frac{X}{\sqrt{\log X}} \qquad \text{as } X \to \infty,$$

where

$$C = \frac{1}{\sqrt{2}} \prod_{p \equiv 3 \mod 4} \left(1 - \frac{1}{p^2}\right)^{-1/2} = \frac{\pi}{4} \prod_{p \equiv 1 \mod 4} \left(1 - \frac{1}{p^2}\right)^{1/2} = 0.764223653 \cdots .$$

As every integer in $D_{2,b}$ is at most $2b^2$ and has the form $u^2 + v^2$ or $1 + u^2 + v^2$ we conclude that

$$\#D_{2,b} < \left(2^{3/2} C + o(1)\right) \frac{b^2}{\sqrt{\log b}} \qquad \text{as } b \to \infty. \tag{9.1}$$

In Example 4 we found that the largest $\#D_{2,b}$ with $b \leq 1000$ was $\#D_{2,816} = 1676$. However

$$2^{3/2}C \cdot \frac{816^2}{\sqrt{\log 816}} = 555859.9 \cdots .$$

This shows that (9.1) is still a very poor upper bound for $\#D_{2,b}$ despite being much better than M.

Now keep $e \geq 2$ fixed. Regarding cycle lengths, it seems reasonable to guess that any cycle for $S_{e,b}$ has length at most $O(b^{e/2})$ as $b \to \infty$. This is what one would expect if $S_{e,b}$ is a 'random walk' on the set $\{1, 2, \ldots, eb^e\}$ which contains the cycles.

References

[1] H. Cohen, Number theory. Vol. I. Tools and Diophantine equations, *Graduate Texts in Math.*, vol. 239, Springer, New York, 2007.

[2] E. El-Sedy and S. Siksek, On happy numbers, *Rocky Mountain J. Math.* 30 (2000), no. 2, 565–570.

[3] H. G. Grundman and E. A. Teeple, Sequences of consecutive happy numbers, *Rocky Mountain J. Math.* 37 (2007), no. 6, 1905–1916.

[4] H. G. Grundman and E. A. Teeple, Iterated sums of fifth powers of digits, *Rocky Mountain J. Math.* 38 (2008), no. 4, 1139–1146.

[5] R. Guy, *Unsolved Problems in Number Theory*, third edition, Problem Books in Mathematics, Springer-Verlag, New York, 2004.

[6] Y. Motohashi, On the number of integers which are sums of two squares, *Acta Arith.* 23 (1973), 401–412.

[7] H. Pan, On Consecutive Happy Numbers, *J. Number Theory* 128 (2008), no. 6, 1646–1654.

[8] N. P. Smart, *The Algorithmic Resolution of Diophantine Equations*, London Mathematical Society Student Texts, vol. 41, Cambridge University Press, Cambridge, 1998.

[9] X. Zhou and T. Cai, On e-Power b-Happy Numbers, *Rocky Mountain J. Math.* 39 (2009), no. 6, 2073–2081.

In: Frontiers of Combinatorics and Number Theory, Volume 2 ISBN: 978-1-61942-918-5
Editor: Zhi-Wei Sun, pp. 241-249

Chapter 17

CONGRUENCE PROPERTIES FOR HYPER m-ARY OVERPARTITIONS

Qing-Lin Lu*
School of Mathematical Sciences, Xuzhou Normal University
Xuzhou, People's Republic of China

Abstract

Let $\overline{h}_m(n)$ be the number of non-increasing sequences of non-negative integral powers of m whose sum is n, and where the first occurrence of power of m may be overlined, such that each power of m is allowed to be used at most m times as a non-overlined part. In this paper we study this hyper m-ary overpartition function $\overline{h}_m(n)$ and obtain some congruence properties. For example, we show that: For all $m \geq 2$, a necessary and sufficient condition for $\overline{h}_m(n) \equiv 1(\mathrm{mod}\ 2)$ is that n can be expressed as $n = (m+1)\sum_{i=0}^{t} \varepsilon_i m^i$, where $\varepsilon_i = 0$ or 1 for each i, and $\varepsilon_t = 1$ if $t > 0$.

2000 Mathematics Subject Classification: Primary 05A17; Secondary 11P83.

1. Introduction

Numerous functions which enumerate partition into powers of a fixed number m (Here m is assumed to be bigger than 1) have been studied by Churchhouse [2], Rødseth [11], Andrews [1], Gupta [8] in the late 1960s, and Dirdal [5-6] in the mid-1970s. For more recent work see [7, 9, 11].

Presently there are a lot of activities in the study of the objects named overpartitions by Corteel and Lovejoy [3]. Rødseth and Sellers [13] discussed divisibility properties of the number of m-ary overpartitions of a natural number. Courtright and Sellers [4] discussed arithmetic properties of hyper m-ary partition functions. The present author [10] gave the conception of hyper m-ary overpartitions of n. A hyper m-ary overpartitions of n is a non-increasing sequence of non-negative integral powers of m whose sum is n, and where the first occurrence of power of m may be overlined, such that each power of m is allowed

Received: August 27, 2010; Accepted: December 3, 2010
*E-mail address: qllu@xznu.edu.cn

to be used at most m times as a non-overlined part. We denoted by $\overline{h}_m(n)$ the number of hyper m-ary overpartitions of n. ($\overline{h}_m(0) = 1$, $\overline{h}_m(n) = 0$ for any negative integer n) In [10], the authors gave the following results which are Theorems 2.2, 2.7 and 2.8 in [10], respectively.

Proposition 1. *For all $n \geq 0$, we have*

$$n \equiv 1, 2 \pmod 3 \iff \overline{h}_2(n) \equiv 0 \pmod 2.$$

Proposition 2. *For all $n \geq 0$, we have*

$$n \equiv 1, 2, 3 \pmod 4 \implies \overline{h}_3(n) \equiv 0 \pmod 2.$$

Proposition 3. *For all $n \geq 0$, $k \geq 0$, we have*

$$\overline{h}_3(4n \cdot 3^k + 2 \cdot 3^k - 1) \equiv 0 \pmod{2^{k+1}},$$

$$\overline{h}_3(4n \cdot 3^k + 3 \cdot 3^k - 1) \equiv 0 \pmod{2^{k+1}},$$

$$\overline{h}_3(4n \cdot 3^k + 4 \cdot 3^k - 1) \equiv 0 \pmod{2^{k+1}}.$$

Note that $\overline{h}_3(n) \equiv 0 \pmod 2$ does not guarantee $n \equiv 1, 2, 3 \pmod 4$. For example, since $8 = 3 + 3 + 1 + 1 = \overline{3} + 3 + 1 + 1 = 3 + 3 + \overline{1} + 1 = \overline{3} + 3 + \overline{1} + 1$ has four representations satisfying the required conditions, we have $\overline{h}_3(8) = 4 \equiv 0 \pmod 2$.

In this paper, we shall give a necessary and sufficient condition for which $\overline{h}_m(n)$ is even /odd.

Theorem 1.1. *For all $m \geq 2$, a necessary and sufficient condition for $\overline{h}_m(n) \equiv 1 \pmod 2$ is that n can be expressed as*

$$n = (m + 1) \sum_{i=0}^{t} \varepsilon_i m^i,$$

where $\varepsilon_i = 0$ or 1 for each i, and $\varepsilon_t = 1$ if $t > 0$.

We then generalize Proposition 3 to a general result.

Theorem 1.2. *For all $n \geq 0$, $k \geq 0$, $m \geq 3$, and $2 \leq l \leq m + 1$, we have*

$$\overline{h}_m((m + 1)nm^k + lm^k - 1) \equiv 0 \pmod{2^{k+1}}.$$

We also give the following congruences which are analogous to those of Churchhouse.

Theorem 1.3. *For all $m \geq 3$, $k \geq 0$, and $n \geq 0$, we have*

$$\overline{h}_m(m^{k+1}n) \equiv \overline{h}_m(m^k n) \pmod{2^{k+1}}$$

and

$$\overline{h}_m(m^{k+2}n + 1) \equiv \overline{h}_m(m^{k+1}n + 1) \pmod{5 \cdot 2^k}.$$

Theorem 1.4. *For all $k \geq 0$, $n \geq 0$, we have*

$$\overline{h}_2(2^{k+1}n-1) \equiv \overline{h}_2(2^k n-1) \pmod{3^k}$$

and

$$\overline{h}_2(2^{k+1}n-2) \equiv \overline{h}_2(2^k n-2) \pmod{3^k}.$$

2. Proofs of the Theorems

Since the overlined parts form an m-ary partition into distinct parts, and the non-overlined parts form a hyper m-ary partition, the generating function for $\overline{h}_m(n)$ is given by

$$\overline{H}_m(q) := \sum_{n\geq 0} \overline{h}_m(n)q^n = \prod_{i\geq 0}(1+q^{m^i})\sum_{k=0}^{m} q^{k\cdot m^i},$$

from which we have

$$\overline{H}_m(q) = (1+q)(1+q+\cdots+q^m)\overline{H}_m(q^m).$$

Comparing the coefficients in the above formula gives the following recurrences:

$$\overline{h}_m(mn) = \overline{h}_m(n) + 2\overline{h}_m(n-1), \tag{2.1}$$

$$\overline{h}_m(mn+1) = 2\overline{h}_m(n) + \overline{h}_m(n-1), \tag{2.2}$$

$$\overline{h}_m(mn+k) = 2\overline{h}_m(n) \ (m \geq 3,\ 2 \leq k \leq m+1). \tag{2.3}$$

We first prove Theorem 1.1. Note that Theorem 1.1 obviously holds for $m = 2$ by Proposition 1. So we need only to prove Theorem 1.1 for $m \geq 3$.

Lemma 2.1. *Let $m \geq 3$ and $n \geq 0$ be integers. If $n \not\equiv 0 \pmod{(m+1)}$, then $\overline{h}_m(n) \equiv 0 \pmod 2$.*

Proof. Let $n = (m+1)l + k,\ 1 \leq k \leq m$. We prove the lemma by induction on l. The lemma obviously holds for $l = 0$, since $\overline{h}_m(k) = 2$ for $1 \leq k \leq m-1$ and $\overline{h}_m(m) = 4$. We assume the lemma is true for all s $(0 \leq s \leq l)$. Now we consider the case $l+1$. Write $l = mt + e, 0 \leq e \leq m-1$. Then

$$n = (m+1)(l+1) + k = m((m+1)t + e + 1) + e + k + 1,$$

where $2 \leq e + k + 1 \leq 2m$.

Case 1. $2 \leq e + k + 1 \leq m - 1$.

In this case, we have

$$\begin{aligned} \overline{h}_m(n) &= \overline{h}_m(m((m+1)t+e+1)+e+k+1) \\ &= 2\overline{h}_m((m+1)t+e+1) \text{ (by (2.3))} \\ &\equiv 0 \pmod 2. \end{aligned}$$

Case 2. $e+k+1=m$.
In this case,

$$n = m((m+1)t+e+1)+m = m((m+1)t+e+2).$$

So, we have by the induction hypothesis

$$\begin{aligned}\overline{h}_m(n) &\equiv \overline{h}_m((m+1)t+e+2) \pmod 2 \text{ (by (2.1))}\\ &\equiv 0 \pmod 2 \text{ (since } t<l \text{ and } 2\le e+2\le m).\end{aligned}$$

Case 3. $e+k+1=m+1$.
In this case,

$$n = m((m+1)t+e+2)+1.$$

So, we have by the induction hypothesis

$$\begin{aligned}\overline{h}_m(n) &\equiv \overline{h}_m((m+1)t+e+1) \pmod 2 \text{ (by (2.2))}\\ &\equiv 0 \pmod 2 \text{ (since } t<l \text{ and } 1\le e+1\le m).\end{aligned}$$

Case 4. $m+2\le e+k+1\le 2m-1$.
In this case,

$$2\le e+k-m+1\le m-1.$$

So we have

$$\begin{aligned}\overline{h}_m(n) &= \overline{h}_m((m(m+1)t+e+2)+e+k-m+1)\\ &= 2\overline{h}_m((m+1)t+e+2) \text{ (by (2.3))}\\ &\equiv 0 \pmod 2.\end{aligned}$$

Case 5. $e+k+1=2m$.
In this case,

$$e=m-1, k=m \text{ and } n=m((m+1)(t+1)+1).$$

So, we have by the induction hypothesis

$$\begin{aligned}\overline{h}_m(n) &= \overline{h}_m(m(m+1)(t+1)+1)\\ &\equiv \overline{h}_m((m+1)(t+1)+1) \pmod 2 \text{ (by (2.1))}\\ &\equiv 0 \pmod 2 \text{ (since } t+1\le l).\end{aligned}$$

This completes the proof. □

Remark 1. Let $m\ge 3$ and $n\ge 0$ be integers. If $\overline{h}_m(n)\equiv 1 \pmod 2$, then

$$n\equiv 0 \pmod{(m+1)}.$$

Lemma 2.2. *For all* $m\ge 3$, $\varepsilon_i=0$ *or* 1 *(*$i=0,1,\cdots,t$*), and* $\varepsilon_t=1$ *if* $t>0$*, we have*

$$\overline{h}((m+1)\sum_{i=0}^{t}\varepsilon_i m^i)\equiv 1 \pmod 2.$$

Proof. We prove the lemma by induction on t. For $t = 0$, we have

$$\overline{h}_m((m+1)\varepsilon_0) = \begin{cases} 1 \pmod 2, & \varepsilon_0 = 0, \\ 5 \equiv 1 \pmod 2, & \varepsilon_0 = 1. \end{cases}$$

Hence, the lemma is true for $t = 0$. We now assume the lemma is true for all s $(0 \leq s \leq t)$. Then we consider the case $t + 1$.

Let $n_1 = \min\{i : \varepsilon_i \neq 0, i = 0, 1, \cdots, t+1\}$. If $n_1 = t + 1$, then we have

$$\begin{aligned} \overline{h}_m(n) &= \overline{h}_m((m+1)m^{t+1}) \\ &\equiv \overline{h}_m(m+1) \pmod 2 \text{ (by (2.1))} \\ &\equiv 1 \pmod 2 \text{ (since } \overline{h}_m(m+1) = 5). \end{aligned}$$

If $n_1 \neq t + 1$, then we let $n_2 = \min\{i : \varepsilon_i \neq 0, i = 0, 1, \cdots, t+1, i \neq n_1\}$. So we have

$$\begin{aligned} &\overline{h}_m((m+1)\textstyle\sum_{i=0}^{t+1} \varepsilon_i m^i) \\ &= \overline{h}_m((m+1)\textstyle\sum_{i=n_1}^{t+1} \varepsilon_i m^i) \\ &= \overline{h}_m((m+1)m^{n_1}\textstyle\sum_{i=n_1}^{t+1} \varepsilon_i m^{i-n_1}) \\ &\equiv \overline{h}_m((m+1)\textstyle\sum_{i=n_1}^{t+1} \varepsilon_i m^{i-n_1}) \pmod 2 \text{ (by (2.1))} \\ &= \overline{h}_m((m+1)m^{n_2-n_1}\textstyle\sum_{i=n_2}^{t+1} \varepsilon_i m^{i-n_2}) \\ &\equiv \overline{h}_m((m+1)\textstyle\sum_{i=n_2}^{t+1} \varepsilon_i m^{i-n_2}) \pmod 2 \text{ (by (2.1))} \\ &\equiv 1 \pmod 2 \text{ (by induction hypothesis and } t+1-n_2 \leq t). \end{aligned}$$

This completes the proof. □

Lemma 2.3. *Let $m \geq 3$, $t \geq 0$ and $n_{t+1} > n_t > \cdots > n_1 > n_0 \geq 0$ be integers. We have*

$$\overline{h}((m+1)(m^{n_{t+1}}n + m^{n_t} + m^{n_{t-1}} + \cdots + m^{n_0})) \equiv \overline{h}((m+1)n) \pmod 2.$$

Proof. We prove this lemma by induction on t. We first consider the case $t = 0$. Since

$$\begin{aligned} &\overline{h}_m((m+1)(m^{n_1}n + m^{n_0})) \\ &\equiv \overline{h}_m((m+1)(m^{n_1-n_0}n + 1)) \pmod 2 \text{ (by (2.1))} \\ &= \overline{h}_m(m((m+1)m^{n_1-n_0-1}n + 1) + 1) \\ &\equiv \overline{h}_m((m+1)m^{n_1-n_0-1}n)) \pmod 2 \text{ (by (2.2))} \\ &\equiv \overline{h}_m((m+1)n) \pmod 2 \text{ (by (2.1))}, \end{aligned}$$

the lemma is true for $t = 0$. Now we suppose that the lemma is true for t. Then we have

$$\begin{aligned} &\overline{h}((m+1)(m^{n_{t+2}}n + m^{n_{t+1}} + m^{n_t} + \cdots + m^{n_0})) \\ &= \overline{h}((m+1)(m^{n_{t+1}}(m^{n_{t+2}-n_{t+1}}n + 1) + m^{n_t} + \cdots + m^{n_0})) \\ &\equiv \overline{h}_m((m+1)(m^{n_{t+2}-n_{t+1}}n + 1)) \pmod 2 \text{ (by induction hypothesis)} \\ &\equiv \overline{h}_m((m+1)n)) \pmod 2. \end{aligned}$$

This completes the proof. □

Lemma 2.4. *Let $m \geq 3$ and $n \geq 0$ be integers. If $\overline{h}_m(n) \equiv 1 \pmod 2$, then there exist $\varepsilon_i = 0$ or 1 ($i = 0, 1, \cdots, t$, and $\varepsilon_t = 1$ if $t > 0$), such that $n = (m+1)\sum_{i=0}^{t} \varepsilon_i m^i$.*

Proof. Since $\overline{h}_m(n) \equiv 1 \pmod 2$, we have $n \equiv 0 (\bmod (m+1))$. Suppose that $n \neq (m+1)\sum_{i=0}^{t} \varepsilon_i m^i$, $\varepsilon_i = 0$ or 1. Then we have three different cases.

Case 1. $n = (m+1)m^{n_0}l_0$, where $n_0 \geq 0, 2 \leq l_0 \leq m-1$.
In this case, we have

$$\begin{aligned}\overline{h}_m(n) &= \overline{h}_m(m^{n_0}(m+1)l_0)\\ &\equiv \overline{h}_m(ml_0+l_0) \pmod 2 \text{ (by (2.1))}\\ &= 2\overline{h}_m(l_0) \text{ (by (2.3))}\\ &\equiv 0 \pmod 2,\end{aligned}$$

a contradiction.

Case 2. $n = (m+1)\sum_{i=0}^{t} l_i m^{n_i}$, where $t \geq 1,\ n_t > n_{t-1} > \cdots > n_0 \geq 0,\ 2 \leq l_0 \leq m-1$, and $1 \leq l_i \leq m-1$ for $i = 1, 2, \cdots, t$.
In this case, we have

$$\begin{aligned}\overline{h}_m(n) &= \overline{h}_m((m+1)m^{n_0}(\textstyle\sum_{i=1}^{t} l_i m^{n_i-n_0}+l_0))\\ &\equiv \overline{h}_m((m+1)(\textstyle\sum_{i=1}^{t} l_i m^{n_i-n_0}+l_0)) \pmod 2 \text{ (by (2.1))}\\ &= \overline{h}_m(m((m+1)m^{n_1-n_0-1}(\textstyle\sum_{i=1}^{t} l_i m^{n_i-n_1}+l_0))+l_0)\\ &\equiv 0 \pmod 2 \text{ (by (2.3))},\end{aligned}$$

a contradiction.

Case 3. $n = (m+1)(\sum_{i=1}^{s} l_i m^{n_{t+i}} + \sum_{i=0}^{t} m^{n_i})$, where $n_{t+s} > n_{t+s-1} > \cdots > n_0$, $2 \leq l_1 \leq m-1$ and $1 \leq l_i \leq m-1$ for $i = 2, 3, \cdots, s$.
In this case, we have

$$\begin{aligned}\overline{h}_m(n) &= \overline{h}_m((m+1)(m^{n_{t+1}}\textstyle\sum_{i=1}^{s} l_i m^{n_{t+i}-n_{t+1}} + \sum_{i=0}^{t} m^{n_i}))\\ &= \overline{h}_m((m+1)\textstyle\sum_{i=1}^{s} l_i m^{n_{t+i}-n_{t+1}}) \text{ (by Lemma 2.3)}\\ &\equiv 0 \pmod 2 \text{ (by Case 1 or Case 2 according as } s=1 \text{ or } s>1).\end{aligned}$$

This contradicts the given conditions and the proof is completed. □

Proof of Theorem 1.1. A proof can be obtained by combining Lemmas 2.1, 2.2 and 2.4. □

Proof of Theorem 1.2. We prove this result by induction on k. First, the theorem obviously holds for the case $k = 0$ by Lemma 2.1. Now we assume the theorem is true for k, i.e.,

$$\overline{h}_m((m+1)nm^k + lm^k - 1) \equiv 0 \pmod{2^{k+1}},$$

where $2 \leq l \leq m+1$. Then we have

$$\begin{aligned}&\overline{h}_m((m+1)nm^{k+1} + lm^{k+1} - 1)\\ &= \overline{h}_m(m((m+1)nm^k + lm^k - 1) + m - 1)\\ &= 2\overline{h}_m((m+1)nm^k + lm^k - 1) \text{ (by (2.3))}\\ &\equiv 0 \pmod{2^{k+2}} \text{ (by induction hypothesis)}.\end{aligned}$$

This completes the proof. □

We now prove Theorems 1.3 and 1.4.

Lemma 2.5 *For all $m \geq 3$, $k \geq 0$ and $n \geq 0$, we have*

$$\overline{h}_m(m^k n) = \overline{h}_m(n) + (2^{k+1} - 2)\overline{h}_m(n-1)$$

and

$$\overline{h}_m(m^{k+1}n+1) = 2\overline{h}_m(n) + (5\cdot 2^k - 4)\overline{h}_m(n-1).$$

Proof. Observe that

$\overline{h}_m(n) = \overline{h}_m(n),$
$\overline{h}_m(mn+1) = 2\overline{h}_m(n) + \overline{h}_m(n-1),$
$\overline{h}_m(mn) = \overline{h}_m(n) + 2\overline{h}_m(n-1),$

$$\begin{aligned}\overline{h}_m(m^2n+1) &= \overline{h}_m(m(mn)+1)\\ &= 2\overline{h}_m(mn) + \overline{h}_m(mn-1)\\ &= 2(\overline{h}_m(n) + 2\overline{h}_m(n-1)) + \overline{h}_m(m(n-1)+m-1)\\ &= 2\overline{h}_m(n) + 4\overline{h}_m(n-1) + 2\overline{h}_m(n-1)\\ &= 2\overline{h}_m(n) + 6\overline{h}_m(n-1).\end{aligned}$$

So we suppose that

$$\overline{h}_m(m^k n) = \overline{h}_m(n) + a_k\overline{h}_m(n-1)$$

and

$$\overline{h}_m(m^{k+1}n+1) = 2\overline{h}_m(n) + b_{k+1}\overline{h}_m(n-1).$$

Then we have

$$\begin{aligned}\overline{h}_m(m^{k+1}n) &= \overline{h}_m(m^k(mn))\\ &= \overline{h}_m(mn) + a_k\overline{h}_m(mn-1)\\ &= \overline{h}_m(n) + 2\overline{h}_m(n-1) + a_k\cdot 2\overline{h}_m(n-1)\\ &= \overline{h}_m(n) + (2a_k+2)\overline{h}_m(n-1),\end{aligned}$$

$$\begin{aligned}\overline{h}_m(m^{k+2}n+1) &= \overline{h}_m(m^{k+1}(mn)+1)\\ &= 2\overline{h}_m(mn) + b_{k+1}\overline{h}_m(mn-1)\\ &= 2(\overline{h}_m(n) + 2\overline{h}_m(n-1)) + b_{k+1}\cdot 2\overline{h}_m(n-1)\\ &= 2\overline{h}_m(n) + (2b_{k+1}+4)\overline{h}_m(n-1).\end{aligned}$$

Hence

$$\begin{cases} a_{k+1} &= 2a_k+2\\ a_0 &= 0\end{cases} \text{ and } \begin{cases} b_{k+1} &= 2b_k+4\\ b_1 &= 1\end{cases}.$$

So we have

$$a_k = 2^{k+1}-2,\ b_{k+1} = 5\cdot 2^k - 4.$$

This complete the proof. □

Proof of Theorem 1.3. A proof can be obtained directly by Lemma 2.5. □

Lemma 2.6. *For all $k \geq 0$, $n \geq 0$, we have*

$$\overline{h}_2(2^k n - 1) = \frac{1}{2}(3^k+1)\overline{h}_2(n-1) + \frac{1}{2}(3^k-1)\overline{h}_2(n-2),$$

$$\overline{h}_2(2^k n - 2) = \frac{1}{2}(3^k-1)\overline{h}_2(n-1) + \frac{1}{2}(3^k+1)\overline{h}_2(n-2).$$

Proof. We prove the lemma by induction on k. We only prove the first formula and the proof of the second one is similar. The first formula obviously holds for $k = 0$. Now we suppose that

$$\overline{h}_2(2^k n - 1) = \frac{1}{2}(3^k + 1)\overline{h}_2(n-1) + \frac{1}{2}(3^k - 1)\overline{h}_2(n-2),$$

i.e.,

$$\sum_{k\geq 0} \overline{h}_2(2^k n - 1)q^n = \left(\frac{1}{2}(3^k + 1)q + \frac{1}{2}(3^k - 1)q^2\right)\overline{H}_2(q).$$

Denote by $[q^n]B(q)$ the coefficient of q^n in $B(q)$. Then we have

$$\begin{aligned}
\overline{h}_2(2^{k+1}n - 1) &= \overline{h}_2(2^k(2n) - 1)\\
&= [q^{2n}](\tfrac{1}{2}(3^k + 1)q + \tfrac{1}{2}(3^k - 1)q^2)\overline{H}_2(q)\\
&= [q^{2n}](\tfrac{1}{2}(3^k + 1)q + \tfrac{1}{2}(3^k - 1)q^2)(1+q)(1+q+q^2)\overline{H}_2(q^2)\\
&= [q^{2n}](\tfrac{1}{2}(3^k + 1)q + \tfrac{1}{2}(3^k - 1)q^2)(1+2q+2q^2+q^3)\overline{H}_2(q^2)\\
&= [q^{2n}](\tfrac{1}{2}(3^{k+1} + 1)q^2 + \tfrac{1}{2}(3^{k+1} - 1)q^4)\overline{H}_2(q^2)\\
&= [q^{n}](\tfrac{1}{2}(3^{k+1} + 1)q + \tfrac{1}{2}(3^{k+1} - 1)q^2)\overline{H}_2(q)\\
&= \tfrac{1}{2}(3^{k+1} + 1)\overline{h}_2(n-1) + \tfrac{1}{2}(3^{k+1} - 1)\overline{h}_2(n-2).
\end{aligned}$$

This completes the proof. □

Proof of Theorem 1.4. A proof can be obtained directly by Lemma 2.6. □

Acknowledgement

Supported by the National Natural Science Foundation (grant 10771100) of People's Republic of China.

References

[1] G. E. Andrews, Congruence properties of the m-ary partition function, *J. Number Theory* 3(1971) 104-110.

[2] R. Churchhouse, Congruence properties of the binary partition function, *Proc. Camb. Philos. Soc.* 66(1969), 371-376.

[3] ES. Corteel and J. Lovejoy, Overpartitions, *Trans. Amer. Math. Soc.* 356(2004) 1623-1635.

[4] K. M. Courtright and J. A. Sellers, Arithmetic properties for hyper m-ary partition functitions, *Integers* 4(2004) #A06.

[5] G. Dirdal, On restricted m-ary partitions, *Math. Scand.* 37(1975) 51-60.

[6] G. Dirdal, Congruences for m-ary partitions, *Math. Scand.* 37(1975) 76-82.

[7] L. L. Dolph, A. Reynolds and J. A. Sellers, Congruences for a restricted m-ary partition function, *Discrete Math.* 219(2000) 265-269.

[8] H. Gupta, On m-ary partitions, *Proc. Cambridge Philos. Soc.* 71(1972) 343-345.

[9] Q. L. Lu, On a restricted m-ary partition function, *Discrete Math*. 275(2004) 347-353.

[10] L. Ma, Q. L. Lu, Congruences for hyper m-ary overpatition functions, *Integers* 10(2010) 1-11.

[11] Ø. Rødseth, Some arithmetic properties of m-ary partitions, *Proc. Cambridge Philos. Soc.* 68(1970) 447-453.

[12] Ø. Rødseth and J. A. Sellers, On m-ary partition function congruences: a fresh look at a past problem, *J. Number Theory* 87(2001) 270-281.

[13] Ø. Rødseth and J. A. Sellers, On m-ary overpartitions, *Ann. Comb.* 9(2005), 345-353.

In: Frontiers of Combinatorics and Number Theory, Volume 2 ISBN: 978-1-61942-918-5
Editor: Zhi-Wei Sun, pp. 251-256

Chapter 18

PURELY PERIODIC NEAREST SQUARE CONTINUED FRACTIONS

***Keith R. Matthews*[1,*] *and John P. Robertson*[2,†]**
[1]Department of Mathematics, University of Queensland
St. Lucia, Brisbane, Australia and
Centre for Mathematics and its Applications, Australian National University,
Canberra, Australia
[2]Actuarial and Economic Services Division
National Council on Compensation Insurance, Boca Raton, FL, US

Abstract

We present a test for determining whether a real quadratic irrational has a purely periodic nearest square continued fraction expansion. This test is somewhat more explicit than the standard test and simplifies the programming of the algorithm.

2010 Mathematics Subject Classification: Primary 11A55; Secondary 11Y65.

1. Introduction

Simple tests have long been known for determining whether a real quadratic irrational $\xi = (P + \sqrt{D})/Q, D > 0$ and not a perfect square, has a purely periodic regular continued fraction expansion, or nearest integer continued fraction expansion. Thus if $\overline{\xi} = (P - \sqrt{D})/Q$, then ξ has a purely periodic regular continued fraction expansion if and only if $\xi > 1$ and $-1 < \overline{\xi} < 0$ (see, e.g., [4, pp. 73–74]). Also ξ has a purely periodic nearest integer continued fraction expansion if and only if $\xi > 2$ and $(1-\sqrt{5})/2 < \overline{\xi} \leq (3-\sqrt{5})/2$ [3].

No test for pure periodicity as simple as these is known for the *nearest square* continued fraction, defined in the next section. Instead, A. A. K. Ayyangar [2, p. 27] gave a definition of *reduced* quadratic irrational and showed that ξ has a purely periodic nearest square continued fraction expansion if and only if ξ is reduced. In this paper, we give a more explicit version of Ayyangar's definition which is useful in detecting the start of a period.

*E-mail address: keithmatt@gmail.com; Website: http://www.numbertheory.org/keith.html
†E-mail address: jpr2718@gmail.com; Website: http://www.jpr2718.org/

2. The Nearest Square Continued Fraction Algorithm

This continued fraction was introduced by A. A. K. Ayyangar in 1938 and 1941 [1, 2]. Let $\xi_0 = \frac{P+\sqrt{D}}{Q}$ be a surd in standard form, i.e.,

(i) P, Q, and D are integers, $Q \neq 0$, D is not a perfect square,

(ii) $(P^2 - D)/Q$ is an integer,

(iii) $\gcd(P, Q, (D - P^2)/Q) = 1$.

Then with $c = \lfloor \xi_0 \rfloor$, the integer part of ξ_0, we can represent ξ_0 in two ways (the *positive* and *negative* representations of ξ_0):

$$\xi_0 = c + \frac{Q'}{P' + \sqrt{D}} = c + 1 - \frac{Q''}{P'' + \sqrt{D}},$$

where $\frac{P'+\sqrt{D}}{Q'} > 1$ and $\frac{P''+\sqrt{D}}{Q''} > 1$ are also in standard form. We choose the partial denominator a_0 and numerator ϵ_1 of the new continued fraction expansion as follows:

(a) $a_0 = c$ and $\epsilon_1 = 1$, if $|Q'| < |Q''|$, or $|Q'| = |Q''|$ and $Q < 0$,

(b) $a_0 = c + 1$ and $\epsilon_1 = -1$, if $|Q'| > |Q''|$, or $|Q'| = |Q''|$ and $Q > 0$.

The term *nearest square* arises on noting that $P'^2 = D - QQ'$ and $P''^2 = D + QQ''$ and restating (a) and (b) using the following equivalence:

$$|Q'| \geq |Q''| \iff |QQ'| \geq |QQ''| \iff |P'^2 - D| \geq |P''^2 - D|.$$

Then $\xi_0 = a_0 + \frac{\epsilon_1}{\xi_1}$, where $|\epsilon_1| = 1, a_0$ is an integer and $\xi_1 = \frac{P_1+\sqrt{D}}{Q_1} > 1$. Also $P_1 = P'$ or P'' and $Q_1 = Q'$ or Q'', according as $\epsilon_1 = 1$ or -1. We call ξ_1 the *successor* of ξ_0. We proceed similarly with ξ_1, and so on. Then the complete quotients ξ_n satisfy

$$\xi_n = a_n + \frac{\epsilon_{n+1}}{\xi_{n+1}} \text{ and } \xi_0 = a_0 + \frac{\epsilon_1|}{|a_1} + \frac{\epsilon_2|}{|a_2} + \cdots + \frac{\epsilon_n|}{|\xi_n}, \tag{2.1}$$

with partial numerator $\epsilon_{i+1} = \pm 1$ and partial denominator $a_i \geq 1$ if $i \geq 1$. This expansion is called the *nearest square* continued fraction (NSCF) expansion.

Analogous relations to those for regular continued fractions also hold for P_n, Q_n and a_n:

$$P_{n+1} + P_n = a_n Q_n, \tag{2.2}$$

$$P_{n+1}^2 + \epsilon_{n+1} Q_n Q_{n+1} = D. \tag{2.3}$$

Ayyangar [2, Theorem II, p. 25] proved that the NSCF expansion is eventually periodic, i.e., the complete quotients ξ_n eventually satisfy $\xi_i = \xi_{i+k}$ for $i \geq i_0$ for some $k \geq 1$. Then $\epsilon_{i+1} = \epsilon_{i+k+1}$ and $a_i = a_{i+k}$ for all $i \geq i_0$.

Our main result is:

Theorem 2.1. *Let* $\xi = (P + \sqrt{D})/Q$ *be in standard form and let* $R = (D - P^2)/Q$. *Then* ξ *has a purely periodic nearest square continued fraction expansion if and only if*

(i) $Q^2 + \frac{1}{4}R^2 \leq D, \quad \frac{1}{4}Q^2 + R^2 \leq D,$

(ii) ξ *is the successor of* $1/\xi$,

(iii) ξ *is not of the form* $\frac{p+q+\sqrt{p^2+q^2}}{2q}, p > 2q > 0.$

3. Reduced NSCF Quadratic Surds

Ayyangar [2, p. 27] gave the following definition of reduced quadratic surd. He first defined a *special* surd ξ_v by the inequalities

$$Q_{v+1}^2 + \tfrac{1}{4}Q_v^2 \leq D, \quad Q_v^2 + \tfrac{1}{4}Q_{v+1}^2 \leq D, \tag{3.1}$$

then defined a *semi-reduced* surd to be the successor of a special surd. Finally a *reduced* surd is defined to be the successor of a semi-reduced surd. Ayyangar [2, p. 28] proved that a semi–reduced surd is a special surd. Consequently a reduced surd is also semi–reduced. That a quadratic surd has a purely periodic NSCF expansion if and only if it is reduced, is proved in [2, pp. 101–102]. To use the Ayyangar characterization to decide if a surd ξ is reduced, one has to determine if there is a special surd whose second successor is ξ; doing this can be combersome.

One example of a reduced surd is the successor of $\sqrt{D}/Q > 1$, where Q divides D (see [2, Theorem XII, p. 102]). Another example that figures prominently in [2] is $\frac{p+q+\sqrt{p^2+q^2}}{p}$, where $p > 2q > 0$.

Lemma 3.1. *If two different semi–reduced surds have the same successor, they have the form* $\frac{p \pm q+\sqrt{D}}{2q}$, *where* $p > 2q > 0$.

Proof. This is [2, Theorem IX, p. 99]. □

Lemma 3.2. ξ *is semi–reduced if and only if* ξ *is reduced, or* $\xi = \frac{p+q+\sqrt{p^2+q^2}}{2q}, p > 2q > 0.$

Proof. (a) Suppose ξ is semi–reduced and let η be its successor. Then η is reduced and has a unique reduced predecessor χ, by [2, Corollary 1, p. 101]. By Lemma 3.1, either $\xi = \chi$, or ξ and χ are equal to $\frac{p \pm q+\sqrt{D}}{2q}$, where $p > 2q > 0$. However by [2, Theorem X, p. 100], $\frac{p+q+\sqrt{p^2+q^2}}{2q}$ has no semi–reduced predecessor and hence is not reduced, so $\chi = \frac{p-q+\sqrt{D}}{2q}$ and hence $\xi = \frac{p+q+\sqrt{p^2+q^2}}{2q}$.

(b) If ξ is reduced, it is the successor of a semi–reduced surd χ and as previously observed, this is special. Hence ξ is semi–reduced. If $\xi = \frac{p+q+\sqrt{p^2+q^2}}{2q}, p > 2q > 0$, in view of the equation

$$\frac{p-q+\sqrt{p^2+q^2}}{p} = 1 + \frac{p}{q+\sqrt{p^2+q^2}} = 2 - \frac{2q}{p+q+\sqrt{p^2+q^2}},$$

as $p > 2q > 0$, we see ξ is the successor of the special surd $\frac{p-q+\sqrt{p^2+q^2}}{p}$ and is hence semi–reduced. □

Corollary 3.1. *ξ is reduced if and only if ξ is semi–reduced and not of the form* $\xi = \frac{p+q+\sqrt{p^2+q^2}}{2q}, p > 2q > 0$.

Ayyangar did not explicitly mention Lemma 3.2 or Corollary 3.1 in his paper [2].

4. Some Lemmas on Successors

Lemma 4.3. *ξ and $-\xi$ have the same successor.*

Proof. This follows from the fact that the positive–negative represention

$$\frac{P+\sqrt{D}}{Q} = c + \frac{Q'}{P'+\sqrt{D}} = c+1-\frac{Q''}{P''+\sqrt{D}},$$

implies the positive–negative represention

$$\frac{P+\sqrt{D}}{-Q} = -c-1+\frac{Q''}{P''+\sqrt{D}} = -c-\frac{Q'}{P'+\sqrt{D}}.$$

Then the conditions defining the successor of ξ also define the same successor of $-\xi$. □

Lemma 4.4. *If ξ is the successor of a quadratic surd, then ξ is the successor of $1/\xi$.*

Proof. Suppose $\xi = \frac{P+\sqrt{D}}{Q}$ is the successor of $\frac{P_0+\sqrt{D}}{Q_0}$. Then the successor equation

$$\frac{P_0+\sqrt{D}}{Q_0} = b + \epsilon\frac{Q}{P+\sqrt{D}} \tag{4.1}$$

gives $P_0 + P = bQ_0$ and $D - P^2 = \epsilon QQ_0$. We also have the positive–negative representation

$$\frac{P_0+\sqrt{D}}{Q_0} = a + \frac{Q'}{P'+\sqrt{D}} = a+1-\frac{Q''}{P''+\sqrt{D}}.$$

Then

$$\begin{aligned}\epsilon/\xi = \frac{\epsilon Q}{P+\sqrt{D}} &= \frac{\epsilon Q(\sqrt{D}-P)}{D-P^2} = \frac{\epsilon Q(\sqrt{D}-P)}{\epsilon QQ_0}\\ &= \frac{-P+\sqrt{D}}{Q_0}\\ &= \frac{P_0-bQ_0+\sqrt{D}}{Q_0}\\ &= -b+\frac{P_0+\sqrt{D}}{Q_0}\\ &= -b+a+\frac{Q'}{P'+\sqrt{D}} = -b+a+1-\frac{Q''}{P''+\sqrt{D}}\end{aligned} \tag{4.2}$$

and this positive–negative representation implies that the successor of ϵ/ξ is ξ. If $\epsilon = -1$, Lemma 4.3 implies that the successor of $1/\xi$ is also ξ. □

Lemma 4.5. *Let* $\xi = \frac{P+\sqrt{D}}{Q}$ *and* $R = (D - P^2)/Q$. *Then* ξ *is semi–reduced if and only if*

(i) $Q^2 + \frac{1}{4}R^2 \le D, \quad \frac{1}{4}Q^2 + R^2 \le D,$

(ii) ξ *is the successor of* $1/\xi$.

Proof. (a) Suppose $\xi = \frac{P+\sqrt{D}}{Q}$ is semi–reduced. Then ξ is the successor of a special surd $\xi_0 = \frac{P_0+\sqrt{D}}{Q_0}$. Then with ϵ as in (4.1), as before, we have $R = (D - P^2)/Q = \epsilon Q_0$ and inequalities $Q_0^2 + \frac{1}{4}Q^2 \le D, \frac{1}{4}Q^2 + Q_0^2 \le D$ become

$$R^2 + \tfrac{1}{4}Q^2 \le D, \quad \tfrac{1}{4}Q^2 + R^2 \le D.$$

Also by Lemma 4.4, ξ is the successor of $1/\xi$.
(b) Suppose (i) and (ii) hold. Then as $1/\xi = \frac{-P+\sqrt{D}}{R}$, (i) and (ii) imply $1/\xi$ is special and that ξ is semi–reduced. □

Our Theorem then follows from Corollary 3.1 and Lemma 4.5.

We also mention the following useful result.

Lemma 4.6. *If* ξ *is semi–reduced, then* ξ *or* $\xi - 1$ *is also reduced in the regular continued fraction sense.*

Proof. By [2, Theorem V, p. 30], we have $\xi > \frac{1+\sqrt{5}}{2} > 1$ and by [2, Theorem III, p. 27], $-1 < \overline{\xi} < 1$. So if $-1 < \overline{\xi} < 0$, ξ is RCF–reduced. If $0 < \overline{\xi}$, let $\xi = \frac{P+\sqrt{D}}{Q}$. Then $0 < \frac{P-\sqrt{D}}{Q}$ and as $0 < Q < \sqrt{D}$ by [2, Theorem 1(iv), p. 22], we deduce

$$2 < \frac{2\sqrt{D}}{Q} < \frac{P+\sqrt{D}}{Q} = \xi.$$

Hence $1 < \xi - 1$ and as $-1 < \overline{\xi - 1} < 0$, it follows that $\xi - 1$ is RCF–reduced. □

We conclude with an example of Ayyangar [2, Theorem XIII, p. 103]. His proof of case (b), when $p > 2q$, involved a complicated discussion of inequalities.

Example. Let $\xi = \frac{P+\sqrt{D}}{Q} = \frac{p+q+\sqrt{p^2+q^2}}{p}, p > 0, q > 0$. Then

(a) ξ is not reduced if $p < 2q$,

(b) ξ is reduced if $p \ge 2q$,

Proof. Here $P = p + q, Q = p, R = (D - P^2)/Q = -2q$ and

$$Q^2 + \tfrac{1}{4}R^2 = p^2 + q^2 = D \text{ and } \tfrac{1}{4}Q^2 + R^2 = \tfrac{1}{4}p^2 + 4q^2.$$

Hence

$$\begin{aligned} \tfrac{1}{4}Q^2 + R^2 \le D &\iff \tfrac{1}{4}p^2 + 4q^2 \le p^2 + q^2 \\ &\iff 3q^2 \le 3p^2/4 \\ &\iff 2q \le p. \end{aligned}$$

Consequently if $p < 2q$, ξ is not reduced. However if $p \geq 2q$, then $\frac{1}{4}Q^2 + R^2 \leq D$. Also the positive–negative representation

$$\begin{aligned} 1/\xi &= \frac{-P+\sqrt{D}}{R} = \frac{-p-q+\sqrt{p^2+q^2}}{-2q} \\ &= 0 + \frac{p}{p+q+\sqrt{p^2+q^2}} = 1 - \frac{p}{p-q+\sqrt{p^2+q^2}}, \end{aligned}$$

shows that ξ is the successor of $1/\xi$. Hence conditions (i) and (ii) of our Theorem are satisfied, so ξ is semi–reduced. Also condition (iii) is satisfied. For assume $\xi = \frac{P+Q+\sqrt{P^2+Q^2}}{2Q}$, $P > 2Q > 0$. Then

$$p + q = P + Q, \quad p = 2Q, \quad p^2 + q^2 = P^2 + Q^2.$$

Hence

$$\begin{aligned} P^2 + Q^2 &= 4Q^2 + q^2 \\ P^2 &= 3Q^2 + q^2 = 3Q^2 + (P-Q)^2 \\ &= 4Q^2 + P^2 - 2PQ \\ 2PQ &= 4Q^2 \\ P &= 2Q. \end{aligned}$$

This contradiction completes the demonstration that ξ is reduced if $p \geq 2q > 0$. □

Acknowledgement

We thank the referee for comments that improved the presentation of the paper.

References

[1] A. A. K. Ayyangar, A new continued fraction, *Current Sci.* 6 (1938), 602–604.

[2] A. A. K. Ayyangar, Theory of the nearest square continued fraction, *J. Mysore Univ. Sect. A.* 1 (1940), 21–32, (1941), 97–117.

[3] K. R. Matthews and J. P. Robertson, *On the definition of nearest–integer reduced quadratic surd*, `http://www.numbertheory.org/pdfs/nicf-reduced.pdf`.

[4] O. Perron, *Die Lehre von den Kettenbrüchen*, Band 1, Teubner, Stuttgart, 1954.

In: Frontiers of Combinatorics and Number Theory, Volume 2 ISBN: 978-1-61942-918-5
Editor: Zhi-Wei Sun, pp. 257-290

Chapter 19

KAC-MOODY FIBONACCI SEQUENCES, HYPERBOLIC GOLDEN RATIOS, AND REAL QUADRATIC FIELDS

***Kasper K.S. Andersen*[1,*], *Lisa Carbone*[†] and *Diego Penta*[2,‡]**
[1]Department of Mathematics Sciences
University of Aarhus
Ny Munkegade, bygn.
Aarhus, Denmark
[2]Department of Mathematics, Hill Center, Busch Campus
Rutgers, The State University of New Jersey
Piscataway, NJ, US

Abstract

Let A be the generalized Cartan matrix of rank 2 Kac-Moody algebra $\mathfrak{g}$. We write $\mathfrak{g} = \mathfrak{g}(a, b)$ when A has non-diagonal entries $-a$ and $-b$. To each such A, its Weyl group and corresponding root lattice, we associate a 'Fibonacci type' integer sequence. These sequences are derived from the coordinates of the real root vectors in the root space. Each element of each sequence can be expressed as a polynomial in the non-diagonal entries of the generalized Cartan matrix, whose coefficients are shallow diagonals of Pascal's triangle. Among the Fibonacci type sequences are the bisected Lucas and Fibonacci sequences, the Kekule numbers for the benzoids, the integers whose squares are triangular numbers, Chebyshev polynomials of the second kind, as well as some other known sequences. Each sequence has an associated 'hyperbolic golden ratio' Ψ which is a unit in a real quadratic field. We show that Ψ can be obtained as the limit of ratios of areas of hyperbolic triangles spanned by a triple of adjacent real roots in the root space. For the bisected Fibonacci sequence that occurs in our setting, Ψ coincides with the classical golden ratio $\psi = 1 + \phi$ for the bisection, where $\phi = 1.61803399\ldots$. It follows that ψ is an eigenvalue for the fundamental endomorphism $w_1 w_2$ in the root system of $\mathfrak{g}(5, 1)$, and ψ^2 is an eigenvalue for $w_1 w_2$ in $\mathfrak{g}(3, 3)$.

*E-mail address: kksa@imf.au.dk
†E-mail address: carbonel@math.rutgers.edu
‡E-mail address: penta@math.rutgers.edu

$\phi = 1.61803399\ldots$. It follows that ψ is an eigenvalue for the fundamental endomorphism $w_1 w_2$ in the root system of $\mathfrak{g}(5,1)$, and ψ^2 is an eigenvalue for $w_1 w_2$ in $\mathfrak{g}(3,3)$. This leads to an infinite family of identities involving continued fractions. We identify the generalized Lucas and Fibonacci sequences with Chebyshev polynomials of the 2nd kind, which in turn can be expressed as products. This leads us to an arithmetic proof that Kitchloo's 'generalized binomial coefficients' occuring in the cohomology of flag varieties of rank 2 Kac-Moody groups over $\mathbb{C}$ are integers.

2000 Mathematics Subject Classification: Primary: 81R10, 11B39; Secondary 14M15.

Keywords: Kac-Moody, Fibonacci and Lucas numbers and polynomials, flag varieties.

1. Introduction

A *Kac-Moody Lie algebra* is the most natural generalization to infinite dimensions of the notion of a finite dimensional semisimple Lie algebra. The data for building a Kac-Moody algebra includes a square integer matrix A called a *generalized Cartan matrix* in which the diagonal entries equal 2, the other entries are nonpositive and A is symmetric in its zeros.

Let A be the 2×2 generalized Cartan matrix

$$A = (a_{ij})_{i,j=1,2} = \begin{pmatrix} 2 & -a \\ -b & 2 \end{pmatrix}$$

with $a, b \in \mathbb{Z}_{>0}$, $ab \geq 4$, with Kac-Moody algebra $\mathfrak{g} = \mathfrak{g}(A)$, root system $\Delta = \Delta(A)$, and Weyl group $W = W(A)$. When $ab = 4$, A is positive semi-definite but not positive definite, and we call A *affine*. When $ab > 4$, A is neither positive semi-definite nor positive definite, but every proper generalized Cartan submatrix is positive definite, and A is *hyperbolic*.

Let A be the generalized Cartan matrix of the rank 2 Kac-Moody algebra $\mathfrak{g}$. We write $\mathfrak{g} = \mathfrak{g}(a, b)$ when A has non-diagonal entries $-a$ and $-b$. To each such A, its Weyl group and corresponding root lattice, we associate a 'Fibonacci type' integer sequence (Section 3).

These sequences are derived from the coordinates of the real root vectors in the root space. Each element of each sequence can be expressed as a polynomial in the non-diagonal entries of the generalized Cartan matrix, whose coefficients are shallow diagonals of Pascal's triangle. Among these sequences are the bisected Lucas and Fibonacci sequences, the Kekule numbers for the benzoids, the integers whose squares are triangular numbers, as well as some other known sequences. For each pair a, b there is a natural pairing of integer sequences. This gives rise to a relation between the entries generalizing the description of elements of the Lucas sequence as a sum of consecutive terms of the Fibonacci sequence. We obtain several other generalizations of Fibonacci type identities, including a generalization of Binet's formula (Proposition 4.3). Each sequence has an associated 'hyperbolic golden ratio' Ψ which is an algebraic integer in a real quadratic field (Section 4). We show that Ψ can be obtained as the limit of ratios of areas of hyperbolic triangles spanned by a triple of adjacent real roots in the root space. This follows from the fact that the distribution of coordinates of adjacent real roots tends towards the 'golden' ratio Ψ. For the bisected Fibonacci sequence that occurs in our setting, Ψ coincides with the classical golden ratio $\psi = 1 + \phi$ for the bisection, where $\phi = 1.61803399\ldots$. It follows that ψ is an eigenvalue for the fundamental endomorphism w_1w_2 in the root system of $\mathfrak{g}(5,1)$, and ψ^2 is an eigenvalue for w_1w_2 in $\mathfrak{g}(3,3)$ (Section 5). This leads to an infinite family of identities involving continued fractions (Section 6).

We also identify the generalized Lucas and Fibonacci sequences with Chebyshev polynomials of the second kind, which in turn can be expressed as products. We then show that each sequence element can be written as products of binomials of the form $k - 4\cos^2(\frac{j\pi}{n})$, where n is the index of the polynomial and j ranges from 1 to $\lfloor\frac{n-1}{2}\rfloor$. Since the resulting polynomials share like terms with our sequence polynomials, we identify the coefficients on both sides, hence indentifying binomial coefficients with elementary symmetric polynomials whose roots are $\cos^2(\frac{j\pi}{n})$ as above (Section 3).

It is interesting to ask if certain ratios of products of consecutive sequence elements are integers. Using topological methods, N. Kitchloo answered this in the affirmative, showing that these integers appear in the cohomology of flag varieties associated to rank 2 Kac-Moody groups over $\mathbb{C}$. Kitchloo also asked if an arithmetic proof was possible. Using properties of Chebyshev polynomials of the second kind, we provide an arithmetic proof (Section 7). We also give a detailed computation of the p-adic valuation of our sequence entries (Section 8).

In Section 9, we give a brief indication of how our methods may be used to study the Weyl group orbits of roots in certain rank 3 hyperbolic root systems.

At the time of submission, there is no description of the infinite family of sequences in the *Online Encyclopedia of Integer Sequences*, though some of the sequences do appear there, as indicated in Table 1 and Table 2.

We remark that our work is within the root system corresponding to a 2×2 generalized Cartan matrix A. In this case, the Cartan subalgebra can be mapped into the Lorentzian space $\mathbb{R}^{(2,1)}$, a *finite dimensional* space. Hence our results do not use the full structure of the Kac-Moody algebra. Furthermore, since we deal only with real roots, the phenomena we describe occur mostly in the 'Euclidean part' of $\mathbb{R}^{(2,1)}$. Hence much of what we describe can be characterized in terms of linear dynamical systems in $\mathbb{R}^2$, though we did not find a suitable reference.

Some of the proofs of our results are elementary, and will be omitted if they are obvious or straightforward. Alex Feingold first discovered the appearance of Fibonacci numbers in a rank 2 Kac-Moody root system when $a = b = 3$ ([4]). This was also observed by Kang and Melville ([7]). Some of the ideas here are implicitly included in the paper of Lepowsky and Moody ([11]) though they do not appear explicitly there. We also suspect that some of the results included here are well known as isolated facts and that our main contribution is to describe them in the framework of rank 2 Kac-Moody root systems. Omission of relevant references in parts of this work is unintentional. We are grateful to Nitu Kitchloo who was a driving force behind Sections 7 and 8.

2. Rank 2 Kac-Moody Root Systems

Let A be the generalized Cartan matrix

$$A = (a_{ij})_{i,j=1,2} = \begin{pmatrix} 2 & -a \\ -b & 2 \end{pmatrix}$$

with $a, b \in \mathbb{Z}_{>0}$, $ab \geq 4$, $a \geq b$, with Kac-Moody algebra $\mathfrak{g} = \mathfrak{g}(A)$, root system $\Delta = \Delta(A)$, and Weyl group $W = W(A)$. Let $\Pi = \{\alpha_1, \alpha_2\}$ be the simple roots and let w_1, w_2 be the simple reflections. We have $W = \langle w_1 \rangle * \langle w_2 \rangle \cong \mathbb{Z}/2\mathbb{Z} * \mathbb{Z}/2\mathbb{Z}$, so every element of W is of the form $w_1^{\epsilon_1}(w_2 w_1)^n w_2^{\epsilon_2}$, where $\epsilon_i \in \{0, 1\}$, $i = 1, 2$ and $n \in \mathbb{Z}_{\geq 0}$. Since W is an infinite group, the number of translates of Π by W is infinite, and $\mathfrak{g}(A)$ is infinite dimensional. However, not all roots Δ are of the form $W\Pi$. We will refer to roots of the form $W\Pi$ as *real roots*. Let

$$B = (b_{ij})_{i,j=1,2} = \begin{pmatrix} 2 & -a \\ -a & 2a/b \end{pmatrix}$$

be a symmetrization of A, so that we have a short root (say α_1) of squared length 2, and a long root (say α_2) of squared length $2a/b$. Let

$$Q(x, y) = 2x^2 - 2axy + (2a/b)y^2$$

be the associated quadratic form. We wish to determine the integer lattice points such that $Q(x, y) = 2$ and $Q(x, y) = 2a/b$ on the corresponding conic section. These correspond to a subset of the root system, namely the roots of positive squared length (real roots). Let $P(x, y)$ be the quadratic form

$$P(x, y) = bx^2 - abxy + ay^2.$$

We have

$$\{(x, y) \in \mathbb{Z} \times \mathbb{Z} \mid Q(x, y) = 2\} = \{(x, y) \in \mathbb{Z} \times \mathbb{Z} \mid P(x, y) = b\},$$

$$\{(x, y) \in \mathbb{Z} \times \mathbb{Z} \mid Q(x, y) = 2a/b\} = \{(x, y) \in \mathbb{Z} \times \mathbb{Z} \mid P(x, y) = a\},$$

so we may characterize the integer solutions of $P(x, y)$ rather than $Q(x, y)$.

We have

$$W\{\alpha_1\} \subseteq \{(x, y) \in \mathbb{Z} \times \mathbb{Z} \mid P(x, y) = b\},$$

$$W\{\alpha_2\} \subseteq \{(x, y) \in \mathbb{Z} \times \mathbb{Z} \mid P(x, y) = a\}.$$

We refer the reader to [6, Exercise 5.25] for a more precise statement of about the relationship between real roots and the quadratic form $P(x, y)$. The real roots then correspond to the set $\Phi = W\{\alpha_1\} \sqcup W\{\alpha_2\}$, and Φ is a proper subset of Δ. When $a = b$, α_1 and α_2 have the same length, as do all translates $W\Pi$.

It is known that we can map Δ into a 2-dimensional Euclidean space E which can be endowed with a Lorentzian metric of signature (1,1) when $ab > 4$. The real roots $\Phi = W\{\alpha_1\} \sqcup W\{\alpha_2\}$ lie on the root lattice $\mathbb{Z}\alpha_1 \oplus \mathbb{Z}\alpha_2 \subset E$. Each root $\beta \in \Phi$ is a linear combination of simple roots with coefficients that are either all positive or all negative. We call β positive or negative accordingly.

3. Kac-Moody Fibonacci Sequences

In this section, we demonstrate that the Weyl group translates of the simple roots α_1 and α_2 give rise to a sequence of polynomials in a and b, where $-a$ and $-b$ are the off diagonal entries of the generalized Cartan matrix. We first tabulate the coefficients of $W\{\alpha_1\} \sqcup W\{\alpha_2\}$ in the root lattice, and then in Lemma 3.1 below, we rewrite these in the language of sequences.

To set up the appropriate notation, we identify the simple roots α_1, α_2 with the vectors $(1, 0)$ and $(0, 1)$, respectively. For every root $\beta \in \Phi$ we write

$$\beta = x(\beta)\alpha_1 + y(\beta)\alpha_2,$$

for $x(\beta), y(\beta) \in \mathbb{Z}$ to give an expression for β in

$$\Phi \cap (\mathbb{Z}\alpha_1 \oplus \mathbb{Z}\alpha_2).$$

Then

$$x(\alpha_1) = 1,\ y(\alpha_1) = 0,$$

$$x(\alpha_2) = 0,\ y(\alpha_2) = 1.$$

For $i = 1, 2$, we write $W^+\{\alpha_i\} = W\{\alpha_i\} \cap \Phi^+$, where Φ denotes the real roots. Then

$$W^+\{\alpha_1\} = \{(w_1 w_2)^n \alpha_1,\ n \geq 0\} \sqcup \{w_2(w_1 w_2)^n \alpha_1,\ n \geq 0\},$$

$$W^+\{\alpha_2\} = \{(w_2 w_1)^n \alpha_2,\ n \geq 0\} \sqcup \{w_1(w_2 w_1)^n \alpha_2,\ n \geq 0\}.$$

Note that for each real root $\alpha \in \Phi$, if we write $\alpha = w\alpha_i$, the word w is reduced and ends in w_{3-i}.

We tabulate $x(\alpha)$ and $y(\alpha)$ for all $\alpha \in \Phi$, where $l(w) = n$ for some $n \geq 0$, and $l(\cdot)$ is the standard length function on W:

$n = l(w)$	$\alpha \in \{(w_1 w_2)^{\lfloor n/2 \rfloor}\alpha_1 \mid n \geq 0\}$	$x(\alpha)$	$y(\alpha)$
0	α_1	1	0
2	$w_1 w_2 \alpha_1$	$ab - 1$	b
4	$w_1 w_2 w_1 w_2 \alpha_1$	$a^2b^2 - 3ab + 1$	$ab^2 - 2b$
$\vdots$	$\vdots$	$\vdots$	$\vdots$

$n = l(w)$	$\alpha \in \{w_2(w_1 w_2)^{\lfloor n/2 \rfloor}\alpha_1 \mid n \geq 0\}$	$x(\alpha)$	$y(\alpha)$
1	$w_2 \alpha_1$	1	b
3	$w_2 w_1 w_2 \alpha_1$	$ab - 1$	$ab^2 - 2b$
5	$w_2(w_1 w_2)^2 \alpha_1$	$a^2b^2 - 3ab + 1$	$a^2b^3 - 4ab^2 + 3b$
$\vdots$	$\vdots$	$\vdots$	$\vdots$

$n = l(w)$	$\alpha \in \{(w_2w_1)^{\lfloor n/2 \rfloor}\alpha_2 \mid n \geq 0\}$	$x(\alpha)$	$y(\alpha)$
0	α_2	0	1
2	$w_2w_1\alpha_2$	a	$ab - 1$
4	$w_2w_1w_2w_1\alpha_2$	$a^2b - 2a$	$a^2b^2 - 3ab + 1$
$\vdots$	$\vdots$	$\vdots$	$\vdots$

$n = l(w)$	$\alpha \in \{w_1(w_2w_1)^{\lfloor n/2 \rfloor}\alpha_2 \mid n \geq 0\}$	$x(\alpha)$	$y(\alpha)$
1	$w_1\alpha_2$	a	1
3	$w_1w_2w_1\alpha_2$	$a^2b - 2a$	$ab - 1$
5	$w_1(w_2w_1)^2\alpha_2$	$a^3b^2 - 4a^2b + 3a$	$a^2b^2 - 3ab + 1$
$\vdots$	$\vdots$	$\vdots$	$\vdots$

We now rewrite the coefficients $x(\alpha)$ and $y(\alpha)$ as tabulated above in the language of sequences. For $i = 1, 2$ and for $\alpha \in W^+\{\alpha_i\}$, we have $\alpha = w\alpha_i$ for some $w \in W$, where $l(w) = n$ for some $n \geq 0$, and $l(\cdot)$ is the standard length function on W. For $i = 1, 2$, we introduce the following notation for the polynomials $x(\alpha)$ and $y(\alpha)$. We set

$$x_{i,n} := x_{i,n}(a, b) = x(\alpha) \text{ where } \alpha = w\alpha_i,\ l(w) = n,$$

$$y_{i,n} := y_{i,n}(a, b) = y(\alpha) \text{ where } \alpha = w\alpha_i,\ l(w) = n.$$

Lemma 3.1. *We have*

$$x_{1,n} = x_{1,n+1},\ n \text{ even},\ n \geq 0,$$

$$y_{1,n} = y_{1,n-1},\ n \text{ even},\ n \geq 2;$$

$$x_{2,n} = x_{2,n-1},\ n \text{ even},\ n \geq 2,$$

$$y_{2,n} = y_{2,n+1},\ n \text{ even},\ n \geq 0.$$

Proof. Let A be the generalized Cartan matrix defined in Section 1 with $ab \geq 4$. Let $\beta = w\alpha_1$ and $\gamma = w'\alpha_2$ where $l(w) = l(w') = n$ for some $n \in 2\mathbb{Z}_{\geq 0}$. That is to say, $w = (w_1w_2)^{n/2}$ and $w' = (w_2w_1)^{n/2}$. The Weyl group generators w_1 and w_2 act on α_1 and α_2 by the general formula:

$$w_i\alpha_j = \alpha_j - A_{ij}\alpha_i.$$

Thus, writing $\beta = x(\beta)\alpha_1 + y(\beta)\alpha_2$ and $\gamma = x(\gamma)\alpha_1 + y(\gamma)\alpha_2$, we obtain:

$$w_2\beta = x(\beta)\alpha_1 + [b \cdot x(\beta) - y(\beta)]\,\alpha_2, \tag{1}$$

$$w_1\gamma = [a \cdot y(\gamma) - x(\gamma)]\,\alpha_1 + y(\gamma)\alpha_2. \tag{2}$$

Note that the Weyl words w_2w and w_1w' associated to $w_2\beta$ and $w_1\gamma$ have lengths $l(w_2(w_1w_2)^{n/2}) = l(w_1(w_2w_1)^{n/2}) = n + 1$. Thus we observe, $y(w_1\gamma) = y(\gamma)$ which implies $y_{2,n+1} = y_{2,n}$ and $x(w_2\beta) = x(\beta)$ proving $x_{1,n+1} = x_{1,n}$.

The two remaining identities in Lemma 3.1 are obtained by the actions of w_1 on β and w_2 on γ, where β and γ are as before, but now $n \geq 2$. Here, we note that the Weyl words associated to the new roots $w_1\beta$ and $w_2\gamma$ begin with w_1^2 and w_2^2, respectively, and so they have length $n-1$:

$$l(w_1 w) = l(w_1(w_1 w_2)^{n/2}) = l(w_2(w_1 w_2)^{\frac{n}{2}-1}) = n-1,$$

$$l(w_2 w') = l(w_2(w_2 w_1)^{n/2}) = l(w_1(w_2 w_1)^{\frac{n}{2}-1}) = n-1.$$

Thus we obtain:

$$w_1\beta = [a \cdot y(\beta) - x(\beta)]\,\alpha_1 + y(\beta)\alpha_2, \tag{3}$$

$$w_2\gamma = x(\gamma)\alpha_1 + [b \cdot x(\gamma) - y(\gamma)]\,\alpha_2, \tag{4}$$

and we observe that $y(w_1\beta) = y(\beta)$ and $x(w_2\gamma) = x(\gamma)$ so that $y_{1,n-1} = y_{1,n}$ and $x_{2,n-1} = x_{2,n}$ completing the proof. □

By Lemma 3.1, we may restrict our attention to the even-indexed coefficients $x_{i,2m}$ and $y_{i,2m}$ ($i = 1, 2$). The values for $m = 0, 1$ and 2 were tabulated earlier in the section, and we note that for each set, they are distinct. The tables also seem to suggest that a pattern might emerge for $m > 2$. Indeed, as the next lemma will show, each set follows a very similar recurrence relation.

Lemma 3.2. *We have the following recursion relations*

(i) $$x_{1,2m} = \begin{cases} 1, & m = 0, \\ ab-1, & m = 1, \\ (ab-2)x_{1,2m-2} - x_{1,2m-4}, & m \geq 2, \end{cases}$$

(ii) $$y_{1,2m} = \begin{cases} 0, & m = 0, \\ b, & m = 1, \\ (ab-2)y_{1,2m-2} - y_{1,2m-4}, & m \geq 2, \end{cases}$$

(iii) $$x_{2,2m} = \begin{cases} 0, & m = 0, \\ a, & m = 1, \\ (ab-2)x_{2,2m-2} - x_{2,2m-4}, & m \geq 2. \end{cases}$$

(iv) $$y_{2,2m} = \begin{cases} 1, & m = 0, \\ ab-1, & m = 1, \\ (ab-2)y_{2,2m-2} - y_{2,2m-4}, & m \geq 2. \end{cases}$$

Proof. We prove part (i) and leave the remaining proofs to the reader, as they involve similar arguments. As already indicated, $x_{1,2m} = x(\alpha)$ where $\alpha = (w_1 w_2)^m \alpha_1$. The starting values, corresponding to reduced Weyl words 1_W and $w_1 w_2$, were tabulated earlier in the section:

$$x_{1,0} = x(\alpha_1) = 1,$$
$$x_{1,2} = x(w_1 w_2 \alpha_1) = ab - 1.$$

Now assume the recursion in (i) is not true for all m, and consider the set of counterexamples, $M \subset \mathbb{N}$. By assumption M is nonempty, and therefore it has a least element, k. Since k is the least element of M, the recursion holds for $k-1$. Thus let $\beta = (w_1w_2)^{k-1}\alpha_1$, so that $x_{1,2k-2} = x(\beta)$ and $x_{1,2k} = x(w_1w_2\beta)$. By equations (1) and (3) above, we have

$$\begin{aligned} w_1w_2\beta &= w_1(x(\beta)\alpha_1 + [b \cdot x(\beta) - y(\beta)]\alpha_2) \\ &= x(\beta)(w_1\alpha_1) + [b \cdot x(\beta) - y(\beta)](w_1\alpha_2) \\ &= -x(\beta) + [b \cdot x(\beta) - y(\beta)](\alpha_2 + a \cdot \alpha_1) \\ &= (a[b \cdot x(\beta) - y(\beta)] - x(\beta))\alpha_1 + [b \cdot x(\beta) - y(\beta)]\alpha_2. \end{aligned}$$

Thus,

$$x_{1,2k} = x(w_1w_2\beta) = (ab-1)x(\beta) - a \cdot y(\beta).$$

The last term expands by equation (3), giving us

$$x_{1,2k} = (ab-1)x(\beta) - (x(w_1\beta) + x(\beta)).$$

We may substitute $x(w_1\beta) = x_{1,2k-4}$, since

$$w_1\beta = w_1(w_1w_2)^{k-1}\alpha_1 = w_2(w_1w_2)^{k-2}\alpha_1,$$

and since the corresponding reduced Weyl word has odd length, by Lemma 3.1 we have

$$x(w_1\beta) = x_{1,2k-3} = x_{1,2k-4}.$$

Hence, we obtain

$$\begin{aligned} x_{1,2k} &= (ab-1)x(\beta) - (x_{1,2k-4} + x(\beta)) \\ &= (ab-2)x(\beta) - x_{1,2k-4} \\ &= (ab-2)x_{1,2k-2} - x_{1,2k-4}, \end{aligned}$$

which is the recursion relation in (i). Thus k cannot be a counterexample, so $k \notin M$. But then M has no least element, contradicting our claim that M is nonempty. Thus, there are no counterexamples, and the recursion holds for all $m \in \mathbb{Z}_{\geq 0}$. □

Similar recursions for rank 2 Kac-Moody root systems are given in [6, Exercises in Ch. 5]. We also refer the reader to
http://www.tanyakhovanova.com/RecursiveSequences/RecursiveSequences.html
for a detailed discussion on recursions $a(n) = d \cdot a(n-1) - a(n-2)$.

We adopt the convention $y_{1,-1} = x_{2,-1} = 0$.

For each pair a, b, where $ab \geq 4$, we obtain the following integer sequences

$$\begin{aligned} X_1(a,b) &:= (x_{1,2m})_{m \in \mathbb{Z}_{\geq 0}} = \{1, ab-1, a^2b^2 - 3ab + 1, \dots\}, \\ Y_1(a,b) &:= (y_{1,2m})_{m \in \mathbb{Z}_{\geq 0}} = \{0, b, ab^2 - 2b, a^2b^3 - 4ab^2 + 3b, \dots\}, \\ X_2(a,b) &:= (x_{2,2m})_{m \in \mathbb{Z}_{\geq 0}} = \{0, a, a^2b - 2a, a^3b^2 - 4a^2b + 3a, \dots\}, \\ Y_2(a,b) &:= (y_{2,2m})_{m \in \mathbb{Z}_{\geq 0}} = \{1, ab-1, a^2b^2 - 3ab + 1, \dots\}. \end{aligned}$$

In Theorem 3.3(i) and (iii) below we prove that $Y_2(a,b) = X_1(a,b)$ and that $X_2(a,a) = Y_1(a,a)$. In particular we can define the sequence:

$$\begin{aligned} X(a) &= \{x_{2,2m}(a,a), y_{2,2m}(a,a)\}_{m\in\mathbb{Z}_{\geq 0}} = \{y_{1,2m}(a,a), x_{1,2m}(a,a)\}_{m\in\mathbb{Z}_{\geq 0}} \\ &= \{0, 1, a, a^2-1, a^3-2a, a^4-3a^2+1, \dots\}. \end{aligned}$$

Note that $X(a)$ is distinct from $X_i(a,a)$ and $Y_i(a,a)$, $i = 1, 2$. In fact for each $a \geq 2$, $X_i(a,a)$ and $Y_i(a,a)$ are each bisections of $X(a)$. It can also be shown that $X(a) = X_2(1, a+2)$.

Theorem 3.3. *For each pair a, b, $ab \geq 4$, we have*

(i) $X_1(a,b) = X_1(b,a) = Y_2(a,b) = Y_2(b,a)$.

(ii) *Let $k \geq 4$. If $k = a_i b_i$ for some $a_i, b_i \in \mathbb{Z}_{\geq 1}$, $i = 1, \dots, t$, then*

$$X_1(a_1,b_1) = X_1(a_2,b_2) = \cdots = X_1(a_t,b_t),$$

and in particular $X_1(a,b) = X_1(b,a) = X_1(ab,1) = X_1(1,ab)$.

(iii) $bX_2(a,b) = aY_1(a,b)$ *and* $X_2(a,b) = Y_1(b,a)$. *In particular* $X_2(a,a) = Y_1(a,a)$.

(iv) $X_2(1,ab) = Y_1(ab,1)$.

Proof. Part (i) follows from Lemma 3.2: Observe that the recursion relations for $x_{1,2m}$ and $y_{2,2m}$ share the same initial values $\{1, ab-1\}$ and recursive formula, namely $d(n) = (ab-2)d(n-1) - d(n-2)$. Since this data completely determines any recursive sequence, we have $X_1(a,b) = Y_2(a,b)$. Moreover, since $ab = ba$, Lemma 3.2 parts (i) and (iv) are invariant under permuting a and b. Thus, $x_{1,2m}(a,b) = x_{1,2m}(b,a)$ and $y_{2,2m}(a,b) = y_{2,2m}(b,a)$, or to summarize in our sequence notation, $X_1(a,b) = X_1(b,a) = Y_2(a,b) = Y_2(b,a)$. Since the polynomial $x_{1,2m}(a,b)$ depends only on the product ab, part (ii) also follows. For part (iii) we use Lemma 3.2 and do induction on m. For $m = 0$ and $m = 1$ we have

$$b \cdot x_{2,0} = b \cdot 0 = 0 = a \cdot 0 = a \cdot y_{1,0}$$

and

$$b \cdot x_{2,2} = ba = ab = a \cdot y_{1,2}.$$

For $m \geq 2$ assume that $b \cdot x_{2,2m} = a \cdot y_{1,2m}$ for all $m < M$ for some $M \in \mathbb{Z}_{\geq 2}$. Then by the induction hypothesis,

$$b \cdot x_{2,2M} = (ab-2)b \cdot x_{2,2M-2} - b \cdot x_{2,2M-4} = (ab-2)a \cdot y_{1,2M-2} - a \cdot y_{1,2M-4} = a \cdot y_{1,2M},$$

and so $bX_2(a,b) = aY_1(a,b)$. The second equality $X_2(a,b) = Y_1(b,a)$ follows from observing that the recursion relations in Lemma 3.2(ii) and (iii) are identical when interchanging a and b. Lastly, (iv) follows from (iii). □

Remark 3.4. The sequences above yield many known sequences as well as some new ones. We highlight the following:

(i) $X_1(1,5) = Y_2(5,1)$ is the bisected Lucas sequence $(L_{2s+1})_{s\geq 0} = \{1, 4, 11, \ldots\}$.

(ii) $X_2(1,5) = Y_1(5,1) = X(3)$ is the bisected Fibonacci sequence $(F_{2s})_{s\geq 0} = \{0, 1, 3, 8, \ldots\}$.

(iii) The sequence of squares of elements of $X(6)$ is a sequence of triangular numbers.

Additional known sequences are given in the tables in Section 10

The following theorem reveals that for each pair a, b, $ab \geq 4$, our sequences are a generalization of the Lucas and Fibonacci sequences and the relationship between them.

Theorem 3.5. *For each pair a, b, $ab \geq 4$ and $n > 0$ we have*

(i) $x_{1,n}(1, ab) = x_{2,n-1}(1, ab) + x_{2,n+1}(1, ab)$,

(ii) $y_{1,n}(1, ab) = x_{1,n-1}(1, ab) + x_{1,n+1}(1, ab)$,

(iii) *When $a = 1$, $b = 5$, the relation in* (i) encodes the relationship $L_{2s+1} = F_{2s} + F_{2s+2}$ that gives the Lucas number L_{2s+1} as the sum of consecutive terms in the bisected Fibonacci sequence F_{2s}.

Proof. The proofs of (i) and (ii) are straightforward. For (iii) we note that Part (i) and Remark 3.4 parts (i) and (ii) give us:

$$L_{2s+1} = x_{1,2s}(1,5) = x_{1,2s+1}(1,5) = x_{2,2s}(1,5) + x_{2,2s+2}(1,5) = F_{2s} + F_{2s+2}. \quad \square$$

Remark 3.6. Since $\beta = x_{1,n}\alpha_1 + y_{1,n}\alpha_2$ is a positive root, we have $x_{1,n} \geq 0$ and $y_{1,n} \geq 0$ for all n. Moreover we have $x_{1,n} = 0$ if and only if $y_{1,n}\alpha_2$ is a positive root, and since α_2 is a real root this is equivalent to $y_{1,n} = 1$. We conclude that $x_{1,n} > 0$ for $n \geq 0$. Similarly we get $y_{1,n} = 0$ if and only if $\beta = \alpha_1$, so $y_{1,n} > 0$ for $n \geq 1$. The same arguments show that $x_{2,n} > 0$ for $n \geq 1$ and $y_{2,n} > 0$ for $n \geq 0$.

For later use, we also prove that the four sequences are increasing. Consider the sequence $\{x_{1,n}\}_{n\geq 0}$. Since $x_{1,2m} = x_{1,2m+1}$ for $m \geq 0$ by Lemma 3.1 it suffices to prove that $x_{1,2m+2} \geq x_{1,2m}$ for $m \geq 0$. By Lemma 3.2(i) we have $x_{1,0} = 1$, $x_{1,2} = ab - 1$ and $x_{1,2m+2} = (ab-2)x_{1,2m} - x_{1,2m-2}$ for $m \geq 1$. Hence the inequality holds for $m = 0$ (since $ab \geq 4$) and if $x_{1,2m} \geq x_{1,2m-2}$ then

$$x_{1,2m+2} = (ab-3)x_{1,2m} + (x_{1,2m} - x_{1,2m-2}) \geq 0,$$

since $ab \geq 4$ and $x_{1,2m} \geq 0$. This proves that $\{x_{1,n}\}_{n\geq 0}$ is increasing. The proof for the three other sequences is identical (again using Lemmas 3.1 and 3.2).

We adopt the terminology used in the literature that Chebyshev polynomials of the first and second kind are denoted by T- and U-polynomials, respectively. These are given by the following recurrence relations:

$$T_n(x) = \begin{cases} 1, & n = 0 \\ x, & n = 1 \\ 2xT_{n-1}(x) - T_{n-2}(x), & n \geq 2, \end{cases}$$

$$U_n(x) = \begin{cases} 1, & n = 0 \\ 2x, & n = 1 \\ 2xU_{n-1}(x) - U_{n-2}(x), & n \geq 2, \end{cases}$$

We also adopt the convention that $U_{-1}(x) = 0$. It is not hard to show from the above recursions that $T_n(\cos(\theta)) = \cos(n\theta)$ and $U_n(\cos(\theta)) = \frac{\sin((n+1)\theta)}{\sin(\theta)}$ for all n and θ (with $\sin(\theta) \neq 0$ in the last case).

Viewing the sequence entries as polynomials in a and b, we show that they are in fact scalar multiples of Chebyshev U-polynomials.

Proposition 3.7. *For each $m \in \mathbb{Z}_{\geq 0}$, $a, b \in \mathbb{Z}_{\geq 1}$, $ab \geq 4$ we have*

$$x_{1,2m}(a,b) = y_{2,2m}(a,b) = U_{2m}\left(\frac{\sqrt{ab}}{2}\right),$$

$$y_{1,2m-1}(a,b) = \sqrt{\frac{b}{a}} \cdot U_{2m-1}\left(\frac{\sqrt{ab}}{2}\right),$$

and

$$x_{2,2m-1}(a,b) = \sqrt{\frac{a}{b}} \cdot U_{2m-1}\left(\frac{\sqrt{ab}}{2}\right),$$

where $U_n\left(\frac{\sqrt{ab}}{2}\right)$ is the Chebyshev U-polynomial $U_n(x)$ evaluated at $x = \frac{\sqrt{ab}}{2}$.

Proof. We first determine the recurrence relations for both the even and odd bisections of the Chebyshev U-polynomials. For initial values, we note that $U_1(x) = 2x$ and $U_2(x) = (2x)(2x) - 1 = 4x^2 - 1$. To find the recurrence formula for $n \geq 3$, we apply the convention $U_{-1}(x) = 0$ to find

$$U_n(x) = (2x)U_{n-1}(x) - U_{n-2}(x) = (2x)[(2x)U_{n-2}(x) - U_{n-3}(x)] - U_{n-2}(x)$$
$$= (4x^2 - 1)U_{n-2}(x) - (2x)U_{n-3}(x) = (4x^2 - 1)U_{n-2}(x) - [U_{n-2}(x) + U_{n-4}(x)]$$
$$= (4x^2 - 2)U_{n-2}(x) - U_{n-4}(x).$$

Thus, evaluating both the even and odd bisections at $x = \frac{\sqrt{ab}}{2}$ gives us the recursion relations

$$U_{2m}\left(\frac{\sqrt{ab}}{2}\right) = \begin{cases} 1, & m = 0, \\ ab - 1, & m = 1, \\ (ab-2)U_{2m-2}(\frac{\sqrt{ab}}{2}) - U_{2m-4}(\frac{\sqrt{ab}}{2}), & m \geq 2, \end{cases}$$

which is $x_{1,2m}$ by Lemma 3.2(i), and

$$U_{2m-1}\left(\frac{\sqrt{ab}}{2}\right) = \begin{cases} 0, & m = 0, \\ \sqrt{ab}, & m = 1, \\ (ab-2)U_{2m-3}(\frac{\sqrt{ab}}{2}) - U_{2m-5}(\frac{\sqrt{ab}}{2}), & m \geq 2, \end{cases}$$

which is $\sqrt{\frac{a}{b}} \cdot y_{1,2m-1}$ by Lemma 3.2(ii), and $\sqrt{\frac{b}{a}} \cdot x_{2,2m-1}$ by Lemma 3.2(iii) (switching to odd indices by Lemma 3.1). □

Corollaries 3.8 and 3.9 below follow immediately from the familiar properties of Chebyshev polynomials, which are found in [14].

Corollary 3.8. *For each $m \geq 1$, $a, b \in \mathbb{Z}_{\geq 1}$, $ab \geq 4$, $x_{1,2m}$ is given by the $2m \times 2m$ determinant*

$$x_{1,2m}(a,b) = \begin{vmatrix} \sqrt{ab} & 1 & 0 & 0 & \dots & 0 & 0 \\ 1 & \sqrt{ab} & 1 & 0 & \dots & 0 & 0 \\ 0 & 1 & \sqrt{ab} & 1 & \dots & 0 & 0 \\ 0 & 0 & 1 & \sqrt{ab} & \dots & 0 & 0 \\ 0 & 0 & 0 & 1 & \dots & 1 & 0 \\ \vdots & \vdots & \vdots & \vdots & \dots & \sqrt{ab} & 1 \\ 0 & 0 & 0 & 0 & \dots & 1 & \sqrt{ab} \end{vmatrix}.$$

In the following, $\mathbb{Z}[ab]$ denotes polynomials in ab with $\mathbb{Z}$-coefficients.

Corollary 3.9. *Let $a, b \in \mathbb{Z}$, $ab \geq 4$. Then, we have:*

(i) *For $s \geq 0$,* $$x_{1,2s+1}(a,b) = y_{2,2s+1}(a,b) = \prod_{j=1}^{s} \left(ab - 4\cos^2\left(\frac{j\pi}{2s+1}\right)\right),$$

(ii) *For $s > 0$,* $$y_{1,2s}(a,b) = b \cdot \prod_{j=1}^{s-1} \left(ab - 4\cos^2\left(\frac{j\pi}{2s}\right)\right),$$

(iii) *For $s > 0$,* $$x_{2,2s}(a,b) = a \cdot \prod_{j=1}^{s-1} \left(ab - 4\cos^2\left(\frac{j\pi}{2s}\right)\right).$$

Moreover, since the sequence entries are polynomials in a and b, the right-hand sides of (i), (ii), and (iii) are polynomials in $\mathbb{Z}[ab]$, $b \cdot \mathbb{Z}[ab]$, and $a \cdot \mathbb{Z}[ab]$, respectively.

Proof. By Proposition 3.7, for n even, $x_{1,n}$ can be written as a Chebyshev U-polynomial. As shown in [14], for $n \geq 0$, the nth Chebyshev U-polynomial can be factored as

$$U_n(x) = 2^n \prod_{j=1}^{n} \left(x - \cos\left(\frac{j\pi}{n+1}\right)\right).$$

Setting $n = 2s$, and bringing 2^{2s} into the product, using that $\cos(\pi - x) = -\cos(x)$ gives

$$x_{1,2s} = x_{1,2s+1} = \prod_{j=1}^{s} \left(\sqrt{ab} - 2\cos\left(\frac{j\pi}{2s+1}\right)\right)\left(\sqrt{ab} + 2\cos\left(\frac{j\pi}{2s+1}\right)\right).$$

We now obtain (i) by pairing up conjugates:

$$x_{1,2s+1}(a,b) = \prod_{j=1}^{s} \left(ab - 4\cos^2\left(\frac{j\pi}{2s+1}\right)\right).$$

Similarly, setting $n = 2s-1$ ($s > 0$) gives us the following identity for $y_{1,2s-1}$:

$$y_{1,2s-1} = y_{1,2s} = \sqrt{\frac{b}{a}} \cdot \prod_{j=1}^{2s-1} \left(\sqrt{ab} - 2\cos\left(\frac{j\pi}{2s}\right)\right)$$

When $j = s$ we have the special factor $\left(\sqrt{ab} - 2\cos(\frac{\pi}{2})\right) = \sqrt{ab}$. Taking this out of the product allows us to pair up conjugates as before:

$$y_{1,2s} = \sqrt{\frac{b}{a}} \cdot \sqrt{ab} \prod_{j=1}^{\lfloor\frac{2s-1}{2}\rfloor} \left(\sqrt{ab} - 2\cos\left(\frac{j\pi}{2s}\right)\right)\left(\sqrt{ab} + 2\cos\left(\frac{j\pi}{2s}\right)\right)$$

$$y_{1,2s}(a,b) = b \cdot \prod_{j=1}^{s-1} \left(ab - 4\cos^2\left(\frac{j\pi}{2s}\right)\right).$$

Lastly, (iii) follows immediately from (ii) by Theorem 3.3(i). □

It is remarkable that the products of $\mathbb{R}$-linear factors shown in Corollary 3.9 will yield polynomials in $\mathbb{Z}[ab]$. Indeed, roots of the type shown in Corollary 3.9, namely $4\cos^2(\frac{j\pi}{n}) \in \mathbb{R}$, have shown similar striking properties in the past. Studies by [13] and [8] revealed that suitable products of these roots yield integers.

We now view the sequence entries as polynomials in a and b. An immediate consequence of Corollary 3.9 is that two such polynomials will share some of the same linear factors if their indices are not coprime. For example, $x_{1,21}$ and $x_{1,7}$ share the linear factor $(ab - 4\cos^2(\frac{\pi}{7}))$, and $y_{1,10}$ and $x_{1,15}$ share the linear factors $(ab - 4\cos^2(\frac{\pi}{5}))$ and $(ab - 4\cos^2(\frac{2\pi}{5}))$. This motivates the question of cancellation: will a ratio of such polynomials sharing a common $\mathbb{R}$-linear factor yield a polynomial in $\mathbb{Z}[ab]$? Indeed, [12] has shown that this is true for Chebyshev U-polynomials. In particular, [12] found that if two U polynomials divide with remainder zero (as in the case of $\frac{x_{1,21}}{x_{1,7}}$ or $\frac{y_{1,18}}{x_{1,9}}$), the resulting product of $\mathbb{R}$-linear factors is a composition of Chebyshev polynomials. It should be noted that striking properties such as this are not uncommon for Chebyshev polynomials, which is why they have found use in so many applications.

The proof of the following theorem uses Chebyshev T-polynomials (defined earlier), which when evaluated at $x = \sqrt{ab}/2$ give us the recurrence relation:

$$T_n(\sqrt{ab}/2) = \begin{cases} 1, & n = 0, \\ \sqrt{ab}/2, & n = 1, \\ \sqrt{ab} \cdot T_{n-1}(\sqrt{ab}/2) - T_{n-2}(\sqrt{ab}/2), & n \geq 2. \end{cases}$$

Theorem 3.10. *We have*

(i) *For all* $n \geq 0$ *and* $s > 0$, $\dfrac{y_{1,2ns}}{y_{1,2s}}, \dfrac{x_{2,2ns}}{x_{2,2s}} \in \mathbb{Z}[\sqrt{ab}/2]$,

(ii) *For odd* $n > 0$ *and* $s \geq 0$, $\dfrac{x_{1,n(2s+1)}}{x_{1,2s+1}} \in \mathbb{Z}[\sqrt{ab}/2]$,

(iii) *For even* $n \geq 0$ *and all* $s \geq 0$, $\dfrac{y_{1,n(2s+1)}}{x_{1,2s+1}}, \dfrac{x_{2,n(2s+1)}}{x_{1,2s+1}} \in \sqrt{\dfrac{b}{a}} \cdot \mathbb{Z}[\sqrt{ab}/2]$.

Proof. (i) Let $s, n \in \mathbb{Z}_{>0}$. By Proposition 3.7, $y_{1,2ns} = \sqrt{\frac{b}{a}} U_{2ns-1}$ and $y_{1,2s} = \sqrt{\frac{b}{a}} U_{2s-1}$. The paper [12] gives the divisibility criteria for two U polynomials $U_q(x)$ and $U_p(x)$ thus: $U_q(x)$ divides $U_p(x)$ if there exists an integer $n > 0$ such that $p = n(q+1) - 1$. Moreover, the U polynomials decompose as follows:

$$\frac{U_p(x)}{U_q(x)} = \frac{U_{n(q+1)-1}(x)}{U_{(q+1)-1}(x)} = U_{n-1}(T_{q+1}(x)).$$

Since the polynomial ring $\mathbb{Z}[x]$ is also a composition ring, composing any two polynomials in $\mathbb{Z}[x]$ will yield another polynomial in $\mathbb{Z}[x]$. Thus we have

$$\frac{y_{1,2ns}}{y_{1,2s}} = \frac{\sqrt{\frac{b}{a}} \cdot U_{n(2s)-1}(\sqrt{ab}/2)}{\sqrt{\frac{b}{a}} \cdot U_{(2s)-1}(\sqrt{ab}/2)} = U_{n-1}\left(T_{2s}\left(\frac{\sqrt{ab}}{2}\right)\right) \in \mathbb{Z}[\sqrt{ab}/2].$$

Similarly, we find that for n odd,

$$\frac{x_{1,n(2s+1)}}{x_{1,2s+1}} = \frac{U_{n(2s+1)-1}(\sqrt{ab}/2)}{U_{(2s)}(\sqrt{ab}/2)} = U_{n-1}\left(T_{2s+1}\left(\frac{\sqrt{ab}}{2}\right)\right) \in \mathbb{Z}[\sqrt{ab}/2],$$

and for n even,

$$\frac{y_{1,n(2s+1)}}{x_{1,2s+1}} = \frac{\sqrt{\frac{b}{a}} \cdot U_{n(2s+1)-1}(\sqrt{ab}/2)}{U_{(2s)}(\sqrt{ab}/2)}$$
$$= \sqrt{\frac{b}{a}} \cdot U_{n-1}\left(T_{2s+1}\left(\frac{\sqrt{ab}}{2}\right)\right) \in \sqrt{\frac{b}{a}} \cdot \mathbb{Z}[\sqrt{ab}/2].$$

Lastly, Theorem 3.3(iii) gives us that the remaining ratios are also in $\sqrt{\frac{b}{a}} \cdot \mathbb{Z}[\sqrt{ab}/2]$. □

It turns out that the expressions in the statement of Theorem 3.10 are integers. Indeed, we will prove that all ratios in Theorem 3.10 are polynomials in ab with coefficients in $\mathbb{Z}, a\mathbb{Z}$, or $b\mathbb{Z}$ (see Corollary 3.14).

We will use the following Lemmas (3.11, 3.12) to prove our Theorem.

Lemma 3.11. *For* $n > 0$ *and* $0 \leq i \leq \frac{n}{2}$ *we have* $\frac{n}{n-i}\binom{n-i}{i} \in \mathbb{Z}$.

Proof. This is clear for $i = 0$. For $i \geq 1$ we have

$$\frac{n}{n-i}\binom{n-i}{i} - \binom{n-i}{i} = \frac{i}{n-i}\binom{n-i}{i} = \frac{i}{n-i} \cdot \frac{(n-i)!}{i! \cdot (n-2i)!}$$
$$= \frac{(n-i-1)!}{(i-1)! \cdot (n-2i)!} = \binom{n-i-1}{i-1}.$$

It follows that $\frac{n}{n-i}\binom{n-i}{i} = \binom{n-i-1}{i-1} + \binom{n-i}{i} \in \mathbb{Z}$. □

Lemma 3.12. *For $m \geq 0$ we have $T_{2m}\left(\frac{x}{2}\right) \in \frac{1}{2}\cdot\mathbb{Z}[x^2]$, $T_{2m+1}\left(\frac{x}{2}\right) \in \frac{x}{2}\cdot\mathbb{Z}[x^2]$, $U_{2m}\left(\frac{x}{2}\right) \in \mathbb{Z}[x^2]$ and $U_{2m+1}\left(\frac{x}{2}\right) \in x\cdot\mathbb{Z}[x^2]$. In particular $U_n\left(\frac{x}{2}\right) \in \mathbb{Z}[x]$ for $n \geq 0$.*

Proof. The claim about T_0 is clear since $T_0(x) = 1$. For $n \geq 1$ we have

$$T_n(x) = \frac{n}{2}\sum_{i=0}^{\lfloor \frac{n}{2} \rfloor} \frac{(-1)^i}{n-i}\binom{n-i}{i}\cdot(2x)^{n-2i}$$

cf. [1, Formula 22.3.6]. This immediately implies that

$$T_n\left(\frac{x}{2}\right) = \frac{1}{2}\sum_{i=0}^{\lfloor \frac{n}{2} \rfloor}(-1)^i\cdot\frac{n}{n-i}\binom{n-i}{i}\cdot x^{n-2i}.$$

By Lemma 3.11 this sum belongs to $\mathbb{Z}[x^2]$ for n even and to $x\cdot\mathbb{Z}[x^2]$ for n odd. The result concerning the Chebyshev U-polynomials follows similarly from the formula

$$U_n(x) = \sum_{i=0}^{\lfloor \frac{n}{2} \rfloor}(-1)^i\binom{n-i}{i}\cdot(2x)^{n-2i} \tag{5}$$

for $n \geq 0$, cf. [1, Formula 22.3.7], except that one does not need Lemma 3.11 in this case. □

The proof of the following theorem follows directly from Lemma 3.12.

Theorem 3.13. *Let $n, s \geq 0$ be integers. Then*

(i) *For $s > 0$ we have $\frac{y_{1,2ns}}{y_{1,2s}} = \frac{x_{2,2ns}}{x_{2,2s}} \in \mathbb{Z}[ab]$.*

(ii) *For n odd, $\frac{x_{1,n(2s+1)}}{x_{1,2s+1}} \in \mathbb{Z}[ab]$.*

(iii) *For n even, $\frac{y_{1,n(2s+1)}}{x_{1,2s+1}} \in b\cdot\mathbb{Z}[ab]$ and $\frac{x_{2,n(2s+1)}}{x_{1,2s+1}} \in a\cdot\mathbb{Z}[ab]$.*

Corollary 3.14. *All ratios in Theorem 3.10 are polynomials in ab with coefficients in $\mathbb{Z}$, $a\mathbb{Z}$, or $b\mathbb{Z}$.*

Theorem 3.15. *For $s \geq 0$ we have closed form expressions*

(i) $x_{1,2s} = y_{2,2s} = x_{1,2s+1} = y_{2,2s+1} = \sum_{i=0}^{s}(-1)^i\binom{2s-i}{i}(ab)^{s-i}$,

and the coefficients in this summation (modulo sign) are the shallow diagonals

$$\binom{2s}{0}, \binom{2s-1}{1}, \binom{2s-2}{2}, \ldots, \binom{s}{s}$$

of Pascal's triangle.

(ii) $y_{1,2s} = y_{1,2s-1} = \sum_{i=0}^{s-1}(-1)^i \binom{2s-1-i}{i} a^{s-1-i} b^{s-i}$, *and*

$$x_{2,2s} = x_{2,2s-1} = \sum_{i=0}^{s-1}(-1)^i \binom{2s-1-i}{i} a^{s-i} b^{s-1-i},$$

and the coefficients in both summations (modulo sign) are the shallow diagonals

$$\binom{2s-1}{0}, \binom{2s-2}{1}, \binom{2s-3}{2}, \ldots, \binom{s}{s-1}$$

of Pascal's triangle.

(iii) *When* $a = b$, $s \geq 0$, *we have*

$$x_{2,2s} = x_{2,2s-1} = y_{1,2s} = y_{1,2s-1} = \sum_{i=0}^{s-1}(-1)^i \binom{2s-1-i}{i} a^{2s-1-2i},$$

$$y_{2,2s} = y_{2,2s+1} = x_{1,2s} = x_{1,2s+1} = \sum_{i=0}^{s}(-1)^i \binom{2s-i}{i} a^{2s-2i}.$$

Proof. By Proposition 3.7 combined with (5) we have

$$x_{1,2s} = U_{2s}(\sqrt{ab}/2) = \sum_{i=0}^{\lfloor 2s/2 \rfloor} (-1)^i \binom{2s-i}{i} \left(2 \cdot \sqrt{ab}/2\right)^{2s-2i}$$
$$= \sum_{i=0}^{s}(-1)^i \binom{2s-i}{i} \cdot (ab)^{s-i}$$

proving (i). The proof of (ii) is similar, and part (iii) follows from (i) and (ii). □

We remark that we do not recognize the expressions involving shallow diagonals in Pascal's triangle, except for the familiar formula for Fibonacci numbers that arises in $x_{1,2s}$ when $ab = 5$. We expect that this sum can be found in the literature, though we did not find a reference.

As we have shown that our sequence polynomials can be written as both products and sums, we now explore the consequence of relating Corollary 3.9 with Theorem 3.15. Denote the root $4\cos^2\left(\frac{j\pi}{n}\right)$ by $4r_j$, and let $k = ab$. Expanding the $x_{1,2s+1}$ product as in Corollary 3.9, we get (by the binomial theorem),

$$k^s - 4\left(\sum_{f=1}^{s} r_f\right) k^{s-1} + 4^2\left(\sum_{f<g}^{s} r_f r_g\right) k^{s-2} - 4^3\left(\sum_{f<g<h}^{s} r_f r_g r_h\right) k^{s-3} + \cdots \pm 4^s \prod_{j=1}^{s} r_j.$$

We may write the polynomials for $y_{1,2s}$ (resp. $x_{2,2s}$) in the same fashion, remembering the extra factor of b (resp. a).

We recognize the ith coefficient as 4^i times the ith elementary symmetric polynomial, $e_i(r_1, r_2, \ldots, r_s)$.

By Theorem 3.15, all x_j and y_j can be written in closed form as polynomials in a and b with coefficients from shallow diagonals of Pascal's Triangle. Since $k = ab$, these closed forms and the polynomials above are identical. Hence, the coefficients above are in fact binomial coefficients, and are therefore integers. This gives us the new identity (in the notation of elementary symmetric polynomials):

Lemma 3.16.

$$\binom{u}{v} = 4^v e_v\left(\cos^2\left(\frac{\pi}{u+v+1}\right), \cos^2\left(\frac{2\pi}{u+v+1}\right), \ldots, \cos^2\left(\frac{\lfloor \frac{u+v}{2} \rfloor \pi}{u+v+1}\right)\right),$$

where $u, v \in \mathbb{Z}_{\geq 0}$.

(See [8], [13] where the number of tilings of a square with dominos is given by a similar expression). The following theorem follows easily from Proposition 3.7.

Theorem 3.17. *Suppose that* $a, b \in \mathbb{Z}_{>0}$ *with* $ab \geq 4$. *Then*

(i) $aX_2(1, ab) = X_2(a, b)$.

(ii) *More generally, let* $q \in \mathbb{Q}_{>0}$. *Then if* $qa, b/q \in \mathbb{Z}$, *we have* $qX_2(a, b) = X_2(qa, b/q)$.

Furthermore, (ii) *allows us to generalize* (i) *as:*

(i)$'$ $qX_2(1, ab) = X_2(u, v)$ *for some* $u, v \in \mathbb{Z}_{\geq 1}$, $uv \geq 4$, *if and only if* q *is a positive integer dividing* ab.

4. Hyperbolic Golden Ratios and Real Quadratic Fields

There are well-documented identities relating Fibonacci numbers to the classical golden ratio, $\phi = \frac{1+\sqrt{5}}{2} = 1.61803399\ldots$, such as Binet's formula which expresses the n-th Fibonacci number, $F(n)$, as the linear combination

$$F(n) = \frac{\phi^n - \phi'^n}{\sqrt{5}},$$

where ϕ' is the conjugate of ϕ. In Theorem 3.5 we showed that, for each $k = ab \geq 4$, sequences $X_1(1, k)$ and $X_2(1, k)$ are generalizations of the bisected Lucas and Fibonacci sequences, respectively. In this section, we investigate our sequences further to determine if there are other generalizations of Fibonacci-type identities. We begin by recognizing that the ratios of consecutive bisected Fibonacci and Lucas numbers both tend to the square of the golden ratio,

$$\lim_{m \longrightarrow \infty} \frac{F(2m+2)}{F(2m)} = \lim_{m \longrightarrow \infty} \frac{L(2m+2)}{L(2m)} = \phi^2 = 2.61803399\ldots$$

This classical number, sometimes referred to by ψ and which we will call the "hyperbolic golden ratio", can indeed be generalized.

Proposition 4.1. *Let $a, b \in \mathbb{Z}_{>0}$ with $k = ab \geq 4$. Then the sequences $\left\{\frac{x_{1,n+2}(a,b)}{x_{1,n}(a,b)}\right\}_{n\geq 0}$, $\left\{\frac{x_{2,n+2}(a,b)}{x_{2,n}(a,b)}\right\}_{n\geq 1}$, $\left\{\frac{y_{1,n+2}(a,b)}{y_{1,n}(a,b)}\right\}_{n\geq 1}$ and $\left\{\frac{y_{2,n+2}(a,b)}{y_{2,n}(a,b)}\right\}_{n\geq 0}$ all converge to $\Psi(k) = \frac{1}{2}[(k-2) + \sqrt{k(k-4)}]$.*

Proof. By Remark 3.6, all numerators and denominators in the fractions above are positive and in particular the fractions are well defined. We first prove that the sequence $\left\{\frac{x_{1,n+2}}{x_{1,n}}\right\}_{n\geq 0}$ is decreasing. Since all entries are positive, this implies that it is convergent. Thus we have to show that for $n \geq 0$ we have $\frac{x_{1,n+2}}{x_{1,n}} \geq \frac{x_{1,n+3}}{x_{1,n+1}}$. For n even the numerators and denominators agree, so it suffices to consider the case $n = 2m-1, m \geq 1$. Since $\frac{x_{1,n+2}}{x_{1,n}} = \frac{x_{1,2m}}{x_{1,2m-2}}$ we have to prove the inequality $\frac{x_{1,2m}}{x_{1,2m-2}} \geq \frac{x_{1,2m+2}}{x_{1,2m}}$ for all $m \geq 1$. We prove the inequality true by induction on m. Since k is positive, we have

$$\frac{x_{1,2}}{x_{1,0}} = \frac{k-1}{1} = \frac{k^2-2k+1}{k-1} \geq \frac{k^2-3k+1}{k-1} = \frac{x_{1,4}}{x_{1,2}},$$

and so the base case holds true. Assume the inequality holds for all $m \leq M$ for some $M \in \mathbb{Z}_{>0}$. Using Lemma 3.2 part (i), we see that

$$\frac{x_{1,2M+2}}{x_{1,2M}} = \frac{(k-2)x_{1,2M} - x_{1,2M-2}}{x_{1,2M}} = k - 2 - \frac{x_{1,2M-2}}{x_{1,2M}}.$$

Multiplying the induction inequality by $-\frac{x_{1,2m-2}}{x_{1,2m+2}}$ (recalling positivity of both numerator and denominator), we see that $-\frac{x_{1,2m}}{x_{1,2m+2}} \leq -\frac{x_{1,2m-2}}{x_{1,2m}}$, which gives us

$$\frac{x_{1,2M+2}}{x_{1,2M}} \geq k - 2 - \frac{x_{1,2M}}{x_{1,2M+2}} = \frac{(k-2)x_{1,2M+2} - x_{1,2M}}{x_{1,2M+2}} = \frac{x_{1,2M+4}}{x_{1,2M+2}},$$

and so the inequality holds true for all $m \in \mathbb{Z}_{>0}$.

This shows that $\left\{\frac{x_{1,n+2}}{x_{1,n}}\right\}_{n\geq 0}$ is convergent. Similarly one proves (using the relevant parts of Lemma 3.2) that the three other sequences are all decreasing and hence convergent.

By Lemma 3.2(i) we have $\frac{x_{1,2m}}{x_{1,2m-2}} = ab - 2 - \left(\frac{x_{1,2m-2}}{x_{1,2m-4}}\right)^{-1}$ and hence $\lambda = \lim_{n\to\infty} \frac{x_{1,n+2}}{x_{1,n}}$ satisfies $\lambda = ab - 2 - \lambda^{-1}$. Since $x_{1,n+2} \geq x_{1,n}$ by Remark 3.6 we get $\lambda \geq 1$ so $\lambda = \Psi(k) = \frac{1}{2}[(k-2) + \sqrt{k(k-4)}]$ as claimed. The proof that the remaining quotients also converge to $\Psi(k)$ is identical. □

We now give two continued fraction expansions of $\Psi(k)$.

Proposition 4.2. *For $k > 4$ we have*

$$\Psi(k) = (k-2) - \cfrac{1}{(k-2) - \cfrac{1}{(k-2) - \cfrac{1}{(k-2) - \cdots}}}$$

$$= (k-3) + \cfrac{1}{1 + \cfrac{1}{(k-4) + \cfrac{1}{1 + \cfrac{1}{(k-4) + \cfrac{1}{1+\cdots}}}}}$$

Proof. This follows directly from [5, Theorem 3.1]. □

The classical hyperbolic golden ratio ψ then corresponds to the special case $k = 5$. Further generalizations of identities involving ψ and ϕ are also found. Set

$$\Psi'(k) = \frac{1}{2}[(k-2) - \sqrt{k(k-4)}].$$

then $\Psi(k)$ and $\Psi'(k)$ are the solutions of the quadratic equation

$$x^2 - (k-2)x + 1 = 0.$$

Note that for $k > 4$ we have $(k-3)^2 < k(k-4) < (k-2)^2$, so $\Psi(k)$ and $\Psi'(k)$ are algebraic integers in the real quadratic field $\mathbb{Q}(\sqrt{d})$, where d denotes the squarefree part of $k(k-4)$. Furthermore $\Psi(k)$ satisfies

$$\Psi(k) = (k-2) - \frac{1}{\Psi(k)}.$$

Multiplying this equation by $\Psi(k)^n$ for any $n \in \mathbb{Z}$ gives the identity

$$\Psi(k)^{n+1} = (k-2)\Psi(k)^n - \Psi(k)^{n-1}. \tag{6}$$

Since $\Psi(k)\Psi'(k) = 1$, $\Psi(k)$ and $\Psi'(k)$ are units in the ring of algebraic integers in $\mathbb{Q}(\sqrt{d})$. Now let $\Phi(k) = \Psi(k) - 1$ (a clear analog of the golden ratio ϕ). Then

$$\Phi(k) = \frac{1}{2}[(k-4) + \sqrt{k(k-4)}].$$

If we set

$$\Phi'(k) = \frac{1}{2}[(k-4) - \sqrt{k(k-4)}],$$

then $\Phi'(k)\Phi(k) = -(k-4)$, $\Psi(k) - \Psi'(k) = \Phi(k) - \Phi'(k) = \sqrt{k(k-4)}$ and $\Phi(k)$ and $\Phi'(k)$ are the solutions of the quadratic equation

$$x^2 - (k-4)x - (k-4) = 0,$$

so $\Phi(k)$ and $\Phi'(k)$ are algebraic integers in the real quadratic field $\mathbb{Q}(\sqrt{d})$ and $\Phi(k)$ is the continued fraction

$$\Phi(k) = (k-4)[1 + \frac{1}{\Phi(k)}].$$

As before, multiplying this equation by $\Phi(k)^n$ for any $n \in \mathbb{Z}$ gives the identity

$$\Phi(k)^{n+1} = (k-4)[\Phi(k)^n + \Phi(k)^{n-1}]. \tag{7}$$

Substituting $k = 5$ recovers the well-known identity relating powers of the classical golden ratio, ϕ.

The following is a generalization of Binet's formula for the Fibonacci numbers.

Proposition 4.3. *Let $a, b \in \mathbb{Z}_{>0}$ with $k = ab > 4$. For $m \geq 1$ we have*

$$\begin{aligned} x_{1,2m-1}(a,b) = y_{2,2m-1}(a,b) &= \frac{1}{\Phi(k)}\Psi(k)^m + \frac{1}{\Phi'(k)}\Psi'(k)^m \\ &= \frac{\Phi(k)^{2m-1} + \Phi'(k)^{2m-1}}{(k-4)^m}, \end{aligned} \tag{8}$$

and for $m \geq 0$ we have

$$\begin{aligned} y_{1,2m}(a,b) &= \sqrt{\frac{b}{a(ab-4)}} \cdot (\Psi(k)^m - \Psi'(k)^m) \\ &= \sqrt{\frac{b}{a(ab-4)}} \cdot \frac{\Phi(k)^{2m} - \Phi'(k)^{2m}}{(k-4)^m}, \end{aligned} \tag{9}$$

$$\begin{aligned} x_{2,2m}(a,b) &= \sqrt{\frac{a}{b(ab-4)}} \cdot (\Psi(k)^m - \Psi'(k)^m) \\ &= \sqrt{\frac{a}{b(ab-4)}} \cdot \frac{\Phi(k)^{2m} - \Phi'(k)^{2m}}{(k-4)^m}. \end{aligned} \tag{10}$$

Proof. From Lemma 3.1 and Lemma 3.2 it follows that the sequence $\{x_{1,2m-1}\}_{m\geq 1} = \{y_{2,2m-1}\}_{m\geq 1}$ satisfies the recurrence $X(m) = (ab-2)X(m-1) - X(m-2)$, $m \geq 3$. The set of solutions to this recurrence is clearly a two dimensional vector space, and if $\lambda^2 = (ab-2)\lambda - 1$, then the sequence $\{\lambda^m\}_{m\geq 1}$ is a solution. Since the solutions to this equation are given by $\lambda = \Psi(k)$ and $\lambda = \Psi'(k)$, it follows that $\{\Psi(k)^m\}_{m\geq 1}$ and $\{\Psi'(k)^m\}_{m\geq 1}$ form a basis for the set of solutions to the recurrence relation. Hence there exists unique constants c_1 and c_2 such that $x_{1,2m-1} = y_{2,2m-1} = c_1\Psi(k)^m + c_2\Psi'(k)^m$ for $m \geq 1$. The values of the constants are easily computed from the values for $m = 1$ and $m = 2$. The second formula in (8) follows from the first using the relation $\Psi(k) = \Phi(k)^2/(k-4)$ and its conjugate $\Psi'(k) = \Phi'(k)^2/(k-4)$. The identities (9) and (10) are proved similarly. □

Now let $a = 1$ and $b = 5$. Then $k = 5$ and

$$\Phi(5) = \frac{1}{2}[1 + \sqrt{5}]$$

is the classical golden ratio $\phi = 1.61803399\ldots$, and as we have already seen, $\Psi(5)$ is the classical golden ratio $\psi = 1 + \phi$ for the bisected Fibonacci sequence. Proposition 4.3

and equations (6), (7), and (10) are generalizations of identities and recurrences of these classical objects.

The following lemma follows directly from Proposition 4.1.

Lemma 4.4. *For* $i = 1, 2$, Ψ *is the limit of the difference of* x *coordinates for adjacent pairs of real roots:*

$$\Psi(k) = \lim_{n \longrightarrow \infty} \frac{x_{i,n+2} - x_{i,n}}{x_{i,n} - x_{i,n-2}}.$$

A similar identity holds for $y_{1,n}$ and $y_{2,n}$. This lemma shows that the limit of the ratio of the differences of x-coordinates of adjacent pairs of real roots is the golden ratio $\Psi(k)$. This can also be used to show that for each $k = ab \geq 4$, $\Psi(k)$ is the limit of ratios of areas of hyperbolic triangles spanned by a triple of adjacent roots in the root system for a rank 2 Kac-Moody algebra. This configuration of hyperbolic triangles is depicted in Fig. 1.

5. Eigenvalues of Weyl Transformations

Let $w \in W$ with $n = l(w)$. If

$$w = \begin{cases} (w_1 w_2)^m, & n = 2m, m \in \mathbb{Z}_{>0}, \\ w_2(w_1 w_2)^m, & n = 2m+1, m \in \mathbb{Z}_{>0}, \end{cases}$$

then w has matrix

$$M_w = \begin{pmatrix} x_{1,n} & -x_{2,n-1} \\ y_{1,n} & -x_{1,n-1} \end{pmatrix}.$$

If

$$w = \begin{cases} (w_2 w_1)^m, & n = 2m, m \in \mathbb{Z}_{>0}, \\ w_1(w_2 w_1)^m, & n = 2m+1, m \in \mathbb{Z}_{>0}, \end{cases}$$

then w has matrix

$$M_w = \begin{pmatrix} -x_{1,n-1} & x_{2,n} \\ -y_{1,n-1} & x_{1,n} \end{pmatrix}.$$

In both cases, the characteristic polynomial is

$$\lambda(n)^2 - (x_{1,n} - x_{1,n-1})\lambda(n) + \det(M_w).$$

Since w is a product of the simple Weyl reflections w_1 and w_2 we obtain

$$\det(M_w) = (-1)^{l(w)} = (-1)^n.$$

Moreover $x_{1,n} - x_{1,n-1} = 0$ when n is odd, so the characteristic polynomial becomes

$$\begin{cases} \lambda(n)^2 - 1, & n \text{ odd}, \\ \lambda(n)^2 - (x_{1,n} - x_{1,n-2})\lambda(n) + 1, & n \text{ even}. \end{cases}$$

It follows that $\lambda(n) = \pm 1$ when n is odd, and when $n = 2$

$$\lambda(2) = \frac{1}{2}[(ab-2) \pm \sqrt{ab(ab-4)}] = \Psi(ab) \text{ or } \Psi'(ab).$$

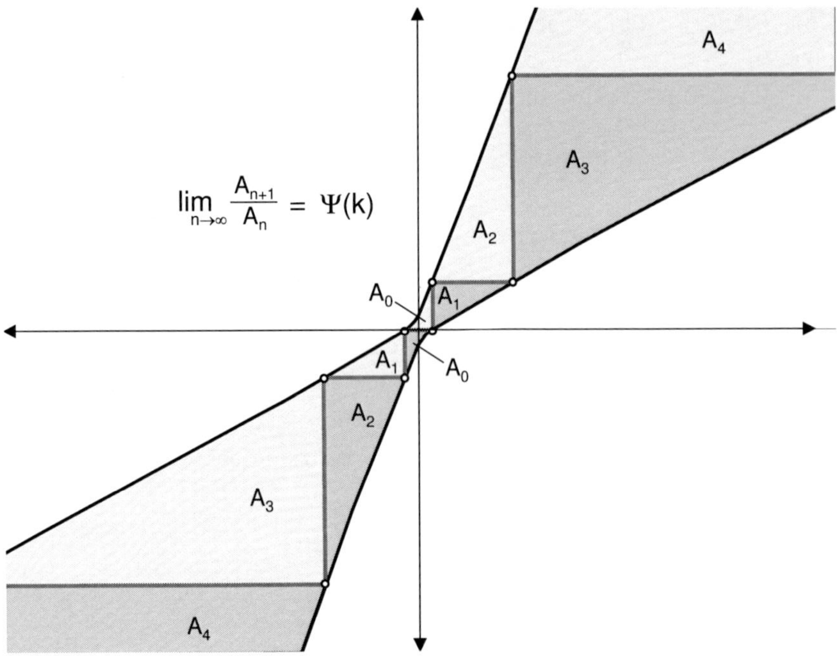

Figure 1.

A basic result from linear algebra shows that for $n = 2m$, $m \in \mathbb{Z}_{>0}$,

$$\lambda(2m) = \Psi(ab)^m \text{ or } \Psi'(ab)^m.$$

Hence for each $k = ab \geq 4$, $m \in \mathbb{Z}_{>0}$, we obtain:

$\Psi(k)^m$ and $\Psi'(k)^m$ are the eigenvalues for $(w_1w_2)^m$ and $(w_2w_1)^m$, respectively.

We believe that for a linear recursion in $\mathbb{R}^2$, it is known that the eigenvalues are quadratic units when the matrix of the recursion has determinant 1 or -1, though we did not find a suitable reference.

6. Identities with Continued Fractions

We now fix a real quadratic field $\mathbb{Q}(\sqrt{d})$, where $d \in \mathbb{Z}_{\geq 2}$ is squarefree. Let $\mathcal{O}$ be the ring of algebraic integers in $\mathbb{Q}(\sqrt{d})$. For $k > 4$, let $\Psi(k) = \frac{1}{2}[(k-2)+\sqrt{k(k-4)}]$ as in Section 4 If $\Psi(k) \in \mathbb{Q}(\sqrt{d})$, we showed that $\Psi(k)$ is a unit in $\mathcal{O}$. By Dirichlet's unit theorem, $\Psi(k)$ must be a power η^n of the fundamental unit η of $\mathbb{Q}(\sqrt{d})$, cf. [3]. The following theorem shows that the only possible values for n are $n = 1$ and 2.

Theorem 6.1. *Let $d \in \mathbb{Z}_{\geq 2}$ be squarefree. Then there exists an integer $k > 4$ such that $\Psi(k) \in \mathbb{Q}(\sqrt{d})$, where $\Psi(k) = \frac{1}{2}[(k-2)+\sqrt{k(k-4)}]$ as above. If k is the smallest such integer, then $\Psi(k) = \eta$ or η^2, where η is the fundamental unit in $\mathbb{Q}(\sqrt{d})$.*

Proof. The fundamental unit of $\mathbb{Q}(\sqrt{d})$ has the form $\eta = x + y\omega(d)$ where $x, y \in \mathbb{Z}_{\geq 0}$ and

$$\omega(d) = \begin{cases} \frac{1}{2}(1+\sqrt{d}) & \text{if } d \equiv 1 \pmod 4, \\ \sqrt{d} & \text{otherwise.} \end{cases}$$

Clearly the norm of η, $N(\eta) = \eta\bar{\eta}$ equals ± 1. Note that $N(\eta) = x^2 - dy^2$ when $d \not\equiv 1 \pmod 4$ and that $N(\eta) = \frac{1}{4}\left((2x+y)^2 - dy^2\right)$ for $d \equiv 1 \pmod 4$. We consider the following cases:

- $N(\eta) = 1$ and $d \not\equiv 1 \pmod 4$: With $k' = 2x + 2$ we have

$$\Psi(k') = \frac{1}{2}\left(2x + \sqrt{(2x+2)(2x-2)}\right) = x + \sqrt{x^2-1} = x + \sqrt{dy^2} = \eta$$

  since $x^2 - dy^2 = 1$.

- $N(\eta) = 1$ and $d \equiv 1 \pmod 4$: A similar computation shows that with $k' = 2x + y + 2$ we have $\Psi(k') = \eta$.

- $N(\eta) = -1$ and $d \not\equiv 1 \pmod 4$: Taking $k' = 4x^2 + 4$ we get

$$\begin{aligned}\Psi(k') &= \frac{1}{2}\left(4x^2 + 2 + \sqrt{(4x^2+4)\cdot 4x^2}\right) \\ &= 2x^2 + 1 + 2x\sqrt{x^2+1} = (x^2 + dy^2) + 2x\sqrt{dy^2} \\ &= (x^2+dy^2) + 2xy\sqrt{d} = \eta^2,\end{aligned}$$

  using that $x^2 - dy^2 = -1$.

- $N(\eta) = -1$ and $d \equiv 1 \pmod 4$: It is easy to verify that with $k' = (2x+y)^2 + 4$ we have $\Psi(k') = \eta^2$.

To sum up we have shown that there exists an integer $k' > 4$ such that

$$\Psi(k') = \begin{cases} \eta & \text{for } N(\eta) = 1, \\ \eta^2 & \text{for } N(\eta) = -1. \end{cases}$$

Now let $k > 4$ be the smallest integer such that $\Psi(k) \in \mathbb{Q}(\sqrt{d})$. Since $\Psi(k)$ is a unit in $\mathcal{O}$ we must have $\Psi(k) = \eta^n$ for some $n \geq 1$, and clearly n must be even if $N(\eta) = -1$. It follows that $\Psi(k) = \Psi(k')^m$ for some $m \geq 1$. By minimality of k, we must have $m = 1$ proving the result. □

Let d be squarefree and let $k_1 > 4$ be the smallest positive integer for which $\Psi(k_1) \in \mathbb{Q}(\sqrt{d})$. Let $S = \{\ldots, \pm\Psi(k_1)^{-3}, \pm\Psi(k_1)^{-2}, \pm\Psi(k_1)^{-1}, \pm 1, \pm\Psi(k_1), \pm\Psi(k_1)^2, \ldots\}$. Then by the proof of Theorem 6.1, S coincides with the set of norm 1 units of $\mathcal{O}$. Moreover, for each $j \geq 1$,

$$\Psi(k_1)^j = \Psi(k_j),$$

where

$$k_1 < k_2 < k_3 < \cdots < k_j < \ldots$$

are the integers $k_j > 4$ such that $\Psi(k_j) \in \mathbb{Q}(\sqrt{d})$.

In $\mathbb{Q}(\sqrt{5})$, the sequence

$$\{\Psi(5),\ \Psi(9), \Psi(20), \Psi(49), \dots\}$$

is the sequence

$$\{\psi,\ \psi^2,\ \psi^3,\ \psi^4, \dots\},$$

for $\psi = 1 + \phi = 2.61803399\dots$, which are eigenvalues for

$$(w_1w_2),\ (w_1w_2)^2,\ (w_1w_2)^3,\ (w_1w_2)^4, \dots$$

in $\mathfrak{g}(1,5)$. Note that in this case $k_j = L_{2j} + 2$ is given in terms of the Lucas numbers L_{2j}. It can also be shown that the bisection

$$\{\Psi(9),\ \Psi(49), \dots\}$$

are

$$\{\psi^2,\ \psi^4, \dots\},$$

eigenvalues for

$$(w_1w_2),\ (w_1w_2)^2,\ \dots$$

in $\mathfrak{g}(3,3)$.

Corollary 6.2. *The relation $\Psi(k_s) = \Psi(k_1)^s$ gives an infinite family of identities expressing the continued fraction for $\Psi(k_s)$ as the s-th power of the continued fraction for $\Psi(k_1)$:*

$$\Psi(k_s) = (k_s - 2) - \cfrac{1}{(k_s - 2) - \cfrac{1}{(k_s - 2) - \cdots}}$$

$$= \left[(k_1 - 2) - \cfrac{1}{(k_1 - 2) - \cfrac{1}{(k_1 - 2) - \cdots}} \right]^s$$

where $4 < k_1 < k_2 < \dots$ are the integers such that $\Psi(k_s) \in \mathbb{Q}(\sqrt{d})$.

7. Generalized Binomial Coefficients

In this section we define a new sequence $(z_n)_{n\geq 0}$ in terms of the sequences $(x_{2,2n})_{n\geq 0}$ and $(x_{1,2n+1})_{n\geq 0}$. We use the sequence $(z_n)_{n\geq 0}$ to define generalized binomial coefficients, which remarkably turn out to be integers. We provide an arithmetic proof of this, established previously by Kitchloo using topological methods ([9]). In [9], Kitchloo also showed that these integers appear in the cohomology of flag varieties associated to rank 2 Kac-Moody groups over $\mathbb{C}$.

Definition 7.1. For natural numbers a and b with $ab \geq 4$ we define the sequence $(z_n)_{n\geq 0}$ by $z_{2n} = x_{2,2n}/a$ and $z_{2n+1} = x_{1,2n+1}$.

Note that by Theorem 3.3 we have $x_{1,2n} = y_{2,2n} = z_{2n+1}$, $x_{2,2n} = az_{2n}$ and $y_{1,2n} = bz_{2n}$ so the sequence $(z_n)_{n\geq 0}$ encodes all the previous sequences.

Proposition 7.2. *The sequence $(z_n)_{n\geq 0}$ is determined by the recursion $z_0 = 0$, $z_1 = 1$, $z_2 = 1$, $z_3 = ab - 1$ and $z_n = (ab-2)z_{n-2} - z_{n-4}$ for $n \geq 4$. In particular $z_n \in \mathbb{Z}[ab]$ for $n \geq 0$. Moreover $z_n > 0$ for $n \geq 1$.*

Proof. The values of z_n, $0 \leq n \leq 3$, are easily checked. For n even, $n \geq 4$, the definition of z_n shows that

$$z_n = \frac{x_{2,n}}{a} = \frac{(ab-2)x_{2,n-2} - x_{2,n-4}}{a} = (ab-2)z_{n-2} - z_{n-4}$$

by Lemma 3.2(iii). Similarly, by definition and Lemma 3.1, $z_{2n+1} = x_{1,2n+1} = x_{1,2n}$, so the recursion formula for n odd, $n \geq 5$, follows from Lemma 3.2(i). This proves the recursion and from this it follows directly that $z_n \in \mathbb{Z}[ab]$ for $n \geq 0$. Finally, Remark 3.6 shows that $z_n > 0$ for $n \geq 1$. □

Note that it follows from Proposition 3.7 and Lemma 3.1 that we have the following formula for z_n in terms of the Chebyshev U-polynomial

$$z_n = \begin{cases} \frac{1}{\sqrt{ab}} \cdot U_{n-1}\left(\frac{\sqrt{ab}}{2}\right) & \text{for } n \text{ even,} \\ U_{n-1}\left(\frac{\sqrt{ab}}{2}\right) & \text{for } n \text{ odd.} \end{cases} \tag{11}$$

Proposition 7.3. *For integers $m \geq 1$ and $n \geq 0$ we have*

$$z_{m+n} = \begin{cases} z_{n+1}z_m - z_{m-1}z_n & \text{if } m \equiv 0 \pmod 2,\ n \equiv 0 \pmod 2, \\ ab \cdot z_{n+1}z_m - z_{m-1}z_n & \text{if } m \equiv 0 \pmod 2,\ n \equiv 1 \pmod 2, \\ z_{n+1}z_m - ab \cdot z_{m-1}z_n & \text{if } m \equiv 1 \pmod 2,\ n \equiv 0 \pmod 2, \\ z_{n+1}z_m - z_{m-1}z_n & \text{if } m \equiv 1 \pmod 2,\ n \equiv 1 \pmod 2. \end{cases}$$

Proof. Note that we have the following trigonometric identity

$$\frac{\sin((m+n)\theta)}{\sin(\theta)} = \frac{\sin((n+1)\theta)}{\sin(\theta)} \cdot \frac{\sin(m\theta)}{\sin(\theta)} - \frac{\sin((m-1)\theta)}{\sin(\theta)} \cdot \frac{\sin(n\theta)}{\sin(\theta)}$$

for $\theta \neq p\pi$, $p \in \mathbb{Z}$. Using the definition of the Chebyshev U-polynomials this translates into the identity

$$U_{m+n-1}(x) = U_n(x) \cdot U_{m-1}(x) - U_{m-2}(x) \cdot U_{n-1}(x).$$

Letting $x = \frac{\sqrt{ab}}{2}$ and using that for $k \geq -1$ we have

$$U_k\left(\frac{\sqrt{ab}}{2}\right) = \begin{cases} z_{k+1} & \text{for } k \text{ even,} \\ \sqrt{ab} \cdot z_{k+1} & \text{for } k \text{ odd,} \end{cases}$$

(cf. (11)) now proves the result. □

Corollary 7.4. *For $m, n \geq 0$ we have $z_{m+n} \in z_m\mathbb{Z}[ab] + z_n\mathbb{Z}[ab]$.*

Proof. This is clear for $m = 0$, and for $m \geq 1$ the claim follows directly from Proposition 7.3 and Proposition 7.2. □

Definition 7.5. Let $m, n \geq 1$. The *generalized binomial coefficient* $Z(m,n)$ is defined by

$$Z(m,n) = \frac{z_{m+n} \cdot z_{m+n-1} \cdot \ldots \cdot z_1}{(z_m \cdot z_{m-1} \cdot \ldots \cdot z_1) \cdot (z_n \cdot z_{n-1} \cdot \ldots \cdot z_1)}.$$

Note that this expression makes sense by Proposition 7.2.

Lemma 7.6. *For $m, n \geq 2$ we have the Pascal triangle type relation*

$$Z(m,n) \in Z(m-1,n)\mathbb{Z}[ab] + Z(m,n-1)\mathbb{Z}[ab].$$

Proof. After cancelling the common factor $z_{m+n-1} \cdot \ldots \cdot z_1$ and multiplying by $(z_m \cdot \ldots \cdot z_1) \cdot (z_n \cdot \ldots \cdot z_1)$ on both sides, we get the equivalent statement $z_{m+n} \in z_m\mathbb{Z}[ab] + z_n\mathbb{Z}[ab]$. This follows directly from Corollary 7.4. □

Theorem 7.7. *For $m, n \geq 1$ we have $Z(m,n) \in \mathbb{Z}[ab]$.*

Proof. Note first that by definition $Z(1,n) = \frac{z_{n+1}}{z_1} = z_{n+1} \in \mathbb{Z}[ab]$. Similarly $Z(m,1) = z_{m+1} \in \mathbb{Z}[ab]$. This proves the result if $m = 1$ or $n = 1$. For $m, n \geq 2$ Lemma 7.6 shows that the result holds for (m,n) if it holds for $(m-1,n)$ and for $(m,n-1)$. Induction on $m+n$ combined with the above now shows that the result holds for all (m,n). □

Remark 7.8. In [9, §10] Kitchloo considers the sequences $(c_n)_{n\geq 0}$ and $(d_n)_{n\geq 0}$ given by $c_0 = d_0 = 0$, $c_1 = d_1 = 1$ and $c_n = ad_{n-1} - c_{n-2}$, $d_n = bc_{n-1} - d_{n-2}$ for $n \geq 2$. By induction it is easy to see that $c_n = az_n$, $d_n = bz_n$ for n even and $c_n = d_n = z_n$ for n odd. Using this it follows that the numbers

$$C(m,n) = \frac{c_{m+n} \cdot \ldots \cdot c_1}{(c_m \cdot \ldots \cdot c_1) \cdot (c_n \cdot \ldots \cdot c_1)}, \qquad D(m,n) = \frac{d_{m+n} \cdot \ldots \cdot d_1}{(d_m \cdot \ldots \cdot d_1) \cdot (d_n \cdot \ldots \cdot d_1)}$$

are given by $C(m,n) = aZ(m,n)$ and $D(m,n) = bZ(m,n)$ if $m \equiv n \equiv 1 \pmod 2$ and $C(m,n) = D(m,n) = Z(m,n)$ otherwise. In particular these numbers are integers. In [9, Thm. 10.3] this is proved by showing that these numbers occur as structure coefficients in the cohomology ring $H^*(G/P;\mathbb{Z})$ where G is the Kac-Moody group corresponding to the generalized Cartan matrix $\begin{pmatrix} 2 & -a \\ -b & 2 \end{pmatrix}$ and P is a maximal parabolic subgroup of K.

8. Arithmetic Properties of the Sequence $(z_n)_{n\geq 0}$

In this section we compute the p-adic valuation of the numbers z_n, see Theorem 8.5. We first need a number of auxiliary results.

Lemma 8.1. *We have $\gcd(z_n, z_{n-1}) = 1$ for $n \geq 1$.*

Proof. For $m \geq 1$, Proposition 7.3 shows that

$$z_{m+1} = \begin{cases} ab \cdot z_m - z_{m-1} & \text{for } m \text{ even, } m \geq 2, \\ z_m - z_{m-1} & \text{for } m \text{ odd, } m \geq 1. \end{cases} \tag{12}$$

It follows that $\gcd(z_{m+1}, z_m) = \gcd(z_m, z_{m-1})$ for all $m \geq 1$. By induction we now get $\gcd(z_n, z_{n-1}) = \ldots = \gcd(z_1, z_0) = \gcd(1, 0) = 1$ for all $n \geq 1$. □

Lemma 8.2. *If p is a prime number with $p \mid ab$ and $p \mid z_n$ then n is even.*

Proof. For $s \geq 0$ Theorem 3.15(i) shows that

$$z_{2s+1} = \sum_{i=0}^{s} (-1)^i \binom{2s-i}{i} (ab)^{s-i} \equiv (-1)^s \binom{2s-s}{s} = (-1)^s \pmod{p},$$

since $p \mid ab$. Hence n must be even. □

Proposition 8.3. *Let p be a prime number. Then there exists a natural number k such that $p \mid z_n$ if and only if $k \mid n$.*

Proof. Let $X = \{n \geq 0 \mid z_n \equiv 0 \pmod{p}\}$. Since $z_0 = 0$ we have $0 \in X$. It is clear from Proposition 7.3 that $i, j \in X \Longrightarrow i + j \in X$. Now suppose $i, j \in X$ with $1 \leq i \leq j$. Setting $m = i$ and $n = j - i$ in Proposition 7.3 shows that $p \mid ab \cdot z_{i-1} z_{j-i}$ if $i \equiv 1 \pmod{2}$ and $j - i \equiv 0 \pmod{2}$ and that $p \mid z_{i-1} z_{j-i}$ otherwise. However by Lemma 8.2, $p \mid ab$ and $p \mid z_i$ implies that i is even so the first case does not occur. Hence $p \mid z_{i-1} z_{j-i}$. By Lemma 8.1 $\gcd(z_{i-1}, z_i) = 1$ so $p \nmid z_{i-1}$. It follows that $p \mid z_{j-i}$ so $j - i \in X$. The above properties show that $X = k\mathbb{Z}_{\geq 0}$ for a unique $k \geq 0$. It remains to show that $k \neq 0$, that is $X \neq \{0\}$. Since there are only p^4 possibilities for $(z_n, z_{n+1}, z_{n+2}, z_{n+3})$ modulo p we can find $0 \leq i < j$ such that $z_j \equiv z_i \pmod{p}$, $z_{j+1} \equiv z_{i+1} \pmod{p}$, $z_{j+2} \equiv z_{i+2} \pmod{p}$ and $z_{j+3} \equiv z_{i+3} \pmod{p}$. Since $z_{n-4} = (ab - 2)z_{n-2} - z_n$ for $n \geq 4$ we see that

$$z_{j-1} \equiv z_{i-1} \pmod{p}, \quad z_{j-2} \equiv z_{i-2} \pmod{p}, \quad \ldots, \quad z_{j-i} \equiv z_0 = 0 \pmod{p}.$$

Hence $X \neq \{0\}$ as desired. □

Remark 8.4. In the language of [10] the preceding proposition says that the sequence $(z_n)_{n \geq 0}$ is *regularly divisible*. By [10, Cor. 2] this implies that the numbers $Z(m, n)$ are integers.

Theorem 8.5. *Let p be a prime number and let k be the smallest natural number with $p \mid z_k$ (cf. Proposition 8.3). For $n \geq 1$, the p-adic valuation of z_n is given by*

$$\nu_2(z_n) = \begin{cases} 0 & \text{if } k \nmid n, \\ \nu_2(z_k) & \text{if } k \mid n \text{ and } 2k \nmid n, \\ \nu_2(z_{2k}) + \nu_2(\frac{n}{2k}) & \text{if } 2k \mid n, \end{cases} \tag{13}$$

for $p = 2$ and

$$\nu_p(z_n) = \begin{cases} 0 & \text{if } k \nmid n, \\ \nu_p(z_k) + \nu_p(\frac{n}{k}) & \text{if } k \mid n. \end{cases} \tag{14}$$

for p odd.

Remark 8.6. This is proved for p odd in [9, Thm. 10.10] by considering $H^*(K;\mathbb{F}_p)$ where K is the Kac-Moody group associated to the generalized Cartan matrix $\begin{pmatrix} 2 & -a \\ -b & 2 \end{pmatrix}$.

Definition 8.7. We define the sequence $(y_n)_{n\geq 0}$ of integers by $y_0 = 2$ and $y_n = z_{n+1} - z_{n-1}$ for $n \geq 1$.

Lemma 8.8. *We have $y_n > 0$ for all $n \geq 0$.*

Proof. By definition we have $y_0 = 2$, $y_1 = 1$, $y_2 = ab-2$ and $y_3 = ab-3$. Moreover from the recursion relation for $(z_n)_{n\geq 0}$ in Proposition 7.2 it follows that $y_n = (ab-2)y_{n-2} - y_{n-4}$ for $n \geq 4$. We now prove by induction that $y_n \geq y_{n-2}$ for $n \geq 2$. This is clear for $n = 2$ and $n = 3$ by the above computations. Suppose that the inequality holds for all n with $2 \leq n \leq N$ for some $N \geq 3$. Then

$$y_{N+1} - y_{N-1} = (ab-3)y_{N-1} - y_{N-3} \geq y_{N-1} - y_{N-3} \geq 0$$

proving the claim. The inequality combined with induction shows immediately that $y_n \geq y_0 = 2$ for n even and $y_n \geq y_1 = 1$ for n odd. This proves the result. □

Proposition 8.9. *The sequence $(y_n)_{n\geq 0}$ may be expressed in terms of Chebyshev T-polynomials as follows*

$$y_n = \begin{cases} 2 \cdot T_n(\sqrt{ab}/2) & \textit{for } n \textit{ even,} \\ \frac{2}{\sqrt{ab}} \cdot T_n(\sqrt{ab}/2) & \textit{for } n \textit{ odd.} \end{cases} \tag{15}$$

Proof. For $n = 0$ this is clear since $T_0(x) = 1$. Assume now that $n \geq 1$. We have the trigonometric identity

$$\frac{\sin((n+1)\theta)}{\sin(\theta)} - \frac{\sin((n-1)\theta)}{\sin(\theta)} = 2\cos(n\theta)$$

for $\theta \neq p\pi$, $p \in \mathbb{Z}$ which translates into the identity $U_n(x) - U_{n-2}(x) = 2 \cdot T_n(x)$. Formula (15) follows from this identity combined with the definition $y_n = z_{n+1} - z_{n-1}$ and (11). □

Lemma 8.10. *We have* $\gcd(z_n, y_n) \in \{1, 2\}$ *for $n \geq 0$.*

Proof. The claim is clear for $n = 0$ since $z_0 = 0$ and $y_0 = 2$. For $n \geq 1$, (12) shows that

$$y_n = z_{n+1} - z_{n-1} = \begin{cases} z_n - 2z_{n-1} & \text{for } n \text{ odd, } n \geq 1, \\ ab \cdot z_n - 2z_{n-1} & \text{for } n \text{ even, } n \geq 2. \end{cases}$$

Hence $\gcd(z_n, y_n) = \gcd(z_n, 2z_{n-1})$, so the result follows from Lemma 8.1. □

Proposition 8.11. *For integers $m \geq 1$ and $n \geq 0$ we have*

$$z_{mn} = \frac{1}{2^{m-1}} \sum_{j=0}^{\lfloor \frac{m-1}{2} \rfloor} \binom{m}{2j+1} z_n^{2j+1} y_n^{m-1-2j} \cdot (ab)^j \cdot (ab-4)^j \qquad \textit{for } n \textit{ even,}$$

$$z_{mn} = \frac{1}{2^{m-1}} \sum_{j=0}^{\lfloor \frac{m-1}{2} \rfloor} \binom{m}{2j+1} z_n^{2j+1} y_n^{m-1-2j} \cdot (ab)^{\lfloor \frac{m-1}{2} \rfloor - j} \cdot (ab-4)^j \quad \textit{for } n \textit{ odd.}$$

Proof. Note that for $\sin(\theta) \neq 0$ we have

$$\sum_{j=0}^{\lfloor \frac{m-1}{2} \rfloor} \binom{m}{2j+1} \cdot \left(\frac{\sin(n\theta)}{\sin(\theta)}\right)^{2j+1} \cdot \cos(n\theta)^{m-1-2j} \cdot (-\sin(\theta)^2)^j$$

$$= \frac{1}{i\sin(\theta)} \sum_{j=0}^{\lfloor \frac{m-1}{2} \rfloor} \binom{m}{2j+1} \cdot (i\sin(n\theta))^{2j+1} \cdot \cos(n\theta)^{m-(2j+1)}$$

$$= \frac{1}{i\sin(\theta)} \cdot \frac{1}{2} \cdot \left((\cos(n\theta) + i\sin(n\theta))^m - (\cos(n\theta) - i\sin(n\theta))^m\right)$$

$$= \frac{1}{2i\sin(\theta)} \cdot (e^{imn\theta} - e^{-imn\theta}) = \frac{\sin(mn\theta)}{\sin(\theta)}.$$

This translates into the following identity

$$U_{mn-1}(x) = \sum_{j=0}^{\lfloor \frac{m-1}{2} \rfloor} \binom{m}{2j+1} \cdot U_{n-1}(x)^{2j+1} \cdot T_n(x)^{m-1-2j} \cdot (x^2-1)^j \tag{16}$$

between Chebyshev polynomials. The formulas of the proposition follow easily from this identity by setting $x = \sqrt{ab}/2$ and using that

$$U_{k-1}\left(\frac{\sqrt{ab}}{2}\right) = \begin{cases} \sqrt{ab} \cdot z_k & \text{for } k \text{ even,} \\ z_k & \text{for } k \text{ odd,} \end{cases} \qquad T_k\left(\frac{\sqrt{ab}}{2}\right) = \begin{cases} \frac{1}{2} \cdot y_k & \text{for } k \text{ even,} \\ \frac{\sqrt{ab}}{2} \cdot y_k & \text{for } k \text{ odd,} \end{cases}$$

for $k \geq 0$ (cf. (11) and Proposition 8.9). □

Proposition 8.12. *Let $n \geq 1$ and $p \neq 2$ be a prime number with $p \mid z_n$. Then $\nu_p(z_{pn}) = \nu_p(z_n) + 1$ and $\nu_p(z_{mn}) = \nu_p(z_n)$ for $m \geq 1$ with $p \nmid m$.*

Proof. Note first that $z_n \neq 0$ since $n \geq 1$. Set $e = \nu_p(z_n) \geq 1$. Since $\gcd(z_n, y_n) \mid 2$ by Lemma 8.10 we must have $p \nmid y_n$. Recall that if $p \mid ab$ then n is even by Lemma 8.2. Now let $m \geq 1$ and consider the formula for z_{mn} from Proposition 8.11. From the above it follows that the term for $j = 0$ in the sum for z_{mn} has p-adic valuation equal to $\nu_p(m) + e$. This equals $e + 1$ for $m = p$ and e for $p \nmid m$. The remaining terms all have p-adic valuations which are $\geq 3e > e + 1$, so $\nu_p(z_{mn})$ is given by $e + 1$ for $m = p$ and by e for $p \nmid m$ as claimed. □

Lemma 8.13. *The values of z_n and y_n modulo 4 are periodic and are given as follows.*

(i) *For $ab \equiv 0 \pmod 4$ we have*

n mod 8	0	1	2	3	4	5	6	7
z_n mod 4	0	1	1	3	2	1	3	3
y_n mod 4	2	1	2	1	2	1	2	1

(ii) *For* $ab \equiv 1 \pmod 4$ *we have*

$n \bmod 6$	0	1	2	3	4	5
$z_n \bmod 4$	0	1	1	0	3	3
$y_n \bmod 4$	2	1	3	2	3	1

(iii) *For* $ab \equiv 2 \pmod 4$ *we have*

$n \bmod 8$	0	1	2	3	4	5	6	7
$z_n \bmod 4$	0	1	1	1	0	3	3	3
$y_n \bmod 4$	2	1	0	3	2	3	0	1

(iv) *For* $ab \equiv 3 \pmod 4$ *we have*

$n \bmod 12$	0	1	2	3	4	5	6	7	8	9	10	11
$z_n \bmod 4$	0	1	1	2	1	1	0	3	3	2	3	3
$y_n \bmod 4$	2	1	1	0	3	3	2	3	3	0	1	1

Proof. This follows easily by direct computation using the values $z_0 = 0$, $z_1 = 1$, $z_2 = 1$, $z_3 = ab - 1$, $y_0 = 2$, $y_1 = 1$, $y_2 = ab - 2$ and $y_3 = ab - 3$ combined with the recursions $z_n = (ab-2)z_{n-2} - z_{n-4}$ and $y_n = (ab-2)y_{n-2} - y_{n-4}$ for $n \geq 4$. □

Proposition 8.14. *Let* $n \geq 1$ *and assume* $2 \,|\, z_n$. *Then* $\nu_2(z_{mn}) = \nu_2(z_n)$ *for* m *odd.*

Proof. Set $e = \nu_2(z_n) \geq 1$ and $f = \nu_2(y_n) \geq 0$ (using that $y_n \neq 0$ by Lemma 8.8). Now let $m \geq 1$ be odd and consider the formula for z_{mn} from Proposition 8.11.

If $2 \mid ab$, Lemma 8.13 shows that $n \equiv 0 \pmod 4$ and $f = 1$. The term for $j = 0$ in the sum for z_{mn} has 2-adic valuation e, and for $1 \leq j \leq \frac{m-1}{2}$ the j'th term in the sum has 2-adic valuation

$$\geq -(m-1) + (2j+1)e + (m-1-2j) + 2j = (2j+1)e \geq 3e > e.$$

It follows that $\nu_2(z_{mn}) = e$ if $2 \,|\, ab$.

Now consider the case $2 \nmid ab$. By Lemma 8.13 it follows that $n \equiv 0 \pmod 3$ and that either $e \geq 2$ and $f = 1$ or $e = 1$ and $f \geq 2$. For $0 \leq j \leq \frac{m-1}{2}$, the j'th term in the sum for z_{mn} has 2-adic valuation equal to

$$\nu_j = -(m-1) + \nu_2\left(\binom{m}{2j+1}\right) + (2j+1)e + (m-1-2j)f. \tag{17}$$

If $e \geq 2$ and $f = 1$ we have $\nu_0 = e$ and $\nu_j \geq (2j+1)e - 2j > e$ for $1 \leq j \leq \frac{m-1}{2}$, so $\nu_2(z_{mn}) = e$ in this case. If $e = 1$ and $f \geq 2$, then $\nu_{(m-1)/2} = 1$ and $\nu_j \geq 2f - 1 > 1$ for $0 \leq j < \frac{m-1}{2}$. It follows that $\nu_2(z_{mn}) = 1 = e$ in this case as well. □

Proposition 8.15. *Let* $k \geq 1$ *be the least integer such that* $2 \,|\, z_k$ *(cf. Proposition 8.3). If* $2k \,|\, n$ *then* $\nu_2(z_{2n}) = \nu_2(z_n) + 1$.

Proof. By Proposition 8.11 we have $z_{2n} = z_n y_n$ for all $n \geq 0$. If ab is even it follows from Lemma 8.13 that $k = 4$, so $n \equiv 0 \pmod 8$ and $\nu_2(y_n) = 1$. If ab is odd, then Lemma 8.13 shows that $k = 3$, so $n \equiv 0 \pmod 6$ and $\nu_2(y_n) = 1$. Thus $\nu_2(z_{2n}) = \nu_2(z_n) + \nu_2(y_n) = \nu_2(z_n) + 1$ as claimed. □

Proof of Theorem 8.5. The case $p = 2$ follows from Propositions 8.14 and 8.15 and the case p odd follows from Proposition 8.12. □

9. Further Directions - Arithmetic in Rank 3 Hyperbolic Root Systems

In this section we give a brief indication of how our methods may extend to certain rank 3 hyperbolic root systems.

Let

$$A = \begin{pmatrix} 2 & -2 & 0 \\ -2 & 2 & -1 \\ 0 & -1 & 2 \end{pmatrix}, \qquad A_0 = \begin{pmatrix} 2 & -2 \\ -2 & 2 \end{pmatrix},$$

so that A_0 is an affine submatrix of A. It follows that A has noncompact hyperbolic type.

We let W denote the Weyl group of A and let $W_0 \subset W$ denote the Weyl group of A_0. Then $W \cong PGL_2(\mathbb{Z})$. Let Φ denote the full real root system of A, let $\Pi = \{\alpha_1, \alpha_2, \alpha_3\}$ denote the simple roots and let

$$\Phi \cap (\mathbb{Z}\alpha_1 \oplus \mathbb{Z}\alpha_2 \oplus \mathbb{Z}\alpha_3)$$

denote the points on the root lattice corresponding to real roots.

The group W_0 acts on $\Phi \cap (\mathbb{Z}\alpha_1 \oplus \mathbb{Z}\alpha_2 \oplus \mathbb{Z}\alpha_3)$. Since W_0 is generated by two simple root reflections w_1 and w_2 (as in Section 3), the coefficient of α_3 in each orbit under the action of W_0 on $\Phi \cap (\mathbb{Z}\alpha_1 \oplus \mathbb{Z}\alpha_2 \oplus \mathbb{Z}\alpha_3)$ is fixed. We denote this coefficient by $z \in \mathbb{Z}$.

The orbit space for the action of W_0 on $\Phi \cap (\mathbb{Z}\alpha_1 \oplus \mathbb{Z}\alpha_2 \oplus \mathbb{Z}\alpha_3)$ consists of an infinite family of (lattice points on) curves indexed by $\mathbb{Z}$. These lie on an infinite family of 'slices' of the hyperboloid of one sheet supporting the real roots Φ, thus each orbit lives in a unique 2 dimensional 'sheet'

$$S_z = \mathbb{Z}\alpha_1 \oplus \mathbb{Z}\alpha_2 \oplus z\alpha_3$$

for a fixed $z \in \mathbb{Z}$. Conversely, for each $z \in \mathbb{Z}$, the sheet S_z contains at least one orbit of W_0.

As in Section 3, every element of W_0 is of the form $w_1^{\epsilon_1}(w_2w_1)^n w_2^{\epsilon_2}$, where $\epsilon_i \in \{0, 1\}$, $i = 1, 2$. We consider the action of $(w_2w_1)^n$ on points $(x, y, z) \in \Phi \cap (\mathbb{Z}\alpha_1 \oplus \mathbb{Z}\alpha_2 \oplus z\alpha_3)$, where z is fixed and we obtain the formula

$$(w_2w_1)^n(x, y, z) = (a_n, b_n, z),$$

where

$$a_0 = x, \qquad a_n = 2T_{n-1}z + 2ny - (2n-1)x, \qquad b_n = n^2 z + (2n+1)y - 2nx,$$

and

$$T_n = \sum_{i=1}^{n} i = \frac{1}{2}(n^2 + n) = \binom{n+1}{2} \text{ is the } n^{\text{th}} \text{ triangular number.}$$

To prove this formula (which was first shown to us by Alex Conway) note that the Jordan decomposition of $w_2 w_1$ is given by $w_2 w_1 = SJS^{-1}$, where

$$S = \begin{pmatrix} 2 & 0 & 0 \\ 2 & 1 & 0 \\ 0 & 0 & 1 \end{pmatrix}, \qquad J = \begin{pmatrix} 1 & 1 & 0 \\ 0 & 1 & 1 \\ 0 & 0 & 1 \end{pmatrix}.$$

Since

$$J^n = \begin{pmatrix} 1 & n & \binom{n}{2} \\ 0 & 1 & n \\ 0 & 0 & 1 \end{pmatrix}$$

by induction, it follows that

$$(w_2 w_1)^n = SJ^nS^{-1} = \begin{pmatrix} -2n+1 & 2n & n^2-n \\ -2n & 2n+1 & n^2 \\ 0 & 0 & 1 \end{pmatrix}$$

which proves the claim.

It is not surprising to find symmetries here relating to Pascal's triangle, though we have not yet understood the full Weyl group action on roots.

We mention the following result which is also of interest. In [2], the authors consider the rank 3 generalized Cartan matrix

$$A = \begin{pmatrix} 2 & -2 & -2 \\ -2 & 2 & -2 \\ -2 & -2 & 2 \end{pmatrix}$$

which is also of noncompact hyperbolic type and contains A_0 as an affine submatrix. The Weyl group W of A is $W \cong \mathbb{Z}/2\mathbb{Z} * \mathbb{Z}/2\mathbb{Z} * \mathbb{Z}/2\mathbb{Z}$.

The authors show that the Weyl group images of twice the fundamental weights give rise to Pythagorean triples, and these in turn correspond to vertices of ideal triangles of the W-tessellation on the perimeter of the Poincaré disk.

We suspect that a detailed investigation of the Weyl group orbits of these rank 3 hyperbolic root systems on their roots and weights will reveal a rich arithmetic structure. We hope to take this up elsewhere.

10. Appendix: Tables of Integer Sequences

Unless otherwise noted, the sequences are not in the Online Encyclopedia of Integer Sequences at the time of publication. In Table 1 we exhibit sequences $X_1(1,k)$, $k = 4, \ldots, 20$. In Table 2, we give a choice of sequence $X_2(a,b)$ for each $k = ab = 4, \ldots, 12$, and in Table 3 we exhibit sequences $X(a)$, $a = 2, \ldots, 12$. We refer the reader to Theorem 3.3 for the properties of these sequences.

Acknowledgement

The second author was supported in part by NSF grant #DMS-0701176.

Table 1. $X_1(1,k)$**,** $k = 4, \ldots, 20$

$X_1(1,4)$	$\{1,3,5,7,9,\ldots\}$	OEIS sequence A005408
$X_1(1,5)$	$\{1,4,11,29,76,\ldots\}$	OEIS sequence A002878 Bisection of the Lucas sequence
$X_1(1,6)$	$\{1,5,19,71,265,\ldots\}$	OEIS sequence A001834
$X_1(1,7)$	$\{1,6,29,139,666,\ldots\}$	OEIS sequence A030221 Chebyshev even indexed U-polynomials evaluated at $\sqrt{7}/2$
$X_1(1,8)$	$\{1,7,41,239,1393,\ldots\}$	OEIS sequence A002315 the Newman-Shanks-Williams numbers
$X_1(1,9)$	$\{1,8,55,377,2584,\ldots\}$	OEIS sequence A033890 Fibonacci(4n+2)
$X_1(1,10)$	$\{1,9,71,559,4401,\ldots\}$	OEIS sequence A057080 Chebyshev even indexed U-polynomials evaluated at $\sqrt{10}/2$ with $\lim_{n\longrightarrow\infty} a(n)/a(n-1) = 4+\sqrt{15}$
$X_1(1,11)$	$\{1,10,89,791,7030,\ldots\}$	OEIS sequence A057081 Chebyshev even indexed U-polynomials evaluated at $\sqrt{11}/2$
$X_1(1,12)$	$\{1,11,109,1079,10681,\ldots\}$	OEIS sequence A054320 Chebyshev even indexed U-polynomials evaluated at $\sqrt{3}$ Squares of entries are star numbers
$X_1(1,13)$	$\{1,12,131,1429,15588,\ldots\}$	OEIS sequence A097783 Chebyshev polynomials with diophantine property
$X_1(1,14)$	$\{1,13,155,1847,22009,\ldots\}$	OEIS sequence A077416 Chebyshev S-sequence with diophantine property
$X_1(1,15)$	$\{1,14,181,2339,30226,\ldots\}$	OEIS sequence A126866
$X_1(1,16)$	$\{1,15,209,2911,40545,\ldots\}$	OEIS sequence A028230 Indices of square numbers which are also octagonal
$X_1(1,17)$	$\{1,16,239,3569,53296,\ldots\}$	
$X_1(1,18)$	$\{1,17,271,4319,68833,\ldots\}$	
$X_1(1,19)$	$\{1,18,305,5167,87534,\ldots\}$	
$X_1(1,20)$	$\{1,19,341,6119,109801,\ldots\}$	OEIS sequence A049629 $a(n) = 1/4[F(6n+4)+F(6n+2)]$ where $F(n)$ is the Fibonacci sequence

Table 2. $X_2(a,b)$, $k = ab = 4, \ldots, 12$

$X_2(2,2)$	$\{0,2,4,6,8,\ldots\}$	OEIS sequence A005843
$X_2(1,5)$	$\{0,1,3,8,21,\ldots\}$	OEIS sequence A001906 Even bisection of the Fibonacci sequence
$X_2(2,3)$	$\{0,2,8,30,112,\ldots\}$	OEIS sequence A052530
$X_2(1,7)$	$\{0,1,5,24,115,\ldots\}$	OEIS sequence A004254
$X_2(4,2)$	$\{0,4,24,140,816,\ldots\}$	OEIS sequence A005319 For any term n, $2n^2+4$ is a perfect square
$X_2(1,9)$	$\{0,1,7,48,329,\ldots\}$	OEIS sequence A004187
$X_2(1,10)$	$\{0,1,8,63,496,\ldots\}$	OEIS sequence A001090
$X_2(1,11)$	$\{0,1,9,80,711,\ldots\}$	OEIS sequence A018913
$X_2(4,3)$	$\{0,4,40,396,3920,\ldots\}$	OEIS sequence A122652 The Kekule numbers for the benzenoids

Table 3. $X(a)$, $a = 2, \ldots, 12$

$X(2)$	$\{0,1,2,3,4,5,\ldots\}$	OEIS sequence A001477
$X(3)$	$\{0,1,3,8,21,\ldots\}$	OEIS sequence A001906 The bisection of the Fibonacci sequence
$X(4)$	$\{0,1,4,15,56,209,\ldots\}$	OEIS sequence A001353 The number of spanning trees in a $2\times n$ grid
$X(5)$	$\{0,1,5,24,115,551,\ldots\}$	OEIS sequence A004254
$X(6)$	$\{0,1,6,35,204,1189,\ldots\}$	OEIS sequence A001109 The squares of entries are triangular numbers
$X(7)$	$\{0,1,7,48,329,2255,\ldots\}$	OEIS sequence A004187
$X(8)$	$\{0,1,8,63,496,3905,\ldots\}$	OEIS sequence A001090
$X(9)$	$\{0,1,9,80,711,6319,\ldots\}$	OEIS sequence A018913
$X(10)$	$\{0,1,10,99,980,\ldots\}$	OEIS sequence A004189 Indices of square numbers which are also generalized pentagonal numbers
$X(11)$	$\{0,1,11,120,\ldots\}$	OEIS sequence A004190
$X(12)$	$\{0,1,12,143,\ldots\}$	OEIS sequence A004191

References

[1] M. Abramowitz and I. A. Stegun. *Handbook of mathematical functions with formulas, graphs, and mathematical tables*, volume 55 of *National Bureau of Standards Applied Mathematics Series*. For sale by the Superintendent of Documents, U.S. Government Printing Office, Washington, D.C., 1964.

[2] G. Benkart and P. Terwilliger. The equitable basis for sl_2. arXiv:0810.2066.

[3] H. Cohn. *Advanced number theory*. Dover Publications Inc., New York, 1980. Reprint of *A second course in number theory*, 1962, Dover Books on Advanced Mathematics.

[4] A. J. Feingold. A hyperbolic GCM Lie algebra and the Fibonacci numbers. *Proc. Amer. Math. Soc.*, 80(3):379–385, 1980.

[5] W. B. Jones and W. J. Thron. *Continued fractions*, volume 11 of *Encyclopedia of Mathematics and its Applications*. Addison-Wesley Publishing Co., Reading, Mass., 1980. Analytic theory and applications, With a foreword by Felix E. Browder, With an introduction by Peter Henrici.

[6] V. G. Kac. *Infinite-dimensional Lie algebras*. Cambridge University Press, Cambridge, third edition, 1990.

[7] S.-J. Kang and D. J. Melville. Rank 2 symmetric hyperbolic Kac-Moody algebras. *Nagoya Math. J.*, 140:41–75, 1995.

[8] P. W. Kasteleyn. The statistics of dimers on a lattice i. the number of dimer arrangements on a quadratic lattice. *Physica*, 27:1209–1225, 1961.

[9] N. Kitchloo. On the topology of Kac-Moody groups. arXiv:0810.0851.

[10] D. E. Knuth and H. S. Wilf. The power of a prime that divides a generalized binomial coefficient. *J. Reine Angew. Math.*, 396:212–219, 1989.

[11] J. Lepowsky and R. V. Moody. Hyperbolic Lie algebras and quasiregular cusps on Hilbert modular surfaces. *Math. Ann.*, 245(1):63–88, 1979.

[12] M. O. Rayes, V. Trevisan, and P. S. Wang. Factorization properties of Chebyshev polynomials. *Comput. Math. Appl.*, 50(8-9):1231–1240, 2005.

[13] H. N. V. Temperley and M. E. Fisher. Dimer problem in statistical mechanics—an exact result. *Philos. Mag. (8)*, 6:1061–1063, 1961.

[14] D. Zwillinger, editor. *CRC standard mathematical tables and formulae*. Chapman & Hall/CRC, Boca Raton, FL, st edition, 2003.

INDEX

A

B

C

D

E

F

G

H

I

J

L

M

N

O

P

Q

R

S

T

U

V

W

Y